Fungal Biology

Series Editors:
Vijai Kumar Gupta, PhD
Molecular Glycobiotechnology Group, Department of Biochemistry,
School of Natural Sciences, National University of Ireland Galway,
Galway, Ireland

Maria G. Tuohy, PhD
Molecular Glycobiotechnology Group, Department of Biochemistry,
School of Natural Sciences, National University of Ireland Galway,
Galway, Ireland

For further volumes:
http://www.springer.com/series/11224

Marco A. van den Berg
Karunakaran Maruthachalam
Editors

Genetic Transformation Systems in Fungi, Volume 2

Springer

Editors
Marco A. van den Berg, Ph.D.
Applied Biochemistry Department
DSM Biotechnology Center
Delft, The Netherlands

Karunakaran Maruthachalam, Ph.D.
Global Marker Technology Lab
 (DuPont-Pioneer)
E.I.DuPont India Pvt Ltd
DuPont Knowledge Center
Hyderabad, Telangana, India

ISSN 2198-7777 ISSN 2198-7785 (electronic)
ISBN 978-3-319-10502-4 ISBN 978-3-319-10503-1 (eBook)
DOI 10.1007/978-3-319-10503-1
Springer Cham Heidelberg New York Dordrecht London

Library of Congress Control Number: 2014952786

© Springer International Publishing Switzerland 2015
This work is subject to copyright. All rights are reserved by the Publisher, whether the whole or part of the material is concerned, specifically the rights of translation, reprinting, reuse of illustrations, recitation, broadcasting, reproduction on microfilms or in any other physical way, and transmission or information storage and retrieval, electronic adaptation, computer software, or by similar or dissimilar methodology now known or hereafter developed. Exempted from this legal reservation are brief excerpts in connection with reviews or scholarly analysis or material supplied specifically for the purpose of being entered and executed on a computer system, for exclusive use by the purchaser of the work. Duplication of this publication or parts thereof is permitted only under the provisions of the Copyright Law of the Publisher's location, in its current version, and permission for use must always be obtained from Springer. Permissions for use may be obtained through RightsLink at the Copyright Clearance Center. Violations are liable to prosecution under the respective Copyright Law.
The use of general descriptive names, registered names, trademarks, service marks, etc. in this publication does not imply, even in the absence of a specific statement, that such names are exempt from the relevant protective laws and regulations and therefore free for general use.
While the advice and information in this book are believed to be true and accurate at the date of publication, neither the authors nor the editors nor the publisher can accept any legal responsibility for any errors or omissions that may be made. The publisher makes no warranty, express or implied, with respect to the material contained herein.

Printed on acid-free paper

Springer is part of Springer Science+Business Media (www.springer.com)

Preface

Fungi are a highly versatile class of microorganisms and their habitats are as diverse. In nature, fungi play a crucial role in a range of degradation processes, enabling recycling of valuable raw materials by wood decaying fungi like the white rot fungus *Phanerochaete chrysosporium*. On the other hand, fungi can be pests to food production like the rice blast fungus *Magnaporthe oryzae*. Furthermore, mankind exploits the enzymatic opportunities of fungi through classical industrial processes as ethanol production by the yeast *Saccharomyces cerevisiae* and heterologous enzyme production by filamentous fungi as *Trichoderma reesei*. All these stimulated an enormous number of studies trying to understand as well as exploit the metabolic capabilities of various fungal species.

One of the game-changing breakthroughs in fungal research was the development of genetic transformation technology. This enabled researchers to efficiently modify the gene content of fungi and study the functional relevance. Interestingly, the first available method (protoplast or spheroplast transformation) evolved from an existing classical method called protoplast fusion, a process which also introduces DNA into a receiving cell however in an uncontrolled way. This publication aims to give an overview of all existing transformation methods used for yeasts and fungi.

Volume I describes in detail the different classical methods as electroporation, protoplast, Agrobacterium mediated, lithium acetate and biolistic transformation as well as more recently developed methods. Transformation methods do not describe the whole story; DNA must enter the cell, the nucleus, and finally integrate in the genome, if required also at predetermined positions. Several chapters will update on the current insights in these processes.

Volume II describes transformation-associated methods and tools as cell fusion, repetitive elements, automation, analysis, markers, and vectors; this volume reflects the many relevant elements at hand for the modern fungal researcher.

This publication is meant not only as reference material for the experienced researcher, but also as introduction for the emerging scientist. Therefore, all methods are supported by several illustrative example protocols from various fungal species and laboratories around the world, which will be a good starting position to develop a working protocol for other fungal species being studied.

Delft, The Netherlands   Marco A. van den Berg
Hyderabad, Telangana, India   Karunakaran Maruthachalam

Contents

Part I Endogenous DNA: Cell Fusion

1 Anastomosis and Heterokaryon Formation 3
Martin Weichert and André Fleißner

2 Induction of the Sexual Cycle in Filamentous Ascomycetes 23
Jos Houbraken and Paul S. Dyer

**3 What Have We Learned by Doing Transformations
in *Neurospora tetrasperma*?** 47
Durgadas P. Kasbekar

Part II Endogenous DNA: Repetitive Elements

**4 Repeat-Induced Point Mutation: A Fungal-Specific,
Endogenous Mutagenesis Process** 55
James K. Hane, Angela H. Williams, Adam P. Taranto,
Peter S. Solomon, and Richard P. Oliver

**5 Calculating RIP Mutation in Fungal Genomes
Using RIPCAL** 69
James K. Hane

6 Fungal Transposable Elements 79
Linda Paun and Frank Kempken

**7 In Vivo Targeted Mutagenesis in Yeast Using
TaGTEAM** 97
Shawn Finney-Manchester and Narendra Maheshri

Part III Endogenous DNA: Gene Expression Control

**8 RNA Silencing in Filamentous Fungi: From Basics
to Applications** 107
Nguyen Bao Quoc and Hitoshi Nakayashiki

Part IV Tools and Applications: Selection Markers and Vectors

9 RNAi-Mediated Gene Silencing in the Beta-Lactam Producer Fungi *Penicillium chrysogenum* and *Acremonium chrysogenum* 125
Carlos García-Estrada and Ricardo V. Ullán

10 Controlling Fungal Gene Expression Using the Doxycycline-Dependent Tet-ON System in *Aspergillus fumigatus* 131
Michaela Dümig and Sven Krappmann

Part IV Tools and Applications: Selection Markers and Vectors

11 Expanding the Repertoire of Selectable Markers for *Aspergillus* Transformation 141
Khyati Dave, V. Lakshmi Prabha, Manmeet Ahuja, Kashyap Dave, S. Tejaswini, and Narayan S. Punekar

12 Arginase (*agaA*) as a Fungal Transformation Marker 155
Kashyap Dave, Manmeet Ahuja, T.N. Jayashri, Rekha Bisht Sirola, Khyati Dave, and Narayan S. Punekar

13 Transformation of Ascomycetous Fungi Using Autonomously Replicating Vectors 161
Satoko Kanematsu and Takeo Shimizu

14 A Recyclable and Bidirectionally Selectable Marker System for Transformation of *Trichoderma* 169
Thiago M. Mello-de-Sousa, Robert L. Mach, and Astrid R. Mach-Aigner

15 Split-Marker-Mediated Transformation and Targeted Gene Disruption in Filamentous Fungi 175
Kuang-Ren Chung and Miin-Huey Lee

Part V Tools and Applications: High Throughput Experimentation

16 Integrated Automation for Continuous High-Throughput Synthetic Chromosome Assembly and Transformation to Identify Improved Yeast Strains for Industrial Production of Biofuels and Bio-based Chemicals 183
Stephen R. Hughes and Steven B. Riedmuller

17 Imaging Flow Cytometry and High-Throughput Microscopy for Automated Macroscopic Morphological Analysis of Filamentous Fungi 201
Aydin Golabgir, Daniela Ehgartner, Lukas Neutsch, Andreas E. Posch, Peter Sagmeister, and Christoph Herwig

Contents

18 Yeast Cell Electroporation in Droplet-Based Microfluidic Chip 211
Qiuxian Cai and Chunxiong Luo

19 Identification of T-DNA Integration Sites: TAIL-PCR and Sequence Analysis 217
Jaehyuk Choi, Junhyun Jeon, and Yong-Hwan Lee

Part VI Tools and Applications: Comprehensive Approaches in Selected Fungi

20 Genetic and Genomic Manipulations in *Aspergillus niger* 225
Adrian Tsang, Annie Bellemare, Corinne Darmond, and Janny Bakhuis

21 Genetic Manipulation of *Meyerozyma guilliermondii* 245
Nicolas Papon, Yuriy R. Boretsky, Vincent Courdavault, Marc Clastre, and Andriy A. Sibirny

Index 263

Contributors

Manmeet Ahuja, Ph.D. Department of Bioscience and Bioengineering, Indian Institute of Technology Bombay, Mumbai, Maharashtra, India

Janny Bakhuis DSM Biotechnology Center, Delft, The Netherlands

Annie Bellemare Centre for Structural and Functional Genomics, Concordia University, Montreal, QC, Canada

Yuriy R. Boretsky, Ph.D. Department of Molecular Genetics and Biotechnology, Institute of Cell Biology, National Academy of Sciences of Ukraine, Lviv, Ukraine

Qiuxian Cai The State Key Laboratory for Artificial Microstructures and Mesoscopic Physics, School of Physics; Center for Qualitative Biology, Academy for Advanced Interdisciplinary StudiesPeking University, Beijing, China

Jaehyuk Choi, Ph.D. Division of Life Sciences, College of Life Sciences and Bioengineering, Incheon National University, Incheon, South Korea

Kuang-Ren Chung, Ph.D. Department of Plant Pathology, National Chung-Hsing University, Taichung, Taiwan

Marc Clastre, Ph.D. Department EA2106 "Biomolécules et Biotechnologies Végétales", Faculté de Pharmacie, Université François-Rabelais de Tours, Tours, France

Vincent Courdavault, Ph.D. Department EA2106 "Biomolécules et Biotechnologies Végétales", Faculté de Pharmacie, Université François-Rabelais de Tours, Tours, France

Corinne Darmond Centre for Structural and Functional Genomics, Concordia University, Montreal, QC, Canada

Kashyap Dave, Ph.D. Department of Bioscience and Bioengineering, Indian Institute of Technology Bombay, Mumbai, Maharashtra, India

Khyati Dave Department of Bioscience and Bioengineering, Indian Institute of Technology Bombay, Mumbai, Maharashtra, India

Michaela Dümig Mikrobiologisches Institut—Klinische Mikrobiologie, Immunologie und Hygiene, Universitätsklinikum Erlangen, Friedrich-Alexander-Universität Erlangen-Nürnberg, Erlangen, Germany

Paul S. Dyer, B.A., M.A., Ph.D. School of Life Sciences, University of Nottingham, Nottingham, UK

Daniela Ehgartner, D.I. M.Sc CD Laboratory for Mechanistic and Physiological Methods for Improved Bioprocesses, Vienna, Austria

Shawn Finney-Manchester, Ph.D. Zymergen Inc, Emeryville, CA, USA

André Fleißner Institut für Genetik, Technische Universität Braunschweig, Braunschweig, Germany

Carlos García-Estrada, D.V.M., Ph.D. INBIOTEC (Instituto de Biotecnología de León), León, Spain

Aydin Golabgir, D.I. M.Sc. Research Division Biochemical Engineering, Vienna University of Technology, Vienna, Austria

James K. Hane Centre for Crop and Disease Management, Curtin University, Perth, WA, Australia

Christoph Herwig, Ph.D. CD Laboratory for Mechanistic and Physiological Methods for Improved Bioprocesses, Vienna, Austria

Jos Houbraken, Ph.D. Department of Applied and Industrial Mycology, CBS-KNAW Fungal Biodiversity Centre, Utrecht, The Netherlands

Stephen R. Hughes, Ph.D. Renewable Product Technology, USDA, ARS, NCAUR, Peoria, IL, USA

T. N. Jayashri, Ph.D. Department of Bioscience and Bioengineering, Indian Institute of Technology Bombay, Mumbai, Maharashtra, India

Junhyun Jeon, Ph.D. Department of Agricultural Biotechnology, Seoul National University, Seoul, Republic of Korea

Satoko Kanematsu, Ph.D. Apple Research Division, NARO, Institute of Fruit Tree Science, Morioka, Japan

Durgadas P. Kasbekar, Ph.D. Centre for DNA Fingerprinting and Diagnostics, Tuljaguda Complex, Hyderabad, India

Frank Kempken Genetics and Molecular Biology in Botany, Botanical Institute, Kiel, Germany

Sven Krappmann, Prof. Dr. rer. nat. Mikrobiologisches Institute—Klinsche Mikrobiolgie, Immunologie und Hygiene, Universitatsklinikum Erlangen, Freiderich-Alexander-Unerverstat Erlangen-Nurnberg, Erlangen, Germany

Yong-Hwan Lee, Ph.D. Center for Fungal Genetic Resources, Seoul National University, Seoul, Republic of Korea

Miin-Huey Lee, Ph.D. Department of Plant Pathology, National Chung-Hsing University, Taichung, Taiwan

Chunxiong Luo, Ph.D. The State Key Laboratory for Artificial Microstructures and Mesoscopic Physics; Center for Quantitative Biology, Academy for Advanced Interdisciplinary Studies, Peking University, Beijing, China

Robert L. Mach, Ph.D. Department for Biotechnology and Microbiology, Institute of Chemical Engineering, Vienna University of Technology, Wien, Austria

Astrid R. Mach-Aigner, Ph.D. Department for Biotechnology and Microbiology, Institute of Chemical Engineering, Vienna University of Technology, Wien, Austria

Narendra Maheshri, Ph.D. Chemical Engineering Department, Massachusetts Institute of Technology/Ginkgo Bioworks, Boston, MA, USA

Thiago M. Mello-de-Sousa, M.Sc. Department for Biotechnology and Microbiology, Institute of Chemical Engineering, Vienna University of Technology, Wien, Austria

Hitoshi Nakayashiki, Ph.D. Graduate School of Agricultural Science, Laboratory of Cell Function and Structure, Kobe University, Kobe, Hyogo, Japan

Lukas Neutsch, Ph.D., M.Pharm.Sci. Department of Biochemical Engineering, Vienna University of Technology, Vienna, Austria

Richard P. Oliver, B.Sc., Ph.D. (Biochem.) Centre for Crop and Disease Management, Curtin University, Bentley, WA, Australia

Nicolas Papon Department EA2106 "Biomolécules et Biotechnologies Végétales", Faculté de Pharmacie, Université François-Rabelais de Tours, Tours, France

Linda Paun Genetics and Molecular Biology in Botany, Botanical Institute, Kiel, Germany

Andreas E. Posch, Ph.D. CD Laboratory for Mechanistic and Physiological Methods for Improved Bioprocesses, Vienna, Austria

V. Lakshmi Prabha, Ph.D. Department of Bioscience and Bioengineering, Indian Institute of Technology Bombay, Mumbai, Maharashtra, India

Narayan S. Punekar, Ph.D. Department of Bioscience and Bioengineering, Indian Institute of Technology Bombay, Mumbai, Maharashtra, India

Nguyen Bao Quoc, Ph.D. Research Institute of Biotechnology and Environment, Nong Lam University, Ho Chi Minh City, Vietnam

Steven B. Riedmuller, B.A. Hudson Robotics, Inc., Springfield, NJ, USA

Peter Sagmeister Vienna University of Technology, Graz, Styria, Austria

Takeo Shimizu, Ph.D. Apple Research Division, NARO, Institute of Fruit Tree Science, Morioka, Japan

Andriy A. Sibirny, Ph.D., Dr.Sc. Department of Molecular Genetics and Biotechnology, Institute of Cell Biology, National Academy of Sciences of Ukraine, Lviv, Ukraine

Department of Biotechnology and Microbiology, University of Rzeszow, Lviv, Ukraine

Rekha Bisht Sirola, Ph.D. Department of Bioscience and Bioengineering, Indian Institute of Technology Bombay, Mumbai, Maharashtra, India

Peter S. Solomon, B.App.Sc. (Hons.), Ph.D. Plant Sciences Division, Research School of Biology, The Australian National University, Canberra, ACT, Australia

Adam P. Taranto, B.Sc. (Hons.), Plant Sciences Division, Research School of Biology, The Australian National University, Canberra, ACT, Australia

S. Tejaswini Department of Bioscience and Bioengineering, Indian Institute of Technology Bombay, Mumbai, Maharashtra, India

Adrian Tsang Centre for Structural and Functional Genomics, Concordia University, Montreal, QC, Canada

Ricardo V. Ullán, Ph.D. INBIOTEC (Instituto de Biotecnología de León), León, Spain

Martin Weichert Institut für Genetik, Technische Universität Braunschweig, Braunschweig, Germany

Angela H. Williams, B.App.Sci. (Hons.) The Institute of Agriculture, The University of Western Australia, Crawley, WA, Australia

Part I
Endogenous DNA: Cell Fusion

Anastomosis and Heterokaryon Formation

1

Martin Weichert and André Fleißner

1.1 Introduction

In filamentous ascomycete fungi, vegetative cell fusion or anastomosis formation constitutes a common growth feature promoting the establishment and expansion of mycelial colonies. Fusion occurs at different developmental stages, where anastomosis might serve various physiological functions (Köhler 1930; Craven et al. 2008; Ruiz-Roldan et al. 2010; Roca et al. 2012). In many fungal species, colony initiation includes the fusion of germinating vegetative spores into supracellular networks, which further develop into the mycelium (Roca et al. 2005a; Fleißner 2012). This merger of numerous individuals into one functional unit appears to provide a competitive advantage and promotes successful habitat colonization (Richard et al. 2012). Within established mycelial colonies, fusion between hyphal branches increases interconnectedness, thereby supporting homeostasis and other vital functions of the mycelium (Hickey et al. 2002).

Recent years have seen a revival of scientific interest in the mechanism and physiological outcomes of anastomosis formation. The vast majority of these studies employed the red bread mold *Neurospora crassa* as a model system (Read et al. 2012). However, other species, including important plant pathogenic fungi, have begun to emerge as experimental objects in fusion research and to make significant contributions to the field (Engh et al. 2007; Craven et al. 2008; Ruiz-Roldan et al. 2010; Roca et al. 2012; Charlton et al. 2012).

While the physiological role of anastomosis is just beginning to surface, the ability of fungal hyphae to fuse has long been experimentally exploited. Examples include genetic mapping in asexual fungi using parasex (Debets et al. 1990, 1993), testing allelism in heterokaryons, or combining strain features by fusing different individuals (Todd et al. 2007; Fleissner et al. 2009b; Roca et al. 2010; Dettmann et al. 2012). On a larger scale, fungal heterokaryons have also been employed in biotechnological applications (Stuart 1997)—a strategy which might still hold much untapped potential for production strain improvement.

1.2 Vegetative Fusion

1.2.1 Germling Fusion

Fusion of genetically identical conidia or conidial germlings during colony initiation has been described in numerous filamentous ascomycete species. A literature survey concluded that fusion at this early developmental stage has been reported in more than 70 species covering more

M. Weichert • A. Fleißner (✉)
Institut für Genetik, Technische Universität
Braunschweig, Braunschweig, Germany
e-mail: a.fleissner@tu-bs.de

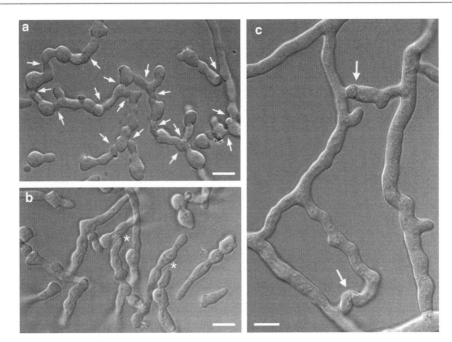

Fig. 1.1 Vegetative cell-to-cell fusion in *Neurospora crassa*. (**a**) Germinating conidia in close proximity readily undergo chemotropic interactions, establish physical contact and finally fuse with each other (*arrows*). As a result, a cellular network is formed, which further develops into the mycelial colony. (**b**) Deletion of the gene *so* (*soft*) completely abolishes intercellular communication. In contrast to wild-type, mutant cells grow in parallel and do not attract each other. Random contact (*asterisks*) of Δ*so* germlings does not result in fusion. (**c**): In the interior part of a mature mycelium, hyphae interconnect via fusion bridges (*arrows*). Hyphal fusion promotes the distribution of nutrients and organelles within the colony

than 20 genera (Roca et al. 2005a). Fusion of spores and/or germlings often involves the formation of specialized hyphal structures termed *c*onidial *a*nastomosis *t*ubes (CATs). These cell protrusions possess characteristic features distinguishing them from germ tubes. They are of significantly thinner diameter, exhibit only limited extension over shorter distances, and do not branch. In contrast to germ tubes, which typically avoid each other, CATs exhibit positive autotropism, resulting in physical contact and fusion (Roca et al. 2005b). Germling fusion is a highly orchestrated multistep process, which comprises a carefully regulated succession of cellular events. These include fusion competence, cell-cell signaling, directed growth, cell adhesion, cell wall breakdown, and finally plasma membrane merger. Formation of the fusion pore allows cytoplasmic mixing and the exchange of organelles, including nuclei, between the two fusion partners. Repeated fusion events within a spore population eventually results in the merger of most individuals into a supracellular network (Fig. 1.1a), which further develops into a mycelial colony.

1.2.2 Hyphal Fusion

Within the mycelium, hyphae fuse via short branches, resulting in hyphal cross-connections, which stimulate the interconnectedness within the fungal colony (Fig. 1.1c). The two processes of germling and hyphal fusion appear to be mechanistically related, since so far all mutant strains affected in conidial fusion were, when tested, also deficient in hyphal fusion. The formation of fusion bridges between hyphae typically occurs in the colony interior, while the periphery of the colony lacks hyphal fusion, indicating differences in fusion competence. Determinants establishing this competence are so far unknown. Fusion-competent hyphal branches actively attract each other via an unknown interhyphal signaling mechanism and re-orientate their growth direction. After establishing physical contact, polar (apical) tip extension turns into non-polar (isotropic) growth, resulting in the swelling of the opposing hyphal tips. Upon growth arrest, the formation of a fusion pore is initiated, which is followed by the exchange of cytoplasm and organelles between the fused hyphal compartments (Hickey et al. 2002). During pore formation, the Spitzenkörper of the two fusion hyphae remain associated with the fusion point. In actively growing hyphae, this organelle serves as a vesicle supply center, which controls the transfer of vesicles to the plasma membrane of the growing tips. Its presence at the forming fusion pore suggests specific functions during the fusion process, such as the controlled delivery of vesicles containing cell wall degrading enzymes or the plasma membrane fusion machinery.

1.3 The Role of Hyphal Fusion in Colony Establishment and Development

It has long been proposed that vegetative fusion and network formation in filamentous fungi promote genetic and nutritional exchange, thereby increasing the fitness and competitiveness of mycelial colonies (Ward 1888; Rayner 1996). Recently, these hypotheses were supported by experimental evidence. In *N. crassa*, significant translocation of nutrients takes place within wild-type mycelia, thereby allowing the expansion of the colony also in heterogeneous environments. In mutants deficient in hyphal fusion, however, translocation of nutrients within individual growing mycelial units is significantly reduced (Simonin et al. 2012). The growth front of these fusion mutants typically appears frayed and is comprised of individual hyphae of various lengths. In contrast, wild-type colonies possess very even boundaries, suggesting an equal resource distribution to individual hyphae (Fleissner et al. 2005; Richard et al. 2012). The reduced translocation capacity of fusion mutants

appears to be not only caused directly by the lack of cross-connection within the colony. In *N. crassa* wild-type hyphae, cytoplasmic flow can occur at high speeds of up to 60 µm/s (Lew 2005). In fusion mutants, the cytoplasmic streaming appears to be restricted (Roper et al. 2013), suggesting that pressure differences between fusing hyphae contribute to the driving forces of cellular flows.

Interestingly, exchange of resources and nuclei between different *N. crassa* individuals via hyphal fusion was only observed in germlings or young undifferentiated colonies, which had not developed clear morphological differences between the colony interior and the periphery, but not between mature mycelia (Simonin et al. 2012). Similarly, in arbuscular mycorrhizal fungi, anastomosis appears to be restricted to certain stages of the life cycle, suggesting that developmental windows of fungal cooperation exist (Purin and Morton 2013). As an outcome of cooperation, fused *N. crassa* germlings gained a competitive advantage, depending on the habitat structure and the initial spore density. While in wild-type strains the number of spores positively correlates with growth rates, in a fusion mutant, increased spore concentrations were neutral or even constituted a disadvantage (Richard et al. 2012). Merging individuals into one functional unit might support coordinated growth and resource consumption, while a lack of fusion creates a population of independent individuals competing for space and nutrients.

In addition to forming functional units, anastomosis between non-clonal individuals results in heterokaryotic mycelia. Such an exchange of genetic material can be followed by parasexual recombination, which is considered to contribute to genetic richness in many fungal species. Genera for which this type of reproduction has been reported include the parasex model *Aspergillus*, but also important plant pathogens, such as *Magnaporthe*, *Alternaria*, or *Colletotrichum* (Stewart et al. 2013; da Silva Franco et al. 2011; Noguchi et al. 2006). During parasex, karyogamy in heterokaryotic hyphae results in instable polyploid nuclei, allowing the recombination of genetical material. During the following mitotic divisions, surplus chromosomes are successively and randomly lost, until the stable, original ploidity state is restored. Parasexual recombination is nowadays considered to be a driving force of fungal evolution, diversity, fitness, and the emergence of new virulent strains (Roper et al. 2011; Baskarathevan et al. 2012; Clay and Schardl 2002). In *A. nidulans*, isogenic diploid strains undergoing parasexual recombination had a significantly higher fitness than haploid isolates (Schoustra et al. 2007), illustrating the potential of parasex to promote competitiveness and adaptation.

Anastomosis formation appears not only to enable parasex, but also to contribute to the maintenance of genetic diversity in heterokaryotic mycelia by mixing flows. In *N. crassa* fusion mutants, different nucleotypes segregate out into colony sectors and the genetic richness is quickly lost, while wild-type colonies maintain their nuclear diversity (Roper et al. 2013).

In nature, however, hyphal fusion between different colonies bears the risk of transmitting infectious agents or resource plundering by aggressive genotypes (Glass and Kaneko 2003). Anastomosis appears to be the main route of mycovirus transmission and has been used experimentally as well as a strategy to control phytopathogenic species (Dawe and Nuss 2013; Ghabrial and Suzuki 2009). An extreme example of anastomosis followed by taking over by one genotype is illustrated in *Fusarium oxysporum*. Here, fusion between germlings results in the quick degradation of the nucleus in the receptive hyphae, thereby extinguishing the recipient's genotype (Ruiz-Roldan et al. 2010).

In order to prevent such deleterious effects of anastomosis formation, many species restrict non-self-fusion by a genetic vegetative incompatibility system. If genetically incompatible hyphae fuse, the merged compartments are quickly sealed and cell death is induced (Glass and Dementhon 2006). This mechanism can efficiently restrict mycovirus transmission, as shown, for example, in the chestnut blight fungus *Cryphonectria parasitica* (Choi et al. 2012). As indicated by the term "vegetative incompatibility", this control mechanism is differentially active at the various developmental stages of the fungal life cycle, thereby providing temporal windows for genetic mixing. In *N. crassa,* for example, no incompatibility is observed during the sexual fusion of mating partners, while vegetative fusion of strains of opposite mating type is quickly followed by an efficient cell death response (Glass and Kuldau 1992). In the plant pathogen *Colletotrichum lindemuthianum*, heterokaryon incompatibility is suppressed during fusion of germinating conidia, allowing a window of cooperation even during vegetative growth. As a result, heterokaryotic strains are formed, whose conidia grow into cultures with phenotypic features different from the parental strains. Temporal suppression of heterokaryon incompatibility in order to allow parasex therefore provides an asexual way of increasing genetic diversity (Ishikawa et al. 2012).

Exceptional temporal suppression of incompatibility during the fusion between conidia or hyphae of different species has also been discussed as a route for horizontal gene transfer (Richards et al. 2011; van der Does and Rep 2007). So far, the contribution of interspecies fusion to the generation of genetic diversity is not assessable. While an early study indicated that fusion between different species is uncommon (Köhler 1930), conidial anastomoses between two different Colletotrichum species have been documented. Strains originating from these hybrids exhibited intermediate phenotypes, suggesting that they were recombinants (Roca et al. 2004). Therefore, further investigations will be essential to determine the significance of interspecies germling and/or hyphal fusion in creating genetic richness in fungi.

In addition to promoting the exchange of nutrients and genetic material, anastomosis formation can serve various other functions during fungal growth and development, such as the formation of specific three-dimensional structures or hyphal repair. For example, in nematode trapping fungi, such as *Arthrobotrys oligospora*, the individual loops of the net-like traps are formed by the fusion of a hyphal branch and a small peg emerging from the same parental hypha (Nordbring-Hertz et al. 1989). In *Trichophyton*,

1 Anastomosis and Heterokaryon Formation

injured and empty hyphal compartments can be repaired by anastomoses between two intrahyphal formed hyphae, emerging from the intact neighboring compartments (Farley et al. 1975). Similar observations were made for various arbuscular mycorrhizal fungi, in which fusion of regenerating hyphae promotes the integrity of the mycorrhizal mycelial network (de la Providencia et al. 2005).

1.4 Molecular Basis of Anastomosis Formation

In recent years, anastomosis formation in filamentous fungi, specifically *N. crassa*, has advanced as one of various model systems for studying the molecular basis of eukaryotic cell fusion (Aguilar et al. 2013). While cell-cell mergers are essential for the growth and development of most eukaryotic organisms, the molecular bases mediating this process are mostly unknown. In genetic approaches, numerous *N. crassa* genes involved in hyphal and germling fusion have been identified (Table 1.1). Further analysis revealed an unusual signaling mechanism mediating communication and tropic growth of fusion germlings. During their interaction, the two fusion germlings appear to coordinately switch between two physiological stages, indicated by the alternating plasma membrane recruitment of two proteins. Strikingly, these switches occur in exact antiphase in the partner cells. This unexpected finding prompted the working hypothesis that the cells coordinately alternate between signal sending and receiving, thereby establishing a kind of cell-cell "dialog" (Fleissner et al. 2009b). Mathematic modeling indicated that by undergoing these alternating switches, genetically identical cells can communicate via a single chemoattractant-receptor pair, while avoiding the otherwise inevitable risk of self-excitation (Goryachev et al. 2012). The two alternating physiological stages are characterized by the presence of either the MAP kinase MAK-2 or the SO protein at the plasma membrane of the cell tips (Fleissner et al. 2009b) (Fig. 1.2). Further analysis of additional fusion mutants suggests an intricate signaling network controlling this intriguing cellular behavior. Emerging principles partaking in this network include two different MAP kinase cascades, reactive oxygen species (ROS) generating systems, cell polarity factors and Ca^{2+} signaling systems (Table 1.1).

1.4.1 The SO Protein

The SO (SOFT) protein is only conserved in filamentous ascomycete fungi (Fleissner et al. 2005). Loss-of-function mutations of the *so* gene result in a pleiotropic phenotype, including shortened aerial hyphae, an altered conidiation pattern, and female sterility. No tropic interactions related to fusion are observed in germlings or hyphae (Fleissner et al. 2005) (Fig. 1.1b). In non-interacting hyphae, SO resides in the cytoplasm. In germlings undergoing tropic reactions related to fusion, the protein is recruited to the plasma membrane of the fusion tips, where it accumulates in complexes of about 300 nm. This membrane recruitment alternates between the two fusion cells, with a phase of 6–12 min (Fig. 1.2). So far, the molecular function of SO remains obscure. The protein contains a WW domain, an amino acid motif associated with protein–protein interactions (Ilsley et al. 2002). Besides mediating germling fusion, SO appears to function also in maintenance of hyphal integrity, since it strongly accumulates at septal plugs, which prevent cytoplasmic loss in damaged, aging, and dying hyphae (Fleissner and Glass 2007).

The function in anastomosis formation appears to be conserved in fungi of different life styles, including the saprophytic dung fungus *Sordaria macrospora*, the phytopathogenic species *Alternaria alternata* and *F. oxysporum,* and the mutualistic endophyte *Epichloe festucae* (Craven et al. 2008; Prados Rosales and Di Pietro 2008; Charlton et al. 2012; Engh et al. 2007). Interestingly, SO appears to be involved in fungus–host interactions. In *A. alternata*, the SO homolog is essential for pathogenesis, while in *E. festucae*, deletion of the *so* gene disturbs the mutualistic fungus–plant interaction and results in a parasitic development. In contrast,

Table 1.1 Genes involved in vegetative cell-to-cell fusion in *N. crassa*

Locus	Gene	Features/functions	References
Cell fusion/fertility pathway			
nrc-1	NCU06182	MAP kinase kinase kinase	Dettmann et al. (2012); Fu et al. (2011); Maerz et al. (2008); Pandey et al. (2004)
mek-2	NCU04612	MAP kinase kinase	Dettmann et al. (2012); Fu et al. (2011); Maerz et al. (2008)
mak-2	NCU02393	MAP kinase	Dettmann et al. (2012); Fleissner et al. (2009b); Fu et al. (2011); Li et al. (2005); Maerz et al. (2008); Pandey et al. (2004)
Cell wall integrity pathway			
mik-1	NCU02234	MAP kinase kinase kinase	Fu et al. (2011); Maerz et al. (2008)
amek-1	NCU06419	MAP kinase kinase	Fu et al. (2011); Maerz et al. (2008)
mak-1	NCU09842	MAP kinase	Fu et al. (2011); Maerz et al. (2008)
Further factors involved in MAPK signaling			
sol/ham-1	NCU02794	WW domain protein[a]	Fleissner et al. (2005, 2009b); Fu et al. (2011)
cot-1	NCU07296	NDR kinase	Seiler et al. (2006); Dettmann et al. (2012); Maerz et al. (2008)
rac-1	NCU02160	Small GTPase	Araujo-Palomares et al. (2011); Fu et al. (2011)
hym-1	NCU03576	NDR scaffolding kinase	Dettmann et al. (2012)
ham-7	NCU00881	GPI-anchored cell wall sensor protein	Fu et al. (2011); Leeder et al. (2013); Maddi et al. (2012)
cdc-42	NCU06454	Rho-type GTPase	Araujo-Palomares et al. (2011); Read et al. (2012)
cdc-24	NCU06067	Guanine exchange factor	Araujo-Palomares et al. (2011); Read et al. (2012)
Transcription factors			
pp-1	NCU00340	Homolog of yeast *STE12*	Leeder et al. (2013); Li et al. (2005)
rcm-1	NCU06842	Homolog of yeast *SSN6*	Aldabbous et al. (2010)
rco-1	NCU06205	Homolog of yeast *TUP1*	Aldabbous et al. (2010); Fu et al. (2011)
adv-1	NCU07392	Clock-controlled gene involved in circadian rhythms	Fu et al. (2011)
ada-3	NCU02896	Central transcription factor	Fu et al. (2011)
snf5	NCU00421	SWI/SNF complex subunit	Fu et al. (2011)
Asm-1	NCU01414	Ascospore maturation protein, transcriptional activator	Leeder et al. (2013)
STRIPAK complex			
ham-2	NCU03727	Homolog of *FAR11* (yeast) and STRIP1/2 (mammals)	Fu et al. (2011); Glass et al. (2004); Xiang et al. (2002); Dettmann et al. (2013)
ham-3	NCU08741	Homolog of *FAR8* (yeast) and striatin (mammals), regulatory subunit B''' of PP2A[b]	Fleissner et al. (2008); Fu et al. (2011); Glass et al. (2004); Simonin et al. (2010); Dettmann et al. (2013)
ham-4	NCU00528	Homolog of *FAR9/10* (yeast) and SLMAP (mammals), coiled-coil and FHA domains[c]	Fleissner et al. (2008); Fu et al. (2011); Glass et al. (2004); Simonin et al. (2010); Dettmann et al. (2013)
mob-3	NCU07674	Phocein (striatin-binding protein, kinase co-activator protein)	Fu et al. (2011; Maerz et al. (2009); Dettmann et al. (2013)

(continued)

1 Anastomosis and Heterokaryon Formation

Table 1.1 (continued)

Locus	Gene	Features/functions	References
ppg-1	NCU06563	Catalytic subunit C of PP2A (PP2A-C)[b]	Fu et al. (2011); Dettmann et al. (2013)
pp2A-A	NCU00488	Regulatory subunit A of PP2A (PP2A-A)[b]	Dettmann et al. (2013)
Redox signaling			
nox-1	NCU02110	NADPH oxidase	Read et al. (2012)
nor-1	NCU07850	NADPH oxidase regulator	Read et al. (2012)
bem1	NCU06593	MAP kinase activator	Schurg et al. (2012)
arg-15	NCU05622	Acetylornithine-glutamate transacetylase	Palma-Guerrero et al. (2013)
lao-1	NCU05113	Laccase (L-ascorbate oxidase-like)	Read et al. (2012)
Calcium signaling			
ham-10	NCU02833	C2 domain-containing protein[d], vesicular trafficking, endocytosis	Fu et al. (2011)
cse-1	NCU04379	Calcium sensor protein	Palma-Guerrero et al. (2013)
pik1	NCU10397	Phosphatidylinositol 4-kinase	Palma-Guerrero et al. (2013)
nfh-2	NCU02806	14-3-3 Protein	Palma-Guerrero et al. (2013)
Plasma membrane fusion			
Prm1	NCU09337	Potential plasma membrane fusogen	Fleissner et al. (2009a)
Further factors			
gpip-1	NCU06663	GPI-anchored protein, cell wall biogenesis/integrity	Bowman et al. (2006)
gpig-1	NCU09757	GPI-anchored protein, cell wall biogenesis/integrity	Bowman et al. (2006)
gpip-2	NCU07999	GPI-anchored protein, cell wall biogenesis/integrity	Bowman et al. (2006)
gpip-3	NCU06508	GPI-anchored protein, cell wall biogenesis/integrity	Bowman et al. (2006)
gpit-1	NCU05644	GPI-anchored protein, cell wall biogenesis/integrity	Bowman et al. (2006)
arp2	NCU07171	Actin-related protein (ARP)	Roca et al. (2010)
arpc3	NCU09572	ARP2/3 protein complex (21 kDa subunit)	Roca et al. (2010)
n.n.	NCU01918	ARP2/3 protein complex (20 kDa subunit)	Roca et al. (2010)
ham-5	NCU01789	WD40 domain protein[a]	Aldabbous et al. (2010); Fu et al. (2011)
ham-6	NCU02767	Transmembrane protein	Fu et al. (2011)
ham-8	NCU02811	Transmembrane protein	Fu et al. (2011)
ham-9	NCU07389	Cytoplasmic protein with PH and SAM domains[a]	Fu et al. (2011)
amph-1	NCU01069	Amphiphysin with BAR domain[c]	Fu et al. (2011)
pkr1	NCU00506	V-ATPase assembly factor in the ER membrane	Fu et al. (2011)
mss-4	NCU02295	Phosphatidylinositol-4-phosphate 5-kinase	Mahs et al. (2012)
bud-6	NCU08468	Actin-interacting protein	Lichius et al. (2012)
spa-2	NCU03115	Polarisome scaffolding protein	Lichius et al. (2012)
bni-1	NCU01431	Formin	Lichius et al. (2012)

(continued)

Table 1.1 (continued)

Locus	Gene	Features/functions	References
sec15	NCU00117	Exocyst complex component	Palma-Guerrero et al. (2013)
sec22	NCU06708	Protein transporter	Palma-Guerrero et al. (2013)
n.n.	NCU06362	GTPase activating protein	Palma-Guerrero et al. (2013)
nik-2	NCU01833	Nonidentical kinase	Palma-Guerrero et al. (2013)
spr-7	NCU07159	Secreted subtilisin-like serine protease	Palma-Guerrero et al. (2013)
ham-11	NCU04732	Transmembrane protein	Leeder et al. (2013)
ham-12	NCU03960	Transmembrane protein	Leeder et al. (2013)

This overview summarizes all the, so far, known loci which have a role during intercellular communication and fusion in *N. crassa*. The loci are listed with their gene-specific numbers and are arranged according to their (proposed) function in different signaling pathways and complexes

n.n. no name

[a]Domains involved in protein–protein interactions: WW, WD40, PH = Pleckstrin homology, SAM = sterile alpha motif

[b]PP2A heterotrimer consisting of PPG-1, PP2A-A and HAM-3

[c]FHA domain: phosphopeptide recognition

[d]C2 domain: calcium-dependent phospholipid binding

[e]BAR = Bin-amphiphysin-Rvs, interaction with lipid bilayers

pathogenicity of the wilt pathogen *F. oxysporum* is not significantly affected in a *so* mutant, suggesting that not the ability to form anastomosis contributes to pathogenicity, but rather additional, unrelated functions of this still obscure protein.

1.4.2 MAP Kinase Signaling

A MAP kinase module homologous to the pheromone response MAP kinase cascade of the unicellular yeast *Saccharomyces cerevisiae* mediates anastomosis formation in filamentous ascomycete species. In *N. crassa*, mutants affected in MAK-2 or its upstream MAP kinase kinase MEK-2 or its MAP kinase kinase kinase NRC-1 display pleiotropic defects, including reduced hyphal growth rate, shortened aerial hyphae, derepressed conidiation, female sterility, and ascospore lethality. Most importantly, the deletion mutants are incapable of undergoing germling or hyphal fusion. In addition, a clear temporal correlation between MAK-2 activation and germling fusion in a spore population exists (Pandey et al. 2004). In *A. nidulans*, the respective MAP kinase MpkB and the MAPKKK SteC are unable to form self and non-self-anastomosis (Wei et al. 2003; Jun et al. 2011). Similarly, a *F. oxysporum* mutant of the MAK-2 homologous

kinase Fmk1 is deficient in hyphal network formation via vegetative fusion (Prados Rosales and Di Pietro 2008).

In *N. crassa*, the three kinases of the MAK-2 module appear to interact at the plasma membrane of fusion germling tips. Complexes of about 300 nm size containing the three kinases accumulate in a dynamic fashion in exact antiphase to the SO protein complexes (see above) (Fig. 1.2). While a physical interaction of these two complexes seems unlikely, a functional relationship exists. In strains carrying a chemically inhibitable variant of MAK-2, SO is not released from the plasma membrane after MAK-2 inhibition and eventually appears in both cells stable but unfocussed at the plasma membrane (Fleissner et al. 2009b).

So far, the upstream factors controlling the MAK-2 module are unknown, including the postulated chemoattractant and its cognate receptor. One target of MAK-2 in *N. crassa* appears to be PP-1, a transcription factor similar to the Fus3 target Ste12 of *S. cerevisiae*. Deletion of the *pp-1* gene causes a phenotype comparable to that of Δ*nrc-1*, Δ*mek-2,* and Δ*mak-2*, including the defects in hyphal fusion (Pandey et al. 2004; Li et al. 2005). Comparison of the transcriptional profiles of Δ*pp-1* and of the mutant carrying the inhibitable MAK-2 variant with the wild-type

1 Anastomosis and Heterokaryon Formation

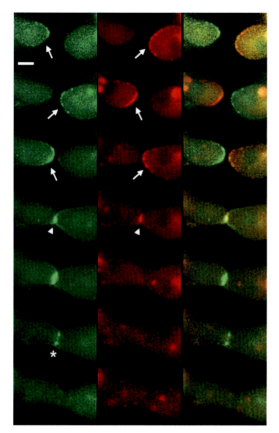

Fig. 1.2 Germling fusion in *Neurospora crassa* involves the oscillatory recruitment of the MAP kinase MAK-2 and the SO protein to interacting cell tips. Conidia of homokaryotic strains expressing either MAK-2-GFP or dsRED-SO were cultured together to form a heterokaryotic mycelium via anastomosis. Spores originating from this heterokaryon were used for live-cell imaging analysis of germling fusion. During tropic cell–cell interactions, MAK-2 and SO are recruited from the cytoplasm to the tips of germlings in a highly dynamic, alternating fashion. MAK-2 and SO localize in perfect anti-phase and do not co-localize at the plasma membrane (0–7 min). After cell–cell contact, both proteins accumulate at the site of future pore formation and transiently co-localize (10 min). While SO readily disappears from the contact point, MAK-2 remains localized there and associates with the fusion pore (25 min)

include *so*, *ham-6*, *ham-7*, *ham-8*, *ham-9*, *Prm-1*, *nox-1*, *nor-1*, *adv-1*, *rco-1*, *mek-1*, *mak-1*, *ppg-1*, *gpip-1*, *gpip-2*, and *mob-3* (Leeder et al. 2013).

The recruitment of MAK-2 to the plasma membrane of fusion tips in *N. crassa* suggests additional functions to its role in transcriptional regulation. One direct target of the yeast homolog Fus3 is the formin Bni1 (Matheos et al. 2004). Formins regulate the polymerization of actin, indicating that the MAP kinase also regulates the cytoskeleton and directed cell growth. A similar function is likely during germling fusion in filamentous fungi. In *N. crassa*, actin is essential for the formation and the directed growth of fusion tips, while in contrast, microtubules are dispensable (Roca et al. 2010). Moreover, the polarisome components BUD-6, SPA-2, and BNI-1, which are known to regulate the actin cytoskeleton during polarized growth in fungi, are strongly recruited to the cell tips of fusing germlings and hyphae (Lichius et al. 2012).

The second MAP kinase cascade essential for anastomosis formation consists of homologs of the yeast cell wall integrity pathway. In *N. crassa*, the three kinases of this module (MIK-1, MEK-1, and MAK-1) are essential for vegetative cell fusion (Pandey et al. 2004; Maerz et al. 2008). These functions appear to be conserved, since the homologous MAP kinase gene *MGV1* of *F. graminearum* is essential for heterokaryon formation (Hou et al. 2002). MAK-1 activity is regulated by the GPI-anchored cell wall sensor protein HAM-7, which is essential for germling fusion and female fertility in *N. crassa* (Fu et al. 2011; Maerz et al. 2008; Maddi et al. 2012).

Phenotypic similarities between the pleiotropic phenotypes of mutants of the MAK-1 and MAK-2 cascades suggest a functional overlap between these two pathways (Maerz et al. 2008; Park et al. 2008). In addition, both MAPK modules appear to be linked to COT-1, a member of the NDR protein kinase family that is, together with the Ste20 protein kinase POD-6, required for hyphal tip extension and coordinated branch formation (Maerz et al. 2008; Yarden et al. 1992; Seiler et al. 2006). Deletion of *mak-2* suppresses the severe growth defects of a temperature-sensitive *cot-1* mutant, which is accompanied by

identified pre-known and novel factors involved in germling fusion. The expression of *mak-2* is decreased in cells treated with the chemical inhibitor 1NM-PP1, suggesting a positive feedback loop, in which MAK-2 promotes its own expression. Other genes involved in germling and hyphal fusion, which appear to be controlled by PP-1,

reduced protein kinase A activity levels in Δ*mak-2*. In return, hyphal fusion is restored in both Δ*nrc-1* and Δ*mak-2* strains, in which COT-1 exhibits reduced activity levels. In addition, COT-1 acts as a potential negative regulator of the MAK-1 MAPK module (Maerz et al. 2008).

Both the COT-1-POD-6 complex and the MAK-2 MAP kinase module share a common scaffolding protein, HYM-1. This scaffold promotes COT-1 activity and appears to be essential for MAK-2 activation by bridging the three kinases of the MAP kinase module. HYM-1 might therefore act as an adaptor to coordinate cell-cell signaling and the control of cell polarity, resulting in directed growth (Dettmann et al. 2012).

1.4.3 ROS Signaling

Based on studies using various fungal species, ROS producing complexes have emerged as additional factors controlling anastomosis formation. In the gray mold, *Botrytis cinerea*, the NADPH oxidase BcNoxA, and the Nox regulator BcNoxR are essential for CAT fusion. Loss of a second NADPH oxidase-encoding gene, *BcNoxB*, results in a reduction of the fusion frequency (Roca et al. 2012). Comparable results were obtained for the homologs in *N. crassa* and *E. festucae* (Read et al. 2012; Takemoto et al. 2011). Interestingly, both ROS generating enzymes of *B. cinerea* are also involved in pathogenic development; however here, BcNoxB is more important. Both processes require BcNoxR, which regulates both NADPH oxidases (Segmuller et al. 2008; Roca et al. 2012). While CAT fusion is readily observed in axenic cultures of the gray mold, it appears to be fully suppressed when spores germinate on plant surfaces, suggesting that in this fungus, pathogenic growth and fusion are two alternative but mutually exclusive developmental choices (Roca et al. 2012; Fleißner 2012).

Recent studies in the plant symbiotic fungus *E. festucae* revealed the scaffolding protein BemA to be part of Nox complexes (Takemoto et al. 2011). In yeast, the homologous protein Bem1 links the pheromone response MAP kinase module to the cell polarity machinery (Leeuw et al. 1995). Together these data suggest that in filamentous fungi Bem1 might integrate ROS and MAP kinase signaling with directed growth. In support of this hypothesis, BEM1 in *N. crassa* is required for efficient MAK-2 signaling during vegetative cell fusion. In the absence of *bem1*, germling interactions become extremely instable. The SO protein strongly mislocalizes in these interactions, in a pattern reminiscent to the one observed after chemical inhibition of MAK-2 (Schurg et al. 2012). In addition, MAK-2 is sensitive to ROS, and the addition of H_2O_2 to axenic cultures readily results in phosphorylation of the kinase (Maerz et al. 2008). Analysis of the Nox complexes in *E. festucae* also identified the small GTPase Rac1 and the Guanine nucleotide exchange factor Cdc24 as components of this protein complex (Takemoto et al. 2011). Interestingly, in *N. crassa* homologs of both proteins are also involved in germling fusion (Read et al. 2012; Fu et al. 2011). Based on these observations, a model can be postulated, in which an extracellular signal results in the activation of ROS generating systems, which in turn activate the MAP kinase cascade. As a result, the polar growth of the fusion tip is redirected towards the partner cell (Fig. 1.3).

A recent study in *N. crassa* further supports the hypothesis that regulated levels of NADPH oxidase activity are crucial for germling communication. Quantitative trait analysis identified the *arg-15* gene as a factor promoting intercellular signaling. The gene encodes an acetylornithine-glutamate transacetylase that potentially participates in the degradation of ROS (Palma-Guerrero et al. 2013).

1.4.4 The STRIPAK Complex

Several recent studies revealed that a number of components of the mammalian STRIPAK complex are also conserved in filamentous fungi (Simonin et al. 2010; Goudreault et al. 2009; Bernhards and Poggeler 2011; Bloemendal et al. 2012; Dettmann et al. 2013). In *S. macrospora* and *N. crassa*, this multiprotein complex controls vegetative cell-cell fusion and fruiting body

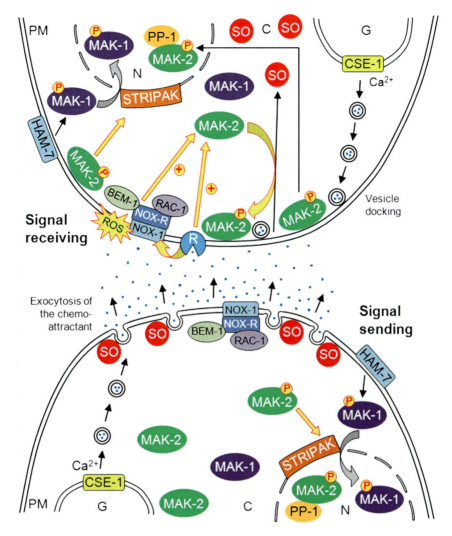

Fig. 1.3 Working model for intercellular communication in filamentous fungi. A model is proposed in which both cells rapidly alternate between signal sending and receiving. Signal emission involves the recruitment of SO from the cytoplasm (C) to the cell tip. Vesicles containing the, so far unknown, chemoattractant are transported from the Golgi apparatus (G) to the plasma membrane (PM). This transport possibly involves calcium signaling (Palma-Guerrero et al. 2013). SO might mediate the fusion of these vesicles with the membrane, so that the signaling molecules are released in a pulse-like manner into the extracellular space (Read et al. 2012). The signal gradient is detected by the opposing cell via a hypothetical receptor protein (R). Signal-receptor binding triggers the activation of MAK-2. This activation involves or is promoted by an NADPH oxidase protein complex consisting of NOX-1, NOR-1, RAC-1, and BEM1. Activation of NOX-1 causes the production of spatially and temporally restricted spikes of reactive oxygen species (ROS), which in turn activate MAK-2. MAK-2 activation occurs at the plasma membrane. Here, the kinase controls polarization of the cytoskeleton, resulting in polarized and directed growth. In addition, the kinase shuttles into the nucleus, where it phosphorylates its downstream targets, including the transcription factor PP-1, which controls transcriptional regulation of genes required for cell–cell communication and fusion. MAK-2 also mediates the nuclear (N) accumulation of the activated MAP kinase MAK-1 via the STRIPAK complex. Apart from its proposed role during cell wall remodeling, MAK-1 might be required for fusion competence. Basal levels of MAK-2 activity are probably also present in the signal sending partner cell. Thus, shuttling of both MAK-1 and MAK-2 into the nucleus might occur here also, to maintain the competence to undergo chemotropic interactions. The switching between signal sending and receiving probably requires intricate positive and negative feedback loops, which are so far not understood

development. In mutants lacking individual components of this complex, hyphal fusion is strongly or even completely disturbed and sexual propagation is blocked.

In *N. crassa*, the complex consists of six proteins: HAM-2/STRIP, HAM-3/striatin, HAM-4/SLMAP, MOB-3/phocein, PPG-1/PP2A-C, and PP2A-A. HAM-2 and HAM-3 are essential for the assembly of the complex, which is located at the nuclear envelope. The STRIPAK complex seems to be directly linked to the MAP kinase cascades involved in anastomosis formation. The accumulation of MAK-1 in the nucleus depends on a functional STRIPAK complex and phosphorylation of MOB-3 by the kinase MAK-2 influences the nuclear accumulation of MAK-1. Although this nuclear accumulation appears to be dispensable for normal germling fusion, these data hint to a clear interconnectedness of the STRIPAK complex and the MAP kinase cascades (Dettmann et al. 2013).

1.4.5 Calcium

In mammalian cells, calcium binds striatin via calmodulin. Therefore, it had been suggested that in filamentous fungi, calcium signaling is required for hyphal fusion, involving the interaction between calmodulin and HAM-3 (Simonin et al. 2010). Further support for a role of calcium came from a recent finding that on growth media reduced in calcium spores of *N. crassa* germinate normally but undergo no interactions and do not form the typical germling network (Palma-Guerrero et al. 2013). The group of identified fusion mutants includes strains affected in calcium-binding proteins. For example, HAM-10 contains a C2-domain, which might link the protein to lipids in a calcium-dependent manner (Fu et al. 2011).

Deletion of *cse-1*, which encodes a homolog of the neuronal calcium sensor protein in vertebrates, strongly reduces intercellular communication between *N. crassa* germlings. Two potential interaction partners of CSE-1 are also required for chemotropic interactions in *N. crassa*. PIK-1 is a phosphatidylinositol kinase involved in secretion from the Golgi to the plasma membrane, while NFH-2 possibly transports CSE-1 from the nucleus to the cytoplasm. Mutations of *pik-1* or *nfh-2* result in phenotypic defects reminiscent of Δ*cse-1*. The three proteins might function together in the secretion pathway, where they could regulate the exocytosis of a chemoattractant and/or receptor in response to calcium signaling (Palma-Guerrero et al. 2013).

1.4.6 Other Factors

The phenomenon of anastomosis is common among many different fungal species, indicating a conserved signaling mechanism. However, the nature of the secreted chemoattractant(s) and the corresponding receptor(s) mediating cell-cell communication remain so far unknown. Cell-to-cell signaling and fusion between different fungi is usually not observed, suggesting that the chemoattractants are species-specific (Köhler 1930). Small secreted peptides could certainly meet this required specificity.

So far, pheromones are the only signaling peptides known to mediate intercellular fungal communication. Mating between *S. cerevisiae* cells requires the secretion of a- and α-pheromones and their binding to their cognate G-protein-coupled membrane receptors (Baskarathevan et al. 2012). Discrimination between equivalent potential mating partners is promoted by the activity of the aspartyl protease Bar1p. This enzyme is secreted by a-cells and degrades the α-pheromone secreted by the opposing α-cell, thereby fine-tuning the cellular interactions (Barkai et al. 1998).

The mating pheromones and their receptors are dispensable for vegetative cell fusion in *N. crassa*, indicating the presence of a different chemoattractant/receptor system (Glass and Fleißner 2006). A recent study suggested that the secreted serine protease SPR-7 degrades the putative signaling peptide during germling fusion (Palma-Guerrero et al. 2013). Intercellular communication between Δ*spr-7* cells was increased compared to wild-type, resulting in networks of multiple germlings interacting with each other. However, when the mutant was confronted with a wild-type cell, normal levels of chemotropism were observed (Palma-Guerrero et al. 2013). Similar to yeast mating, activity of this secreted protease might sharpen the signaling gradient and promote fusion partner selection.

Germlings of several fusion mutants of *N. crassa* also undergo no interactions with wild-type cells, suggesting that cell-cell signaling requires an interdependent coordination of fusion partner behavior (Pandey et al. 2004; Fleissner et al. 2005; Simonin et al. 2010; Palma-Guerrero et al. 2013; Leeder et al. 2013). In contrast, the Δ*ham-11* mutant, which is also blind-to-self, still communicates with wild-type germlings. Interactions of such mixed pairs involve normal dynamics of MAK-2 and SO oscillation, suggesting that the protein might be required for the initiation of intercellular communication (Leeder et al. 2013). The *ham-11* gene encodes a transmembrane protein of unknown molecular function conserved among Sordariomycetes.

1.4.7 Plasma Membrane Merger

So far, the vast majority of isolated fusion mutants in filamentous fungi is affected in developmental stages preceding the actual membrane fusion, such as fusion competence, cell–cell communication, or directed growth. The molecular process of plasma membrane merger, however, remains one of the great mysteries in modern life sciences. Only very few proteins, which directly mediate plasma membrane merger, have been isolated in non-fungal model systems (Aguilar et al. 2013). Interestingly, these factors all appear to be species-specific, raising questions concerning the evolution of plasma membrane fusion.

The only known factor, which acts after cell-cell contact and cell wall break down in hyphal fusion, is the integral membrane protein PRM1. Its function has first been described in yeast mating, where about 50 % of Δ*prm-1* mating pairs failed to fuse. Electron micrographs revealed that in these pairs the cell wall is successfully degraded, but the subsequent membrane merger fails (Heiman and Walter 2000). A respective *N. crassa* mutant displays a comparable pheno-type during germling fusion (Fleissner et al. 2009a). Interestingly, sexual mating partner fusion is also deficient in these isolates, suggesting that PRM-1 is part of a general membrane fusion machinery. *N. crassa* is a heterothallic fungus, in which fertilization occurs after the fusion of a female fruiting body born receptive hypha with a male conidium. In Δ*prm-1* pairings, also 50 % of these fusion events fail. In addition, the mutant is completely sterile as female or male, suggesting additional PRM-1 functions after fertilization. In yeast, additional mutants exhibiting similar phenotypes to Δ*prm-1* have been isolated. Analysis of these factors in filamentous fungi is still awaiting.

1.5 Applications of Anastomosis Formation and Heterokaryons

1.5.1 Genetic Analysis

The observation of parasexual recombination in some asexual fungi opened the field of classical genetic analysis also for this group of organisms.

During parasex two individuals undergo vegetative fusion resulting in the formation of a heterokaryon. Within this heterokaryon, nuclei can fuse giving rise to unstable diploid nuclei, in which mitotic crossing over occurs. During induced haploidization, the surplus chromosomes are lost in a random manner until a stable haploid stage is restored, resulting in segregation of whole chromosomes (Pontecorvo 1956).

Parasexual recombination was employed to establish a genetic map in *Aspergillus niger*, one of the most important production organism in biotechnological applications. By random mutation and UV irradiation, more than 100 independent mutants were isolated, including strains carrying auxotrophies, defects in nitrogen assimilation, or fungicide resistances. Using these mutations as markers in genetic analysis revealed eight linkage groups. Strains carrying different markers for all chromosomes were constructed, serving as master strains in mapping studies (Debets et al. 1990, 1993).

Besides enabling genetic mapping, heterokaryons allow further genetic analysis, including complementation and dominance tests (Todd et al. 2007). Complementation tests determine if two independent mutants are affected in the same gene. In this case, formation of a heterokaryon of these two strains will not result in complementation of the phenotype. If however two different genomic loci are affected, the two strains will complement each other's defects, resulting in a wild-type appearance. Most filamentous ascomycete fungi are usually haploid, allowing the direct observation of defects caused by a mutation. To test, however, if a mutant is dominant requires the presence of an additional wild-type gene copy, which can be provided in a heterokaryon.

1.5.2 Combination of Strain Features in Heterokaryons

Vegetative hyphal fusion also provides a valuable tool for combining certain strain features in one test culture. For example, the subcellular localization of a protein can be determined by its fusion to a fluorescent marker. The combination of strains expressing differentially tagged candidate proteins in a heterokaryon allows easy co-localization tests, without the need to construct additional strains. This strategy has been proven successful even in unbalanced heterokaryons, with no pressure for maintenance of the heterokaryotic stage (Fleissner et al. 2009b; Roca et al. 2010; Dettmann et al. 2012) (Fig. 1.2). If stable heterokaryons are desired, strains carrying different auxotrophic markers can be mixed (Wada et al. 2013). Subsequent cultivation on minimal medium will prevent the formation of homokaryotic sectors. By choosing specific combinations of auxotrophic markers, the heterokaryon can be balanced and the ratio of the two types of nuclei can be controlled. Another example of his experimental approach is the combination of differentially tagged proteins in a heterokaryon for protein–protein interaction studies, such as co-immunoprecipitation experiments (Maerz et al. 2009).

A great advantage of filamentous fungi as experimental organisms is the relative ease with which gene targeting approaches can be conducted. This strategy is however hampered by the limited number of selectable markers. Therefore, strategies of marker recycling have been adapted. A common system is the bacteriophage Cre-*lox*P system, in which expression of the Cre recombinase leads to the excision of DNA fragments flanked by the *lox*P sites. A major drawback of this strategy in filamentous fungi is the lack of tightly regulated promoters to control expression of the recombinase gene. A recent study solved this problem by simply inducing excision through the delivery of Cre via anastomosis (Zhang et al. 2013). Transformants of the chestnut blight fungus *C. parasitica* carrying *lox*P-flanked marker genes were fused with isolates constitutively expressing the Cre recombinase. As a result the markers were efficiently removed. The same strategy was successfully applied to the insect pathogen *Metarhizium robertsii*, proving its general potential (Zhang et al. 2013).

1.5.3 Heterokaryon Formation in Biotechnological Applications

The ability of filamentous fungi to form heterokaryons by anastomosis has not only been exploited in basic research, but is also applied in biotechnological processes. For example, a patent describes the heterologous expression of dimeric proteins by a heterokaryotic production strain, in which the two types of nuclei present in the host strain encode the two subunits of the heterologous protein (Stuart 1997). Stability of the heterokaryon is ensured by the use of auxotrophic markers. This methodology allows the simple combination of various different subunits, without the need to construct a new production strain for each combination. Once a collection of strains expressing the various individual subunits is established, all possible combinations can be produced by simply picking and combining of the two desired strains.

Parasexual crosses have been successfully applied to improve strains by combining desirable features of individual haploid isolates. Example applications involve the production of various enzymes or citric acid in *A. niger*. For example, invertase production by *A. niger* increased 5–18 times in diploid strains obtained from parasexual crosses of selected haploid isolates (Montiel-Gonzalez et al. 2002). Similarly, xylanase productivity of the same fungus could be significantly increased by the parasexual formation of diploids from haploid mutants (Loera and Cordova 2003). Combination of independent mutations, which both led to increased chymosin production in *A. niger*, by parasexual recombination resulted in progeny with even higher productivity of the heterologous enzyme (Bodie et al. 1994).

Parasexual recombination was also used to improve bioremediation of the insecticide dichlorodiphenyltrichloroethane (DDT) by *Fusarium solani*. Natural soil isolates of this fungus produced different DDT degrading enzymes with different bioremediation abilities. Mixed cultures of these different isolates degraded the toxins more efficiently, suggesting synergistic functions of the different enzyme sets of the individual strains. Parasexual crosses of these isolates resulted in progeny carrying these beneficial features combined in individual strains (Mitra et al. 2001).

Hyphae of the entomopathogenic fungus *Beauvaria bassiana* frequently fuse during colonization of host insects (Guerri-Agullo et al. 2010). Heterokaryosis was observed when compatible mutant strains were mixed and cultured on insect cadavers (Castrillo et al. 2004). This ability for heterokaryon formation via anastomosis has been recently exploited for strain improvement. Co-cultivation of two different isolates on agar plates resulted in strains exhibiting new beneficial phenotypic features including increased thermotolerance (Kim et al. 2011).

1.6 Conclusion

While we are just at the beginning of understanding the role and molecular basis of anastomosis, it is already becoming apparent that hyphal fusion constitutes a fascinating unique biological mechanism. Studying hyphal fusion will promote our knowledge on cell–cell signaling, cell fusion, the functioning of biological networks, and fungal biology in general.

Exploiting anastomosis and heterokaryon formation in experimental set-ups still holds much potential in research and applications.

Acknowledgments Work in our group is supported by funding from the German Research Foundation (FL 706/1-2) and the European Union [PITN-GA-2013-607963].

We thank David Havlik, Stephanie Herzog, and Marcel Schumann for critical reading of the manuscript.

References

Aguilar PS, Baylies MK, Fleißner A, Helming L, Inoue N, Podbilewicz B, Wang H, Wong M (2013) Genetic basis of cell-cell fusion mechanisms. Trends Genet 29(7):427–437. doi:10.1016/j.tig.2013.01.011

Aldabbous MS, Roca MG, Stout A, Huang IC, Read ND, Free SJ (2010) The ham-5, rcm-1 and rco-1 genes regulate hyphal fusion in Neurospora crassa. Microbiology 156(Pt 9):2621–2629. doi:10.1099/mic.0.040147-0

Araujo-Palomares CL, Richthammer C, Seiler S, Castro-Longoria E (2011) Functional characterization and cellular dynamics of the CDC-42 - RAC - CDC-24 module in Neurospora crassa. PLoS One 6(11):e27148. doi:10.1371/journal.pone.0027148

Barkai N, Rose MD, Wingreen NS (1998) Protease helps yeast find mating partners. Nature 396(6710):422–423. doi:10.1038/24760

Baskarathevan J, Jaspers MV, Jones EE, Cruickshank RH, Ridgway HJ (2012) Genetic and pathogenic diversity of Neofusicoccum parvum in New Zealand vineyards. Fung Biol 116(2):276–288. doi:10.1016/j.funbio.2011.11.010, S1878-6146(11)00234-0 [pii]

Bernhards Y, Poggeler S (2011) The phocein homologue SmMOB3 is essential for vegetative cell fusion and sexual development in the filamentous ascomycete *Sordaria macrospora*. Curr Genet 57:133–149. doi:10.1007/s00294-010-0333-z

Bloemendal S, Bernhards Y, Bartho K, Dettmann A, Voigt O, Teichert I, Seiler S, Wolters DA, Poggeler S, Kuck U (2012) A homologue of the human STRIPAK complex controls sexual development in fungi. Mol Microbiol 84(2):310–323. doi:10.1111/j.1365-2958.2012.08024.x

Bodie EA, Armstrong GL, Dunn-Coleman NS (1994) Strain improvement of chymosin-producing strains of Aspergillus niger var. awamori using parasexual recombination. Enzym Microb Technol 16(5):376–382

Bowman SM, Piwowar A, Al Dabbous M, Vierula J, Free SJ (2006) Mutational analysis of the glycosylphosphatidylinositol (GPI) anchor pathway demonstrates that GPI-anchored proteins are required for cell wall biogenesis and normal hyphal growth in Neurospora crassa. Eukaryot Cell 5(3):587–600

Castrillo LA, Griggs MH, Vandenberg JD (2004) Vegetative compatibility groups in indigenous and mass-released strains of the entomopathogenic fungus Beauveria bassiana: likelihood of recombination in the field. J Invertebr Pathol 86(1–2):26–37. doi:10.1016/j.jip.2004.03.009, S0022201104000485 [pii]

Charlton ND, Shoji JY, Ghimire SR, Nakashima J, Craven KD (2012) Deletion of the fungal gene soft disrupts mutualistic symbiosis between the grass endophyte Epichloe festucae and the host plant. Eukaryot Cell 11(12):1463–1471. doi:10.1128/EC.00191-12, EC.00191-12 [pii]

Choi GH, Dawe AL, Churbanov A, Smith ML, Milgroom MG, Nuss DL (2012) Molecular characterization of vegetative incompatibility genes that restrict hypovirus transmission in the chestnut blight fungus Cryphonectria parasitica. Genetics 190(1):113–127. doi:10.1534/genetics.111.133983, genetics.111.133983 [pii]

Clay K, Schardl C (2002) Evolutionary origins and ecological consequences of endophyte symbiosis with grasses. Am Nat 160(Suppl 4):S99–S127. doi:10.1086/342161

Craven KD, Velez H, Cho Y, Lawrence CB, Mitchell TK (2008) Anastomosis is required for virulence of the fungal necrotroph Alternaria brassicicola. Eukaryot Cell 7(4):675–683. doi:10.1128/EC.00423-07, EC.00423-07 [pii]

da Silva Franco CC, de Sant' Anna JR, Rosada LJ, Kaneshima EN, Stangarlin JR, De Castro-Prado MA (2011) Vegetative compatibility groups and parasexual segregation in Colletotrichum acutatum isolates infecting different hosts. Phytopathology 101(8):923–928. doi:10.1094/PHYTO-12-10-0327

Dawe AL, Nuss DL (2013) Hypovirus molecular biology: from Koch's postulates to host self-recognition genes that restrict virus transmission. Adv Virus Res 86:109–147. doi:10.1016/B978-0-12-394315-6.00005-2, B978-0-12-394315-6.00005-2 [pii]

de la Providencia IE, de Souza FA, Fernandez F, Delmas NS, Declerck S (2005) Arbuscular mycorrhizal fungi reveal distinct patterns of anastomosis formation and hyphal healing mechanisms between different phylogenic groups. New Phytol 165(1):261–271. doi:10.1111/j.1469-8137.2004.01236.x, NPH1236 [pii]

Debets AJ, Swart K, Bos CJ (1990) Genetic analysis of Aspergillus niger: isolation of chlorate resistance mutants, their use in mitotic mapping and evidence for an eighth linkage group. Mol Gen Genet 221(3):453–458

Debets F, Swart K, Hoekstra RF, Bos CJ (1993) Genetic maps of eight linkage groups of Aspergillus niger based on mitotic mapping. Curr Genet 23(1):47–53

Dettmann A, Illgen J, März S, Schürg T, Fleißner A, Seiler S (2012) The NDR kinase scaffold HYM1/MO25 is essential for MAK2 MAP kinase signaling in Neurospora crassa. PLoS Genet 8(9):e1002950

Dettmann A, Heilig Y, Ludwig S, Schmitt K, Illgen J, Fleissner A, Valerius O, Seiler S (2013) HAM-2 and HAM-3 are central for the assembly of the Neurospora STRIPAK complex at the nuclear envelope and regulate nuclear accumulation of the MAP kinase MAK-1 in a MAK-2-dependent manner. Mol Microbiol 90(4):796–812. doi:10.1111/mmi.12399

Engh I, Wurtz C, Witzel-Schlomp K, Zhang HY, Hoff B, Nowrousian M, Rottensteiner H, Kuck U (2007) The WW domain protein PRO40 is required for fungal fertility and associates with Woronin bodies. Eukaryot Cell 6(5):831–843. doi:10.1128/EC.00269-06, EC.00269-06 [pii]

Farley JF, Jersild RA, Niederpruem DJ (1975) Origin and ultrastructure of intra-hyphal hyphae in Trichophyton terrestre and T. rubrum. Arch Microbiol 106(3):195–200

Fleißner A (2012) Hyphal fusion. In: Perez-Martin J, Di Pietro A (eds) Morphogenesis and pathogenicity in fungi, vol 22. Springer, Berlin, pp 43–59

Fleissner A, Glass NL (2007) SO, a protein involved in hyphal fusion in Neurospora crassa, localizes to septal plugs. Eukaryot Cell 6(1):84–94. doi:10.1128/EC.00268-06, EC.00268-06 [pii]

Fleissner A, Sarkar S, Jacobson DJ, Roca MG, Read ND, Glass NL (2005) The so locus is required for vegetative cell fusion and postfertilization events in Neurospora crassa. Eukaryot Cell 4(5):920–930. doi:10.1128/EC.4.5.920-930.2005, 4/5/920 [pii]

Fleissner A, Simonin AR, Glass NL (2008) Cell fusion in the filamentous fungus, Neurospora crassa. Methods Mol Biol 475:21–38. doi:10.1007/978-1-59745-250-2_2

Fleissner A, Diamond S, Glass NL (2009a) The Saccharomyces cerevisiae PRM1 homolog in Neurospora crassa is involved in vegetative and sexual cell fusion events but also has postfertilization functions. Genetics 181(2):497–510. doi:10.1534/genetics.108.096149, genetics.108.096149 [pii]

Fleissner A, Leeder AC, Roca MG, Read ND, Glass NL (2009b) Oscillatory recruitment of signaling proteins to cell tips promotes coordinated behavior during cell fusion. Proc Natl Acad Sci U S A 106(46):19387–19392. doi:10.1073/pnas.0907039106, 0907039106 [pii]

Fu C, Iyer P, Herkal A, Abdullah J, Stout A, Free SJ (2011) Identification and characterization of genes required for cell-to-cell fusion in Neurospora crassa. Eukaryot Cell 10(8):1100–1109. doi:10.1128/EC.05003-11, EC.05003-11 [pii]

Ghabrial SA, Suzuki N (2009) Viruses of plant pathogenic fungi. Annu Rev Phytopathol 47:353–384. doi:10.1146/annurev-phyto-080508-081932

Glass NL, Dementhon K (2006) Non-self recognition and programmed cell death in filamentous fungi. Curr Opin Microbiol 9(6):553–558. doi:10.1016/j.mib.2006.09.001, S1369-5274(06)00150-0 [pii]

Glass N, Fleißner A (2006) Re-wiring the network: understanding the mechanism and function of anastomosis in filamentous ascomycete fungi. In: Kues U, Fischer R (eds) The mycota. I. Growth, differentiation and sexuality. Springer, Berlin, pp 123–139

Glass NL, Kaneko I (2003) Fatal attraction: nonself recognition and heterokaryon incompatibility in filamentous fungi. Eukaryot Cell 2(1):1–8

Glass NL, Kuldau GA (1992) Mating type and vegetative incompatibility in filamentous ascomycetes. Annu

Rev Phytopathol 30:201–224. doi:10.1146/annurev.py.30.090192.001221

Glass NL, Rasmussen C, Roca MG, Read ND (2004) Hyphal homing, fusion and mycelial interconnectedness. Trends Microbiol 12(3):135–141

Goryachev AB, Lichius A, Wright GD, Read ND (2012) Excitable behavior can explain the "ping-pong" mode of communication between cells using the same chemoattractant. Bioessays 34:259–266. doi:10.1002/bies.201100135

Goudreault M, D'Ambrosio LM, Kean MJ, Mullin MJ, Larsen BG, Sanchez A, Chaudhry S, Chen GI, Sicheri F, Nesvizhskii AI, Aebersold R, Raught B, Gingras AC (2009) A PP2A phosphatase high density interaction network identifies a novel striatin-interacting phosphatase and kinase complex linked to the cerebral cavernous malformation 3 (CCM3) protein. Mol Cell Proteomics 8(1):157–171. doi:10.1074/mcp.M800266-MCP200, M800266-MCP200 [pii]

Guerri-Agullo B, Gomez-Vidal S, Asensio L, Barranco P, Lopez-Llorca LV (2010) Infection of the red palm weevil (Rhynchophorus ferrugineus) by the entomopathogenic fungus Beauveria bassiana: a SEM study. Microsc Res Tech 73(7):714–725. doi:10.1002/jemt.20812

Heiman MG, Walter P (2000) Prm1p, a pheromone-regulated multispanning membrane protein, facilitates plasma membrane fusion during yeast mating. J Cell Biol 151(3):719–730

Hickey PC, Jacobson D, Read ND, Louise Glass NL (2002) Live-cell imaging of vegetative hyphal fusion in Neurospora crassa. Fungal Genet Biol 37(1):109–119, S108718450200035X [pii]

Hou Z, Xue C, Peng Y, Katan T, Kistler HC, Xu JR (2002) A mitogen-activated protein kinase gene (MGV1) in *Fusarium graminearum* is required for female fertility, heterokaryon formation, and plant infection. Mol Plant Microbe Interact 15(11):1119–1127. doi:10.1094/MPMI.2002.15.11.1119

Ilsley JL, Sudol M, Winder SJ (2002) The WW domain: linking cell signalling to the membrane cytoskeleton. Cell Signal 14(3):183–189, S0898656801002364 [pii]

Ishikawa FH, Souza EA, Shoji JY, Connolly L, Freitag M, Read ND, Roca MG (2012) Heterokaryon incompatibility is suppressed following conidial anastomosis tube fusion in a fungal plant pathogen. PLoS One 7(2):e31175. doi:10.1371/journal.pone.0031175, PONE-D-11-17981 [pii]

Jun SC, Lee SJ, Park HJ, Kang JY, Leem YE, Yang TH, Chang MH, Kim JM, Jang SH, Kim HG, Han DM, Chae KS, Jahng KY (2011) The MpkB MAP kinase plays a role in post-karyogamy processes as well as in hyphal anastomosis during sexual development in Aspergillus nidulans. J Microbiol 49(3):418–430. doi:10.1007/s12275-011-0193-3

Kim JS, Skinner M, Gouli S, Parker BL (2011) Generating thermotolerant colonies by pairing Beauveria bassiana isolates. FEMS Microbiol Lett 324(2):165–172. doi:10.1111/j.1574-6968.2011.02400.x

Köhler E (1930) Zur Kenntnis der vegetativen Anastomosen der Pilze (II. Mitteilung). Planta 10:495–522

Leeder AC, Jonkers W, Li J, Glass NL (2013) Early colony establishment in neurospora crassa requires a MAP kinase regulatory network. Genetics 195(3):883–898. doi:10.1534/genetics.113.156984, genetics.113.156984 [pii]

Leeuw T, Fourest-Lieuvin A, Wu C, Chenevert J, Clark K, Whiteway M, Thomas DY, Leberer E (1995) Pheromone response in yeast: association of Bem1p with proteins of the MAP kinase cascade and actin. Science 270(5239):1210–1213

Lew RR (2005) Mass flow and pressure-driven hyphal extension in Neurospora crassa. Microbiology 151(Pt 8):2685–2692. doi:10.1099/mic.0.27947-0, 151/8/2685 [pii]

Li D, Bobrowicz P, Wilkinson HH, Ebbole DJ (2005) A mitogen-activated protein kinase pathway essential for mating and contributing to vegetative growth in Neurospora crassa. Genetics 170(3):1091–1104. doi:10.1534/genetics.104.036772, genetics.104.036772 [pii]

Lichius A, Yanez-Gutierrez ME, Read ND, Castro-Longoria E (2012) Comparative live-cell imaging analyses of SPA-2, BUD-6 and BNI-1 in Neurospora crassa reveal novel features of the filamentous fungal polarisome. PLoS One 7(1):e30372. doi:10.1371/journal.pone.0030372, PONE-D-11-16914 [pii]

Loera O, Cordova J (2003) Improvement of xylanase production by a parasexual cross between Aspergillus niger strains. Braz Arch Biol Technol 46(2):177–181

Maddi A, Dettman A, Fu C, Seiler S, Free SJ (2012) WSC-1 and HAM-7 are MAK-1 MAP kinase pathway sensors required for cell wall integrity and hyphal fusion in Neurospora crassa. PLoS One 7(8):e42374. doi:10.1371/journal.pone.0042374, PONE-D-12-11850 [pii]

Maerz S, Ziv C, Vogt N, Helmstaedt K, Cohen N, Gorovits R, Yarden O, Seiler S (2008) The nuclear Dbf2-related kinase COT1 and the mitogen-activated protein kinases MAK1 and MAK2 genetically interact to regulate filamentous growth, hyphal fusion and sexual development in *Neurospora crassa*. Genetics 179(3):1313–1325. doi:10.1534/genetics.108.089425, genetics.108.089425 [pii]

Maerz S, Dettmann A, Ziv C, Liu Y, Valerius O, Yarden O, Seiler S (2009) Two NDR kinase-MOB complexes function as distinct modules during septum formation and tip extension in *Neurospora crassa*. Mol Microbiol 74(3):707–723. doi:10.1111/j.1365-2958.2009.06896.x, MMI6896 [pii]

Mahs A, Ischebeck T, Heilig Y, Stenzel I, Hempel F, Seiler S, Heilmann I (2012) The essential phosphoinositide kinase MSS-4 is required for polar hyphal morphogenesis, localizing to sites of growth and cell fusion in Neurospora crassa. PLoS One 7(12):e51454. doi:10.1371/journal.pone.0051454

Matheos D, Metodiev M, Muller E, Stone D, Rose MD (2004) Pheromone-induced polarization is dependent on the Fus3p MAPK acting through the formin Bni1p.

J Cell Biol 165(1):99–109. doi:10.1083/jcb.200309089, jcb.200309089 [pii]

Mitra J, Mukherjee PK, Kale SP, Murthy NB (2001) Bioremediation of DDT in soil by genetically improved strains of soil fungus Fusarium solani. Biodegradation 12(4):235–245

Montiel-Gonzalez AM, Fernandez FJ, Viniegra-Gonzalez G, Loera O (2002) Invertase production on solid-state fermentation by Aspergillus niger strains improved by parasexual recombination. Appl Biochem Biotechnol 102–103(1–6):63–70

Noguchi MT, Yasuda N, Fujita Y (2006) Evidence of genetic exchange by parasexual recombination and genetic analysis of pathogenicity and mating type of parasexual recombinants in rice blast fungus, magnaporthe oryzae. Phytopathology 96(7):746–750. doi:10.1094/PHYTO-96-0746

Nordbring-Hertz B, Friman E, Veenhuis M (1989) Hyphal fusion during initial stages of trap formation in Arthrobotrys oligospora. Antonie Van Leeuwenhoek 55(3):237–244

Palma-Guerrero J, Hall CR, Kowbel D, Welch J, Taylor JW, Brem RB, Glass NL (2013) Genome wide association identifies novel loci involved in fungal communication. PLoS Genet 9(8):e1003669. doi:10.1371/journal.pgen.1003669, PGENETICS-D-13-00985 [pii]

Pandey A, Roca MG, Read ND, Glass NL (2004) Role of a mitogen-activated protein kinase pathway during conidial germination and hyphal fusion in *Neurospora crassa*. Eukaryot Cell 3(2):348–358

Park G, Pan S, Borkovich KA (2008) Mitogen-activated protein kinase cascade required for regulation of development and secondary metabolism in Neurospora crassa. Eukaryot Cell 7(12):2113–2122. doi:10.1128/EC.00466-07, EC.00466-07 [pii]

Pontecorvo G (1956) The parasexual cycle in fungi. Annu Rev Microbiol 10:393–400. doi:10.1146/annurev.mi.10.100156.002141

Prados Rosales RC, Di Pietro A (2008) Vegetative hyphal fusion is not essential for plant infection by *Fusarium oxysporum*. Eukaryot Cell 7(1):162–171. doi:10.1128/EC.00258-07, EC.00258-07 [pii]

Purin S, Morton JB (2013) Anastomosis behavior differs between asymbiotic and symbiotic hyphae of Rhizophagus clarus. Mycologia 105(3):589–602. doi:10.3852/12-135, 12-135 [pii]

Rayner ADM (1996) Interconnectedness and individualism in fungal mycelia. A century of mycology. Cambridge University Press, Cambridge

Read ND, Goryachev AB, Lichius A (2012) The mechanistic basis of self-fusion between conidial anastomosis tubes during fungal colony initiation. Fung Biol Rev 26:1–11

Richard F, Glass NL, Pringle A (2012) Cooperation among germinating spores facilitates the growth of the fungus, Neurospora crassa. Biol Lett 8(3):419–422. doi:10.1098/rsbl.2011.1141, rsbl.2011.1141 [pii]

Richards TA, Leonard G, Soanes DM, Talbot NJ (2011) Gene transfer into the fungi. Fung Biol Rev 25:98–110

Roca MG, Davide LC, Davide LM, Mendes-Costa MC, Schwan RF, Wheals AE (2004) Conidial anastomosis fusion between *Colletotrichum* species. Mycol Res 108(Pt 11):1320–1326

Roca M, Read ND, Wheals AE (2005a) Conidial anastomosis tubes in filamentous fungi. FEMS Microbiol Lett 249(2):191–198. doi:10.1016/j.femsle.2005.06.048, S0378-1097(05)00438-6 [pii]

Roca MG, Arlt J, Jeffree CE, Read ND (2005b) Cell biology of conidial anastomosis tubes in *Neurospora crassa*. Eukaryot Cell 4(5):911–919. doi:10.1128/EC.4.5.911-919.2005, 4/5/911 [pii]

Roca MG, Kuo HC, Lichius A, Freitag M, Read ND (2010) Nuclear dynamics, mitosis, and the cytoskeleton during the early stages of colony initiation in *Neurospora crassa*. Eukaryot Cell 9(8):1171–1183. doi:10.1128/EC.00329-09, EC.00329-09 [pii]

Roca MG, Weichert M, Siegmund U, Tudzynski P, Fleissner A (2012) Germling fusion via conidial anastomosis tubes in the grey mould Botrytis cinerea requires NADPH oxidase activity. Fung Biol 116(3):379–387. doi:10.1016/j.funbio.2011.12.007, S1878-6146(11)00252-2 [pii]

Roper M, Ellison C, Taylor JW, Glass NL (2011) Nuclear and genome dynamics in multinucleate ascomycete fungi. Curr Biol 21(18):R786–R793. doi:10.1016/j.cub.2011.06.042, S0960-9822(11)00715-9 [pii]

Roper M, Simonin A, Hickey PC, Leeder A, Glass NL (2013) Nuclear dynamics in a fungal chimera. Proc Natl Acad Sci U S A 110(32):12875–12880. doi:10.1073/pnas.1220842110, 1220842110 [pii]

Ruiz-Roldan MC, Kohli M, Roncero MI, Philippsen P, Di Pietro A, Espeso EA (2010) Nuclear dynamics during germination, conidiation, and hyphal fusion of *Fusarium oxysporum*. Eukaryot Cell 9(8):1216–1224. doi:10.1128/EC.00040-10, EC.00040-10 [pii]

Schoustra SE, Debets AJ, Slakhorst M, Hoekstra RF (2007) Mitotic recombination accelerates adaptation in the fungus Aspergillus nidulans. PLoS Genet 3(4):e68.doi:10.1371/journal.pgen.0030068,06-PLGE-RA-0505R2 [pii]

Schurg T, Brandt U, Adis C, Fleissner A (2012) The Saccharomyces cerevisiae BEM1 homologue in Neurospora crassa promotes co-ordinated cell behaviour resulting in cell fusion. Mol Microbiol 86(2):349–366. doi:10.1111/j.1365-2958.2012.08197.x

Segmuller N, Kokkelink L, Giesbert S, Odinius D, van Kan J, Tudzynski P (2008) NADPH oxidases are involved in differentiation and pathogenicity in Botrytis cinerea. Mol Plant Microbe Interact 21(6):808–819. doi:10.1094/MPMI-21-6-0808

Seiler S, Vogt N, Ziv C, Gorovits R, Yarden O (2006) The STE20/germinal center kinase POD6 interacts with the NDR kinase COT1 and is involved in polar tip extension in Neurospora crassa. Mol Biol Cell 17(9):4080–4092. doi:10.1091/mbc.E06-01-0072, E06-01-0072 [pii]

Simonin AR, Rasmussen CG, Yang M, Glass NL (2010) Genes encoding a striatin-like protein (*ham-3*) and a forkhead associated protein (*ham-4*) are required for hyphal fusion in *Neurospora crassa*. Fungal Genet Biol 47(10):855–868. doi:10.1016/j.fgb.2010.06.010, S1087-1845(10)00122-2 [pii]

Simonin A, Palma-Guerrero J, Fricker M, Glass NL (2012) Physiological significance of network organization in fungi. Eukaryot Cell 11(11):1345–1352. doi:10.1128/EC.00213-12, EC.00213-12 [pii]

Stewart JE, Thomas KA, Lawrence CB, Dang H, Pryor BM, Timmer LM, Peever TL (2013) Signatures of recombination in clonal lineages of the citrus brown spot pathogen, Alternaria alternata sensu lato. Phytopathology 103(7):741–749. doi:10.1094/PHYTO-08-12-0211-R

Stuart WD (1997) Heterologous dimeric proteins produced in heterokaryons.

Takemoto D, Kamakura S, Saikia S, Becker Y, Wrenn R, Tanaka A, Sumimoto H, Scott B (2011) Polarity proteins Bem1 and Cdc24 are components of the filamentous fungal NADPH oxidase complex. Proc Natl Acad Sci U S A 108(7):2861–2866. doi:10.1073/pnas.1017309108, 1017309108 [pii]

Todd RB, Davis MA, Hynes MJ (2007) Genetic manipulation of Aspergillus nidulans: heterokaryons and diploids for dominance, complementation and haploidization analyses. Nat Protoc 2(4):822–830. doi:10.1038/nprot.2007.113, nprot.2007.113 [pii]

van der Does HC, Rep M (2007) Virulence genes and the evolution of host specificity in plant-pathogenic fungi. Mol Plant Microbe Interact 20(10):1175–1182. doi:10.1094/MPMI-20-10-1175

Wada R, Jin FJ, Koyama Y, Maruyama JI, Kitamoto K (2013) Efficient formation of heterokaryotic sclerotia in the filamentous fungus Aspergillus oryzae. Appl Microbiol Biotechnol 98:325. doi:10.1007/s00253-013-5314-y

Ward H (1888) A lily disease. Ann Bot 2(7):319–382

Wei H, Requena N, Fischer R (2003) The MAPKK kinase SteC regulates conidiophore morphology and is essential for heterokaryon formation and sexual development in the homothallic fungus Aspergillus nidulans. Mol Microbiol 47(6):1577–1588, 3405 [pii]

Xiang Q, Rasmussen C, Glass NL (2002) The ham-2 locus, encoding a putative transmembrane protein, is required for hyphal fusion in Neurospora crassa. Genetics 160(1):169–180

Yarden O, Plamann M, Ebbole DJ, Yanofsky C (1992) cot-1, a gene required for hyphal elongation in Neurospora crassa, encodes a protein kinase. EMBO J 11(6):2159–2166

Zhang DX, Lu HL, Liao X, St Leger RJ, Nuss DL (2013) Simple and efficient recycling of fungal selectable marker genes with the Cre-loxP recombination system via anastomosis. Fungal Genet Biol 61:1–8. doi:10.1016/j.fgb.2013.08.013, S1087-1845(13)00153-9 [pii]

Induction of the Sexual Cycle in Filamentous Ascomycetes

2

Jos Houbraken and Paul S. Dyer

2.1 Introduction

The vast majority of fungal species are able to undergo sexual reproduction involving the formation of sexual spores via meiosis. This form of reproduction is thought to have many evolutionary advantages, and where present offers a valuable laboratory tool for experimental genetic analysis (Dyer and Paoletti 2005; Aanen and Hoekstra 2007; Lee et al. 2010). However, a surprisingly high minority of fungal species (approximately 20 %) are only known to reproduce by asexual (mitotic) means. This includes many species of industrial importance, notably several *Aspergillus* and *Penicillium* species (Dyer and Paoletti 2005; Dyer and O'Gorman 2011). The aim of this chapter is twofold. Firstly, to describe how sexual cycles may be induced in filamentous fungi including 'fastidious' species (Kwon-Chung and Sugui 2009), which require very specific conditions, and sexually 'recalcitrant' species, where a sexual cycle might not yet have been reported. Secondly,

to describe methods by which sexual progeny can be isolated, with further possibilities suggested for progeny analysis. The chapter will focus on filamentous ascomycete species (*Pezizomycotina*), which represent one of the largest groups in the fungal kingdom, and in particular on members of industrial importance. A final section is included to briefly describe ways in which the sexual cycle can be exploited for purposes including gene identification and localization, strain improvement, and gene complementation.

2.2 Sexual Reproduction and Breeding Systems in Filamentous Ascomycete Fungi

In 1820, microscopic sexual structures in fungi were reported for the first time in a culture of *Syzygites megalocarpus* (*Zygomycota, Mucorales*) (Ehrenberg 1820; Idnurm 2011). In 1904, Blakeslee showed that *S. megalocarpus* is a self-fertile species and also observed the existence of different "sexes" ('mating types') in *Rhizopus stolonifer* (reported as *Rhizopus nigricans*). Based on his findings he introduced the terms 'homothallism' for self-fertile (or self-compatible) and 'heterothallism' for self-incompatible (or obligate outcrossing) individuals (Blakeslee 1904). Thus, by definition, individuals of homothallic fungal species can complete the sexual cycle without the

J. Houbraken, Ph.D. (✉)
Department of Applied and Industrial Mycology,
CBS-KNAW Fungal Biodiversity Centre,
Uppsalalaan 8, 3584 CT Utrecht, The Netherlands
e-mail: j.houbraken@cbs.knaw.nl

P.S. Dyer, B.A., M.A., Ph.D.
School of Life Sciences, University of Nottingham,
Nottingham, NG7 2RD, UK
e-mail: Paul.Dyer@nottingham.ac.uk

need for a mating partner, whereas individuals of heterothallic species require a mating partner of compatible mating type for sexual reproduction to occur. However, it is important to note that homothallic species are not restricted to self-fertility, as individuals normally retain the ability to out-cross under suitable conditions (Dyer et al. 1992; Burnett 2003; Cavindera and Trail 2012). In the case of heterothallic pezizomycete species there are normally only two mating types present. By convention these are now termed *MAT1-1* and *MAT1-2*, although for some species alternative established terminology such as matA and mat*a* (e.g. in *Neurospora*) or plus '+' and minus '−' (e.g. in *Podospora*) are used (Dyer et al. 1992; Turgeon and Yoder 2000). Also some heterothallic species have an additional layer of sexual compatibility superimposed on the mating type. Individuals can either be male (M), female (F), or hermaphrodites (MF) with respect to their ability to form sexual mating structures such as ascogonia, protoperithecia, microconidia, and spermatia (Debuchy et al. 2010). For a cross to be successful not only must isolates of opposite mating type be present, but also one mating partner must be able to act as a male and the other as a female. However, this system appears to be restricted to certain taxonomic groupings such as *Fusarium* and *Magnaporthe* species (Gordon 1961; Takan et al. 2012). A third reproductive strategy occurring in filamentous fungi was later described, called 'pseudohomothallism' (or secondary homothallism) (Dodge 1957). Ascomycetous pseudohomothallic species (e.g. *Neurospora tetrasperma*, *Podospora anserina*) develop four spored asci in which most ascospores contain two nuclei, one of each mating type (Raju and Perkins 1994). A typical binucleate ascospore germinates to form a self-fertile mycelium due to the fact that the arising heterokaryic hyphae contains both matA and mat*a* nuclei (Pöggeler 2001).

In heterothallic pezizomycete species sexually compatible haploid strains are normally morphologically indistinguishable and not differentiated into male and female sexes, the mating partners are instead distinguished only by their mating type (Bistis 1998). Extensive studies over the past 25 years have revealed that both sexual identity and later stages of sexual development are controlled in fungi by so called 'mating-type' (*MAT*) genes (Debuchy et al. 2010). In heterothallic pezizomycete fungi there is usually only one *MAT* locus at which between one and three *MAT* genes may be present. The DNA sequence of the *MAT* genes of isolates of opposite mating type, together with other non-coding sequence present at the *MAT* locus, is highly dissimilar although the regions flanking the *MAT* locus are highly conserved (Debuchy and Turgeon 2006). Because of the dissimilarity in sequence between the opposite *MAT* loci, they are referred to as 'idiomorphs' instead of alleles to emphasize that the sequences at the same locus are highly dissimilar (Metzenberg and Glass 1990). By convention, *MAT1-1* mating-type isolates contain a *MAT* idiomorph which includes a *MAT1-1* gene encoding a protein with a motif called the alpha box, whereas *MAT1-2* mating-type isolates contain a *MAT* idiomorph which includes a *MAT1-2* gene encoding a regulatory protein with a DNA-binding domain of the high mobility group (HMG) family. These two idiomorphs are designated *MAT1-1* and *MAT1-2*, respectively. Where more than one *MAT* gene is present in an idiomorph, each gene within an idiomorph is indicated by the idiomorph symbol followed by a dash and a number, e.g., *MAT1-2-1* (Turgeon and Yoder 2000). The organization of *MAT* genes differs in homothallic species, where both alpha box *MAT1-1* and HMG domain *MAT1-2* genes are normally present in the same individual; this co-occurrence appears to confer the ability to self-fertilize. The alpha and HMG genes can be tightly linked at a single *MAT* locus or be present at two distinct *MAT* loci within the genome (Paoletti et al. 2007; Debuchy et al. 2010).

2.2.1 Use of *MAT* Genes as Diagnostic Tools for Induction of a Sexual Cycle

The discovery and characterization of mating-type genes from a diverse range of fungi has provided a major advance in the ability to induce sexual cycles of chosen fungal species in vitro.

Fungal *MAT* genes were first identified using molecular methods from the yeast *Saccharomyces cerevisiae* (Astell et al. 1981) and since then have been characterized from numerous filamentous ascomycete species (Debuchy and Turgeon 2006; Debuchy et al. 2010).

Although the *MAT* genes and idiomorphs show considerable sequence divergence overall, it has nevertheless been possible to identify partly conserved alpha box and HMG domain regions of the *MAT1-1* and *MAT1-2* genes, respectively, encoding homologous 65–80 amino acid regions of the MAT proteins. This has allowed the design of degenerate PCR primers [which can include the alternative base inosine (I) to avoid too high a rate of degeneracy] that can be used to amplify these regions of the *MAT* genes from species where genome sequence data is lacking. Due to sequence divergence within the Pezizomycotina, degenerate primers often need to be designed for groups of related fungal species because it has proved difficult to design all encompassing pezizomycete *MAT* degenerate primer sets (Dyer et al. 1995; Arie et al. 1997; Singh et al. 1999; Table 2.1). But once *MAT* amplicons have been obtained they can then be sequenced to confirm homology to known *MAT* genes, and if desired the arising sequence can be used to design specific primers (which are likely to be less prone to PCR artefacts than degenerate primers) for use as a *MAT* diagnostic tool to determine the mating type of isolates. It is noted that the *MAT1-2* gene sequence tends to be conserved to a higher extent than *MAT1-1* gene sequence, so it might be necessary to obtain *MAT1-1* idiomorph sequence via a chromosome walking approach based on inwards sequencing from the *SLA* and *APC* genes found in the conserved region bordering the *MAT* loci (Eagle 2009, C Eagle and PS Dyer unpublished results) or use of TAIL-PCR to amplify whole *MAT* idiomorph regions (Arie et al. 1997). More recently it has also been possible to use whole genome sequence to identify *MAT* genes by BLAST analysis and thereby design *MAT* diagnostic primers set directly. Such *MAT* diagnostic tests may use different primer pairs for the *MAT1-1* or *MAT1-2* genes, necessitating two rounds of PCR (see examples in Table 2.1).

Alternatively, multiplex PCR-based tests have been designed to allow mating type to be determined using a single PCR (Table 2.1). The latter diagnostic relies on the use of one primer binding to conserved sequence in the flanking regions of isolates of both mating type, together with two further primers that bind to sequence either present in the *MAT1-1* or *MAT1-2* idiomorph. The location of the latter two primers is designed such that differential size PCR products are generated according to whether isolates are of *MAT1-1* or *MAT1-2* genotype, thereby allowing the rapid and efficient determination of mating type [for further explanation of rationale see Dyer et al. (2001b) and Paoletti et al. (2005)].

The availability of a PCR-based mating-type detection method is a major aid when dealing with induction of sexual reproduction in heterothallic species. Before the introduction of these *MAT* diagnostic techniques, isolates had to be crossed in all combinations with each other and when successful mating occurred, mating-type tester strains were selected. With these tester strains, larger sets of isolates were then examined in order to determine their fertility and mating type. Whereas now with the use of *MAT*-specific primers, the mating type of isolates can be determined prior to crossing on agar media. It is therefore possible to set up, in a more efficient manner, directed crossings between isolates known to be of the opposite mating type. This significantly reduces the amount of effort needed to study for example the presence of a sexual cycle in presumed asexual species or the sexual fertility within a population. Indeed, this approach has led to major breakthroughs over recent years with the induction of a sexual cycle in a series of previously considered 'asexual' species when crosses were set up between known *MAT1-1* and *MAT1-2* isolates. A model example concerns the opportunistic human pathogen *Aspergillus fumigatus*. This species was described by Fresenius in 1863 and for more than 140 years was only known as an asexual organism (Samson et al. 2009). However, there was accumulating evidence for the presence of a cryptic (i.e., hidden, so far undescribed) sexual cycle. Firstly, studies of the population structure provided evidence for sexual

Table 2.1 Degenerate and specific PCR primers used in mating-type diagnostic tests for a selection of heterothallic ascomycete species, especially in the *Eurotiales* and *Hypocreales*

Taxonomic grouping/ species	Primers (MAT1-1; 5′–3′)	Primers (MAT1-2; 5′–3′)	Remarks	Reference
Leotiomycete and eurotiomycete fungi	N/A	MAT 5-3 (F) AARIIICCIMGICCIMYIAAT MAT 3-1 (R) CKIGGIIIRTAIYKRTAIIINGG	Degenerate primer pair used successfully in conjunction with hot start PCR to amplify *MAT1-2* sequence from leotiomycete and eurotiomycetes species.	Dyer et al. (1995), Singh et al. (1999), Paoletti et al. (2005)
Eurotiomycete and lecanoromycete fungi	N/A	MAT5-4 (F) AARRTICCIMGICCICCIAAYGC MAT3-2 (R) TTNCKIGGIGTRTAITGRTARTCNGG	Degenerate primer pair used successfully with hot start PCR to amplify *MAT1-2* sequence from eurotiomycete and lecanoromycete species.	Seymour et al. (2005)
Eurotiomycete fungi	MAT5-6 (F) GIMGICCIYTIAAYWSITTYATHGC MAT3-4 (R) ARRAAICKIARIATICCISWYTT	MAT5-7 (F) THSCIMGICCICCIAAYKSITTYAT MAT3-5 (R) TTICKIGGIKKRWAIYKRTARTYNGG	Degenerate PCR primer pairs used successfully with hot-start PCR to amplify *MAT* sequence from a range of *Aspergillus* and *Penicillium* species.	Eagle (2009), C Eagle and PS Dyer (unpublished results)
Loculoascomycete fungi	N/A	ChHMG1 (F) AAGGCNCCNCGYCCNATGAAC ChHMG2 (R) CTNGGNGTGTAYTTGTAATTNGG	Degenerate PCR primer pairs used successfully to amplify *MAT1-2* sequence from a range of loculoascomycete species.	Arie et al. (1997)

Pyrenomycete fungi	N/A	NcHMG1 (F) CCYCGYCCYCCYAAYGCNTAYAT NcHMG2 (R) CGNGGRTTRTARCGRTARTNRGG	Degenerate PCR primer pairs used successfully to amplify *MAT1-2* sequence from a range of pyrenomycete species.	Arie et al. (1997)
Aspergillus felis	AFM1_F65655 (F) CCTYGACGMGATGGGITGG MAT1_R6215 (R) TGTCAAAGARTCCAAAAGGAGG	MAT2_F6086 (F) TCGACAAGATCAAAWCYCGTC MAT2_R6580 (R) CTTYTTGARCTCTTCYGCTAG	Annealing temperature 48 °C.	Barrs et al. (2013)
Aspergillus flavus, A. *parasiticus*, A. *nomius*	M1F (F) ATTGCCCATTTGGCCTTGAA M1R (R) TTGATGACCATGCCACCAGA	M2F (F) GCATTCATCCTTTATCGTCAGC M2R (R) GCTTCTTTTCGGATGGCTTGCG	Multiplex.	Ramirez-Prado et al. (2008), Horn et al. (2011)
Aspergillus fumigatus, A. *lentulus*	AFM1 (F) CCTTGACGCGATGGGGTGG AFM3 (R) CGGAAATCTGATGTCGCCACG	AFM2 (F) CGCTCCTCATCAGAACAACTCG AFM3 (R) CGGAAATCTGATGTCGCCACG	Multiplex; AFM3 is "common" primer and binds in the flanking region bordering both idiomorphs.	Paoletti et al. (2005), Swilaiman et al. (2013)
Aspergillus terreus	AteM1F (F) GCGAGGCAGACACATTCAGGAT AteM1R (R) CGAGGATGCCAATAAAACCAGC	AteM2F (F) TCTATCGCCAGCACCATCATCC AteM2R (R) CTTGTTGTGGTGGTGGTCGTTCT	Annealing temperature MAT1-1: 53.5 °C; MAT1-2: 55 °C.	Eagle (2009); C Eagle and P.S. Dyer (unpublished results)
Aspergillus tubingensis	M1F_Anig (F) GGTCATCGCGAATGATGGAG M1R_Anig (R) CAGCGTGCTTTCAACGCATTC	MAT5-4 (F) AARRTICCIMGICCICCIAAYGC MAT3-2 (R) TTNCKIGGIGTRTAITGRTARTCNGG	Annealing temperature MAT1-1: 62 °C; MAT1-2: 55 °C.	Horn et al. (2013), Rydholm et al. (2007)
Aspergillus felis, A. *udagawae*, A. *wyomingensis*	alpha1 (F) CTGGAGGAGCTTCTGCAGTAC alpha2 (R) GGAGTACGCCTTCGCGAG	HMG1 (F) CTCTTGTGGCAGGATGCTCT HMG2 (R) TTGCTGGTAGAGGGCAGTCT		Sugui et al. (2010)

(continued)

Table 2.1 (continued)

Taxonomic grouping/ species	Primers (MAT1-1; 5′–3′)	Primers (MAT1-2; 5′–3′)	Remarks	Reference
Paecilomyces variotii	MAT1-F1-VarSp (F) TATGCCTCCTGGTGAGCTGG MAT1-R2-VarMar (R) GATCCCRGAYTTSGYCTTCTG	MAT2-F1Paec (F) AYCAYCAYCCKATYGTCAAAGC MAT2-R1Paec (R) GYTTGCGYTTTATCTSCTCYGC	Multiplex; annealing 58 °C.	Houbraken et al. (2008)
Penicillium rubens (reported as *P. chrysogenum*)	MAT-1-f (F) CTTCGTCCATTGAACTCTTTTATG MAT-1-r (R) ATCCCAAC CAGCCATCCTGAGAT	MAT-2-f (F) CCAAGT CTATCCACGAGGCTG MAT-2-r (R) GCAGGCAGTTGGCACGGGAAC		Hoff et al. (2008)
Talaromyces derxii	MAT 1-1b (F) CCACGTATAACGGGGCATC (R) CGGCTTGCCAMAGGTCTT	MAT 1-2b (F) GTGATAATGCTTSCGATAGAGAATG (R) GTTGGAGAGGAGGCGTTGAC	Touchdown 60–50 °C.	López-Villavicencio et al. (2010), Y. Yilmaz (personal communication)
Trichoderma reesei	MatA2-fw (F) CTCGAGAGGGATATACACCAG MatA2-rv (R) CTTCCTACACGGATGCCAGA	Mat2fw (F) CAACACGTATGAAAGAGAGATG Mat2rv (R) ATTGGAACGGATCACCTTCTTG		Seidl et al. (2009)
Fusarium keratoplasticum	FS3MAT1-1 (F) ATGGCTTTCCGCAGTAAGGA FS3MAT1-1 (R) CATGATAGGGCAGCAAAGAG	FS3MAT1-2 (F) GGGAATCTGAGAAAGATACGTAC FS3MAT1-2 (R) CGGTACTGGTAGTCGGGAT		Short et al. (2013)
Fusarium culmorum	FusALPHAfor (F) CGCCCTCTKAAYGSCTTCATG FusALPHArev (R) GGARTARACYTTAGCAATYAGGGC	FusHMGfor (F) CGACCTCCCAAYGCYTACAT FusHMGrev (R) TGGGCGGTACTGGTARTCRGG	Multiplex; annealing 55 °C.	Kerényi et al. (2004)
Fusarium fujikuroi	GFmat1a (F) GTTCATCAAAGGGCAAGCG GFmat1b (R) TAAGCGCCCTCTTAACGCCTTC	GFmat2d (F) CTACGTTGAGAGCTGTACAG Gfmat2c (R) AGCGTCATTATTCGATCAAG		Steenkamp et al. (2000)

recombination based on the analysis of the association of alleles of five loci (Pringle et al. 2005) and sequence present at three intragenic regions (Paoletti et al. 2005). Secondly, it was possible to identify *MAT1-1* and *MAT1-2* isolates of *A. fumigatus* using genomic BLAST searching and degenerate PCR approaches, together with the presence of a series of genes related to sex within the genome (Galagan et al. 2005; Paoletti et al. 2005). A multiplex PCR-based *MAT* diagnostic test was then developed, and analysis of 290 worldwide clinical and environmental isolates revealed the presence of *MAT1-1* and *MAT1-2* genotypes in similar proportions (43 % and 57 %, respectively) (Paoletti et al. 2005). The presence of the two mating types in equal frequencies within a population is an indication of sexual reproduction (Milgroom 1996). In a subsequent study, analysis of a population of *A. fumigatus* strains from five locations in Dublin, Ireland, revealed an almost exact 1:1 ratio of *MAT1-1:MAT1-2* isolates (O'Gorman et al. 2009). Furthermore, 88 out of 91 isolates were genetically unique according to a RAPD-PCR DNA fingerprinting study, and a phylogenetic analysis demonstrated that the *MAT-1* and *MAT1-2* isolates were interleaved when represented on a phylogenetic tree. This provided strong evidence for recent or extant sexual reproduction and led O'Gorman et al. (2009) to set up directed crosses between known *MAT1-1* and *MAT1-2* isolates on a range of media under a variety of different growth conditions. An exciting result was then obtained when it was found that a sexual cycle, leading to production of cleistothecia containing recombinant ascospores, could be induced when cultures were crossed on oatmeal agar in Parafilm-sealed Petri-dishes which were incubated at 30 °C in darkness for 6–12 months (O'Gorman et al. 2009). More recently, a "supermater" pair of *A. fumigatus* isolates have been identified which produce abundant cleistothecia containing viable ascospores after only 4 weeks incubation under the same conditions (Sugui et al. 2011). There have subsequently been further examples of the discovery of sexual states in other supposedly 'asexual' fungal species using similar directed crosses between known *MAT1-1*

and *MAT1-2* isolates [reviewed by Dyer and O'Gorman (2012)]. These have included notably the description of a sexual state for the industrial workhorse *Trichoderma reesei* (Seidl et al. 2009), the penicillin producer *Penicillium chrysogenum* (Böhm et al. 2013), the opportunistic pathogen *Aspergillus lentulus* (Swilaiman et al. 2013), the aflatoxin producers *Aspergillus flavus* and *Aspergillus parasiticus* (Moore 2014), and the starter culture of blue veined cheese *Penicillium roqueforti* (Ropars et al. 2014; S Swilaiman, J Houbraken, J Frisvad, R Samson and PS Dyer, unpublished results).

It should be cautioned that the presence of the *MAT* genes alone is insufficient to prove that a sexual stage exists. Given that several hundred other genes are also likely to be required for a functional sexual cycle to occur, it is possible that loss of function of any of these genes could result in reduced fertility or asexuality. For example, a series of over 75 genes have been identified which are required for sexual reproduction in the aspergilli, encompassing processes such as environmental sensing, mating, fruit body formation, and ascospore production. Such 'sex-related' genes might be predicted to accumulate deleterious mutations or even be lost in purely asexual species, in which there was no functional constraint on their conservation (Dyer 2007; Dyer and O'Gorman 2012).

2.3 Methods to Induce Sexual Reproduction in Filamentous Fungi

In this next section methods will be described to induce sexual reproduction in filamentous ascomycete species based on procedures that have been used to obtain successful mating in a variety of fungal species. The first step involves the selection of suitable strains; this is followed by *MAT* analysis; then selection of suitable agar media, inoculation procedures, and incubation conditions to induce a sexual cycle. Finally after formation of a sexual state, single ascospore isolates should be obtained and these should be examined for evidence of recombination. It is noted that sexual reproduction for many homo-thallic species can be achieved fairly readily as there is no need for any mating step, so most of the following discussion will apply to heterothallic species which can be more demanding in their sexual requirements.

2.4 Materials

2.4.1 Solutions

Sterile distilled water or tap water.
Tween 80: 0.5 g/L, ddH$_2$O to 1 L.

2.4.2 Agar Media Inducing Sexual Reproduction

Carrot agar (CA): Fresh washed, peeled, diced carrots (400 g) in 400 mL ddH$_2$O. Autoclave at 121 °C for 15 min. After autoclaving, blend the carrots and add additional 500 mL ddH$_2$O. Add ZnSO$_4$·7H$_2$O (0.01 g/L), CuSO$_4$·5H$_2$O (0.005 g/L), agar (20 g/L). Mix well and autoclave at 121 °C for 15 min.

Mixed cereal agar (MCA): Gerber mixed grain cereal (Gerber Products Co., Freemont, Michigan) (50 g/L), agar (20 g/L), in 1 L ddH$_2$O. Mix well and autoclave at 121 °C for 15 min (McAlpin and Wicklow 2005).

Oatmeal agar (OA)[1]:

Version (1): Blend 30 g of oats and add 1 L ddH$_2$O. Boil and let it stand for 1 h. Add ZnSO$_4$·7H$_2$O (0.01 g/L), CuSO$_4$·5H$_2$O (0.005 g/L), agar (20 g/L). Autoclave for 15 min, at 121 °C (Samson et al. 2010).

Version (2): Add 40 g of oats to 1 L tap water. Bring to the boil then lower the heat to just below boiling point (i.e., bubbling gently) for a further 45 min. Then filter through two layers of cheese cloth and restore the volume of the solution to 1 L with tap water and mix thoroughly. Add agar (20 g/L). Autoclave for 15 min at 121 °C with slow cool down and slow release of pressure to prevent media loss (O'Gorman et al. 2009).

Potato dextrose agar (PDA): 200 g sliced potatoes are boiled in 1 L of ddH$_2$O and sieved, add glucose (20 g/L), agar (20 g/L), ZnSO$_4$·7H$_2$O (0.01 g/L), CuSO$_4$·5H$_2$O (0.005 g/L). pH is approximately 5.6. Autoclave for 15 min, at 121 °C (Samson et al. 2010).

Tap water agar (TWA)[2]: Bacto™ agar (15 g/L) in 1 L tap water. Autoclave for 15 min, at 121 °C. Supplement with appropriate natural growth substrate (Dyer et al. 1993).

V8 agar (V8): V8® vegetable juice (Campbell) (175 mL/L), CaCO$_3$ (3 g/L), ZnSO$_4$·7H$_2$O (0.01 g/L), CuSO$_4$·5H$_2$O (0.005 g/L), agar (20 g/L) in 1 L ddH$_2$O. Mix well and autoclave at 121 °C for 15 min (Samson et al. 2010).

2.5 Methods

2.5.1 Strain Selection

2.5.1.1 Identification

The first step in inducing a sexual cycle is the requirement for the correct identification of isolates of the same biological species. In the past, fungal taxonomy was primary based on phenotypic and physiological characters. Nowadays molecular techniques like DNA sequencing are commonly applied for identification purposes. Such data has shown that many well-known 'species' are actually species complexes composed of closely related species that might be sexually fertile when crossing within a species, but sexually sterile when attempts are made to cross different species. For example, *P. chrysogenum* is a complex of five species, namely *P. chrysogenum sensu stricto*, *Penicillium rubens*, *Penicillium*

---

[1] Commercially made oatmeal agar (OA) is available from certain manufacturers, but in our experience this is not able to induce sex in demanding species. Instead, it is best to prepare OA in house. Different brands of oats can be used and these have can have an effect on the mating. Commonly used brands are Pinhead oatmeal (Odlums, Ireland) and Quaker Oats. For *Sordaria* and *Chaetomium* species and *Penicillium rubens* (*P. chrysogenum*) this medium needs to be supplemented with biotin (6.4 µg/L) to induce sex (Böhm et al. 2013).

[2] In our experience Bacto™ agar is less prone to condensation problems than some cheaper, less pure, commercial agars.

allii-sativi, Penicillium tardochrysogenum, and *Penicillium vanluykii* (Henk et al. 2011; Houbraken et al. 2012). Using these current taxonomic insights, the main penicillin producer is named *P. rubens* and in the study of Böhm et al. (2013) all *P. chrysogenum* strains that were able to reproduce sexually are actually *P. rubens*. The induction of the sexual cycle would have probably been much more difficult or even impossible if attempts had been made to cross different members of the *P. chrysogenum sensu lato* complex and might explain why the sexual stage had remained undiscovered up to that point. Similarly, it was discovered that the 'single' anamorphic species *Pseudocercosporella herpotrichoides* is composed of two closely related, but intersterile, species *Tapesia (Oculimacula) yallundae* and *Tapesia acuformis* (Dyer et al. 1996). Again, attempts to induce a sexual cycle by erroneously crossing isolates of the different species would have failed in this case.

Therefore, it is essential to verify that isolates to be used in crossing experiments are of the same biological species. The internal transcribed spacer regions (ITS) of the ribosomal gene have been used in many taxonomic studies for species identification and have been selected as fungal barcodes (Schoch et al. 2012). The main advantages of using the ITS locus for identification are the ease of amplification by PCR and the presence of a high number of ITS sequences in the public databases. However, the resolution of this locus is insufficient for species identification of all fungi. For example, closely related species belonging to the industrially important genera *Aspergillus, Penicillium, Fusarium,* and *Trichoderma* can share the same ITS barcode. In those cases, it is recommended to sequence other (protein coding) genes. There is no consensus which region to sequence and the choice largely depends on the genus/species being dealt with. Details on molecular and phylogenetic identification methods can be found in Crous et al. (2009).

2.5.1.2 Origin of Strains

Freshly isolated strains are generally more fertile than strains maintained for longer periods in culture collections, which can be prone to a 'slow decline' in fertility following prolonged subculture (Dyer and Paoletti 2005). For example, *Paecilomyces variotii* strains isolated from heat-treated products proved to be fertile, while older isolates from a culture collection were unable to mate (Houbraken et al. 2008). Similar observations were found in the heterothallic *Histoplasma capsulatum*. This species lost fertility during laboratory passage and it was suggested that selective pressures may serve to maintain fertility in the environment (Kwong-Chung et al. 1974; Fraser et al. 2007). Furthermore, even when obtaining isolates from the field it is important to be aware that such isolates can exhibit a range of fertility due to various physiological and genetic factors (Dyer et al. 1992). Indeed the same 'slow decline' in sexual fertility observed during in vitro culture may be occurring in vivo in natural populations subject to strong selection pressure favoring asexual propagation (Dyer and Paoletti 2005). For example, Sugui et al. (2010) found that most attempted crosses involving the emerging agent of aspergillosis *Aspergillus (Neosartorya) udagawae* either failed to produce cleistothecia or produced ascospores which did not germinate. Similarly, Swilaiman et al. (2013) found that many clinical isolates of the opportunistic pathogen *A. lentulus* exhibited low fertility or were sterile in crosses. However, in both of these cases it was possible to detect isolates that successfully crossed to form cleistothecia with ascospores. This illustrates the fact that it is very important to select a number of representative field isolates for crossing studies to ensure that at least some representatives will exhibit sexual fertility if it is present.

2.5.1.3 Strain Typing

Before the start of the mating experiments, the isolates should ideally be typed using methods such as AFLP (Amplified Fragment Length Polymorphism), SSR (microsatellites or simple sequence repeats), RFLP (Restriction Fragment Length Polymorphism), RAPD (Random Amplified Polymorphic DNA), or MLST (MultiLocus Sequence Typing) DNA fingerprinting. The use of such typing methods allows the selection of independent (non-clonal) strains for crossing purposes, avoiding the error of setting up repeated crosses with different isolates of the same field strain. Typing can also generate insights as to whether there is evidence of recombination among the strains.

2.5.2 Detection of MAT Genes

In order to increase the likelihood of success and reduce the number of crosses that need to be set up, it is recommended that the mating type of test strains be determined prior to crossing efforts. This enables directed crosses to be set up on agar media (see below) between *MAT1-1* and *MAT1-2* mating partners that are known to be potentially sexually compatible. For certain species or species groups, mating type-specific primers have been developed that amplify part of either the *MAT1-1* or *MAT1-2* gene. A selection of primers pairs already published for some important pezizomycete species is given in Table 2.1 together with details of some degenerate primers pairs that should be more broadly applicable to wide groups of species (Dyer et al. 1995; Arie et al. 1997; Singh et al. 1999; Paoletti et al. 2005; Seymour et al. 2005). Especially note that the use of hot-start PCR can greatly increase the chances of success; for example, Singh et al. (1999) were unable to amplify a *MAT1-2* region from *T. yallundae* using standard PCR, but found that the use of hot-start PCR gave very strong amplification of the required product. In cases where no *MAT* amplicons are obtained, it is recommended that new degenerate primers are designed based on known *MAT* gene sequence of species that are phylogenetically closely related to the test species.

2.5.3 Agar Media

A large variety of agar media have been used for the induction of fungal sexual cycles. Generally, media based on natural substrates are more effective than synthetic media and which agar to use largely depends on the species or genus. Some species are fastidious and need specific nutrients, which often mimic their natural growth substrate.

For example, the sexual cycle of the cereal pathogen *T. yallundae* occurs in nature on straw stubble left after harvest (Dyer et al. 2001a) and attempts to induce the sexual cycle in vitro on a range of synthetic media failed. However, it was possible to induce the sexual cycle when *MAT1-1* and *MAT1-2* isolates were inoculated onto straw segments (especially those with nodes), which were kept moist by being placed on TWA (Dyer et al. 1993). Similarly, it was only possible to induce sex in *Thermomyces dupontii* (=*Talaromyces thermophilus*) on natural oat grains rather than synthetic agar media (Pitt 1979; Houbraken et al. 2014), and three dermatophytic *Trichophyton* species required growth on sterilized baby or rabbit hair (placed on agar) to induce sexual reproduction (Kawasaki et al. 2010). By contrast, for species such as *Neurospora crassa* it is possible to induce sex on fully synthetic media, one reason why this is used as a model organism (Perkins 1986). The diversity in nutrient requirements for sexual reproduction is illustrated well by members of the genus *Aspergillus*. Species such as the homothallic *Aspergillus nidulans* and *Aspergillus fischeri* reproduce sexually on a fairly wide range of media, including fully synthetic complete media and oatmeal agar (OA) (Paoletti et al. 2007). This contrasts with some heterothallic species which have more exacting demands. Members of the section *Flavi* (e.g. *A. flavus*, *A. parasiticus*) have only been successfully crossed on mixed cereal agar (MCA), while crosses of members of the section *Fumigati* (*A. fumigatus*, *A. lentulus*) have only proved fruitful on OA. In contrast to these two high water activity media, species belonging to section *Aspergillus* (Eurotium-type ascomata) require a low water activity medium (e.g., malt extract agar with 40 % sucrose) for fruiting body formation (Dyer and O'Gorman 2012). In *Fusarium*, the standard medium to induce fruiting bodies, called perithecia, is carrot agar. In contrast, attempts with *Fusarium keratoplasticum* to use this standard medium were unsuccessful while crosses on V8 agar induced the sexual cycle in this species (Covert et al. 2007; Short et al. 2013). In the case of homothallic *Giberella zeae* sex can be induced by the gentle removal of surface mycelia, followed by treatment with detergent solution (Cavindera and Trail 2012). One intriguing example is that of different *Cryptococcus* species. Nielsen et al. (2007) found that it was possible to induce the sexual cycle of *Cryptococcus neoformans* on pigeon guano media, but that this was not possible for the related species *Cryptococcus gattii*. It was suggested that the ability to undergo sexual reproduction on pigeon guano represented an evolutionary adaptation that allowed ancestral strains of *C. neoformans* to sweep the globe (Nielsen et al. 2007; Heitman et al. 2014). It is important to note that when making agar media there can be difference among ingredients of different suppliers. For example, different brands of yeast extracts are available and these can have a strong influence on the phenotype of the culture. Furthermore, agar media based on natural ingredients can vary between manufactures and labs, and even within one lab batch to batch differences can occur.

Some ascomycetes may require exogenous vitamins, minerals, or other natural materials for ascomata (ascocarp) production, and these are often not present in synthetic media. This might be one of the explanations why sexual reproduction is more often found on media made from natural substrates such as oatmeal, (mixed) cereals, and cornmeal agar. A fractionation study of V8 juice agar revealed that no single factor was responsible for its utility in inducing sex in *C. neoformans*, but rather the unique composition of V8 juice provided sustenance for sex, especially the copper content (Kent et al. 2008). Other studies have found it necessary to add compounds to a standard medium to induce sex; for example biotin was added to OA to stimulate mating in *P. rubens* (Böhm et al. 2013). Meanwhile, certain nutritional auxotrophs of *A. nidulans* can require supplementation of media to ensure sexual development, e.g., tryptophan, arginine, and *riboB2* mutants are self-sterile (Dyer and O'Gorman 2012) and heterologous expression of *pyrG* can result in reduced fertility (C Scazzocchio pers. comm.; Robellet et al. 2010). An overview of agar media used in mating experiments of selected heterothallic species belonging to either the *Eurotiales* or *Hypocreales* as representative groupings is given in Table 2.2.

Thus, the best strategy when attempting to induce sexual reproduction in vitro is to trial a range of agar media, which should include some basal media supplemented with the natural growth substrate of the species in question.

2.5.4 Incubation Conditions

Besides the nutrient availability, various other environmental factors such as light, temperature, and oxygen determine the success of mating experiments. These factors are genus and in some cases also species-specific. There are numerous reports in the literature of how different environmental conditions influence fungal sexual reproduction, and in the present chapter only some representative examples can be given. For example, *Fusarium* perithecia are formed in abundance under alternating 12 h/dark and 12 h/light cycles (with both fluorescent and near ultra violet light), whereas perithecia are absent when incubated in darkness (Table 2.2). Similarly the sexual cycle of *T. acuformis* can be induced under near UV or white light, but not in darkness (Dyer et al. 1996). By contrast, it is necessary to incubate *Aspergillus* species in darkness to trigger sexual reproduction because light preferentially induces asexual sporulation, reflecting the natural ecology of many *Aspergillus* species (Mooney and Yager 1990; Han et al. 2003; Dyer and O'Gorman 2012). Incubation temperature also has a strong influence on sexual fertility. For example, Choi et al. (2009) showed that their *Fusarium fuijikuroi* strains only produced perithecia at 23 °C and none were formed at 18, 26, and 28 °C. However, in *Fusarium graminearum*, the optimal temperature was 28.5 °C and *Fusarium circinatum* perithecia were more abundantly produced at 20 °C than at 25 °C (Table 2.2) (Tschanz et al. 1976; Covert et al. 1999). Meanwhile, in *Aspergillus* and *Penicillium*, oxygen limitation can induce sexual reproduction (Dyer and O'Gorman 2012), which can be achieved by sealing Petri dishes with Parafilm (Table 2.2).

Table 2.2 Details of media and incubation conditions required to induce sexual reproduction in a selection of heterothallic fungal species

Species	Structure ascomata	Agar medium	Incubation temperature and time	Inoculation method	Additional conditions	Reference
Aspergillus felis	Neosartorya	Oatmeal agar	30 °C, 2–4 weeks	Barrage zone	Darkness, Petri dishes	Barrs et al. (2013)
Aspergillus flavus	Petromyces	Mixed cereal agar	30 °C, 6–11 months	Mixed culture	Slants, in sealed plastic bags	Horn et al. (2009a)
Aspergillus fumigatus	Neosartorya	Oatmeal agar	30 °C, 6 months	Barrage zone	Darkness, Parafilm-sealed Petri dishes	O'Gorman et al. (2009)
Aspergillus heterothallicus	Emericella	Oatmeal agar	Room temperature, 2–3 weeks	Barrage zone		Raper and Fennell (1965)
Aspergillus lentulus	Neosartorya	Oatmeal agar	28–30 °C, 3–7 weeks	Barrage zone	Darkness, Parafilm-sealed Petri dishes	Swilaiman et al. (2013)
Aspergillus nomius	Petromyces	Mixed cereal agar	30 °C, 5–11 months	Mixed culture	Slants, in sealed plastic bags	Horn et al. (2011)
Aspergillus parasiticus	Petromyces	Mixed cereal agar	30 °C, 6–9 months	Mixed culture	Slants, in sealed plastic bags	Horn et al. (2009b, c)
Aspergillus sclerotiicarbonarius	Petromyces	Oatmeal agar	25 °C, 6–11 months	Barrage zone	Darkness, Parafilm-sealed Petri dishes	Darbyshir et al. (2013)
Aspergillus terreus	Fennellia	Mixed cereal agar	37 °C, up to 6 months	Barrage zone	Darkness, 56 mm Petri dishes	Arabatzis and Velegraki (2013)
Aspergillus tubingensis	Petromyces	Mixed cereal agar, Czapek agar	30 °C, 5–6 months	Mixed culture	Slants, in sealed plastic bags	Horn et al. (2013)
Aspergillus wyomingensis, A. udagawae	Neosartorya	Oatmeal agar	25–30 °C, 4–5 weeks	Barrage zone		Nováková et al. (2014)
Paecilomyces variotii	Byssochlamys	Potato dextrose agar	30 °C, 6–9 weeks	Barrage zone	Darkness, Petri dishes	Houbraken et al. (2008)
Penicillium rubens (reported as *P. chrysogenum*)	Eupenicillium	Oatmeal agar supplemented with biotin (6.4 µg/L)	20 °C; 5 weeks	Barrage zone	Darkness, Parafilm-sealed Petri dishes	Böhm et al. (2013)
Talaromyces derxii	Talaromyces	Oatmeal agar, malt extract agar	37 °C, 7–10 days	Barrage zone	Darkness	Takada and Udagawa (1988)
Trichoderma reesei	Hypocrea	Optimal: Malt extract agar (3 % wt/vol; Merck); also potato dextrose agar	Optimal 20–22 °C; 7–10 days	Barrage zone	12-h light–dark cycle or daylight	Seidl et al. (2009)

Species						
Fusarium keratoplasticum	Nectria	V-8 agar	Room temperature; 8 weeks	Barrage zone	Parafilm sealed; under 12 h cycles of direct fluorescent and UV light from 120 V bulbs	Short et al. (2013)
Fusarium tucumaniae	Nectria	Carrot agar	18 °C, 3–4 weeks	Barrage zone	Cool white, fluorescent bulbs	Covert et al. (2007)
Fusarium fujikuroi	Gibberella	V-8 agar, Carrot agar	23 °C	Fertilization	Alternating 12 h/light and 12 h/dark cycles with both FL/NUV and NUV light[a]	Choi et al. (2009)

A diverse range of conditions are evident even though species belonging to the genera *Aspergillus*, *Penicillium*, *Paecilomyces* and *Talaromyces* (members of *Eurotiales*), and *Fusarium* and *Trichoderma* species (both order *Hypocreales*), are phylogenetically related within their respective groupings

[a]Fluorescent (FL) and near ultra violet (NUV) light

Thus, the best strategy when attempting to induce sexual reproduction in vitro is to trial a range of growth conditions, which ideally might mimic those encountered in the wild when sexual reproduction occurs. This approach was used to induce sex in plant pathogenic *Tapesia* species which were known to sexually reproduce in the field in early spring in the UK on exposed straw stubble. It was found that incubation in vitro at low temperatures between 7 and 10 °C under white light could induce sex, but the sexual cycle was inhibited above these temperatures (Dyer et al. 1996).

2.5.5 Inoculations

For homothallic species, self fertilization can be induced using either point inoculation or spore spread methods (e.g. Paoletti et al. 2007; Todd et al. 2007; Cavindera and Trail 2012). In the case of heterothallic species, several methods have been described in the literature for crossing on agar media. We have summarized three different methods below. In the first, the "barrage zone" method, strains of opposite mating type are inoculated in close proximity on an agar medium. During incubation, the strains grow towards each other. Fruiting body formation then mainly occurs in the barrage zone (Fig. 2.1a, b), but sometimes also towards the centre or on the opposite periphery of the colony (Fig. 2.1c). Interestingly, this method can also be used to promote outcrossing in homothallic species, especially when using complementary auxotrophic strains [see Todd et al. (2007) and Cavindera and Trail (2012) for further details]. In the second, the "mixed culture" method, spore suspensions of the same concentration are prepared for isolates of opposite mating type. These suspensions are mixed together and then used to inoculate the agar medium. This results in intermingled growth of both partners from an early stage allowing close sexual interaction. Finally, in *Fusarium, Neurospora,* and other pyrenomycete fungi, a third "fertilization" crossing technique is also sometimes applied. In this method, a strain of one mating type is cultured on an agar medium to allow development of protoperithecia. After growth, the culture is then fertilized (so-called 'spermatization') with a spore-suspension of the opposite mating partner. This fertilization method can also be used in heterothallic *Botrytis* species that produce sclerotia, which can be spermatized by being soaked with conidia of the opposite mating type (Faretra et al. 1988). Crosses with strains of the same mating type should be used as controls.

2.5.5.1 Barrage Zone Method

Cultures of opposite mating types are crossed in Petri dishes containing an agar medium inducing recombination. The following protocol is based on O'Gorman et al. (2009).

1. Prepare single spore isolates of each isolate and incubate under conditions inducing sporulation.
2. Harvest the spores of each isolate in sterile water containing 0.05 % Tween 80.
3. Inoculate 1.0–2.5 µL of each spore suspension (e.g., containing 500 spores) onto the agar surface about 4 cm apart and perpendicular to aliquots of spore-suspensions of the opposite mating type. This configuration created four interaction/barrage zones as colonies grew.
4. Seal, if required, with Petri dishes with Parafilm and incubate at conditions inducing recombination.
5. Regularly check the cultures on the production of fruiting bodies.

2.5.5.2 Mixed Culture Method

Agar slants are used in the protocol mentioned below and this method is useful when the agar medium needs be incubated for a long time. However, this method can be adopted for agar plates as well. In that case the agar media can be sealed with Parafilm in order to induce fruiting body formation and prevent drying out of the plates. The following protocol is based on Horn et al. (2009c).

1. Grow the fungal strains under conditions inducing sporulation.

2 Induction of the Sexual Cycle in Filamentous Ascomycetes

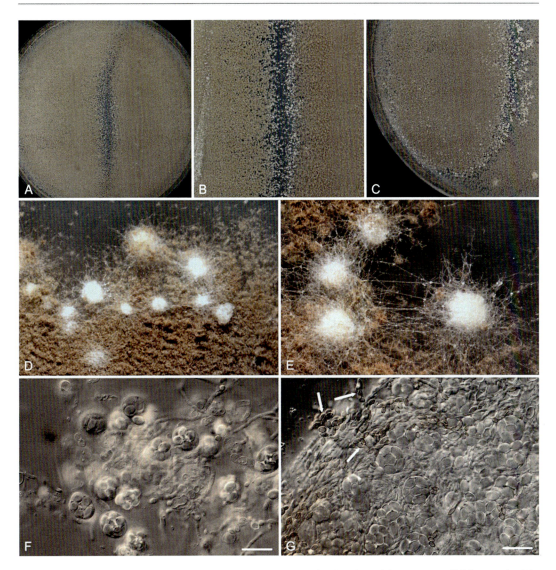

Fig. 2.1 Mating experiments between *P. variotii* strains. (**a–c**) Detail image of a potato dextrose agar plate where two *P. variotii* strains of opposite mating types were inoculated on each side of the Petri-dish. In **a** and **b**, ascomata are formed in the *middle*, in **c**, ascomata are also present on the opposite side of the colony. (**d**, **e**) Higher-magnification view of the ascomata. (**f**) Micrograph of the ascomata showing asci with ascospores. (**g**) Similar as **f**, but also showing the presence of heat-sensitive conidia (*arrows*); isolation of the ascospores, which are heat-resistant, can be achieved by applying the "heat treatment method". Scale bars = 10 μm

2. Harvest the spores in sterile water containing 0.05 % Tween 80.
3. Dilute or concentrate (e.g., by centrifugation) the spore suspensions until a concentration of 5×10^5 spores per mL is obtained.
4. Mix the spore suspensions of strains of opposite mating types. Spore suspensions of single isolates can be used as negative controls.
5. Inoculate the slants with 10 μL spore suspensions on a medium inducing a sexual cycle (production of sclerotia).
6. Incubate the slants with loose caps until sclerotia are produced.
7. Enclose the caps of the slants and enclose the slants in sealed plastic bags to prevent desiccation.

8. Allow prolonged incubation and regularly check the cultures for the production of fruiting bodies.

2.5.5.3 Fertilization Method

The fertilization method below is derived from standard protocols used for *Fusarium* species (Klittich and Leslie 1988).

1. Grow the fungal strains of one mating type under conditions inducing sporulation.
2. Harvest the spores in sterile water containing 0.05 % Tween 80.
3. Spread the cultures of one mating type on a suitable agar media.
4. Incubate the agar plates until the agar is covered with fungal growth. Plates can be checked for evidence of ascogonia and protoperithecial formation.
5. Fertilize the cultures by dispensing 1 mL of spore-suspension in 0.05 % Tween 80 carrying at least 5×10^5 conidia from the opposite strains.
6. Work the spore suspension into the mycelia with a glass rod until the suspension is absorbed. Self-fertilizations can be made by substituting sterilized 0.05 % Tween 80 solution for the spore suspension. Any excess conidial suspension can be removed using a sterile Pasteur pipette.
7. Seal, if required, with Petri dishes with Parafilm and incubate at conditions inducing recombination.
8. Regularly check the cultures for the production of fruiting bodies.

2.5.6 Single Ascospore Cultures

In heterothallic ascomycete species, a successful mating experiment will lead to production of ascomata (ascocarps) (Fig. 2.1d, e) containing asci and ascospores (Fig. 2.1f, g). Single ascospores cultures then need to be obtained for further analysis. The method of isolation of the ascospores depends on the way the ascospores are produced (e.g., in a closed ascoma or "open" perithecium) and other features of the species in question (e.g., degree of heat resistance). Four commonly used methods are described below.

2.5.6.1 Direct Isolation

The easiest method to obtain single ascospores cultures is by transferring ascospores directly from the fruiting body. This method can be applied in, e.g. *Fusarium*, where ascospores are produced in a perithecium and ooze out to form a prominent spore mass. This method can also be adapted for closed fruiting bodies as described by Todd et al. (2007) for *A. nidulans*.

1. Examine the agar plates using a stereomicroscope and isolate ascospores direct from a spore mass if present with a sharp needle. If necessary suspend the ascospores in 0.01 % Tween 20 or 0.05 % Tween 80 solution. In the case of closed fruiting bodies, these should be individually picked up and rolled across the surface of 4 % TWA plates to remove adhering conidia and hyphae [see Todd et al. (2007) for a detailed description]. The latter ascomata can then be transferred to a microfuge tube and crushed to release the ascospores. These should then be suspended in 0.01 % Tween 20 or 0.05 % Tween 80 solution.
2. Transfer the ascospores onto a clear agar medium (e.g. water agar), spreading or using a decimal dilution series if a spore suspension is used.
3. Incubate overnight (or longer as need be) at a temperature allowing germination of the ascospores.
4. Locate single ascospores using a dissecting microscope and transfer these or hyphal tips from germinating ascospores to separate agar plates. Transfers can be efficiently made using a fine, flattened platinum wire mounted on the end of glass tubing. Such wires can be flame-sterilized and cool down rapidly.

2.5.6.2 Heat Treatment

The second method is based on a heat treatment of the ascospores. In some genera, ascospores are markedly more heat-resistant than the (heat-sensitive) asexual spores. This feature can be used to generate single ascospore isolates free of contaminating conidia and hyphae. For some species this heat treatment step is also required to trigger ascospore germination; without this ascospores do not germinate at standard growth

2 Induction of the Sexual Cycle in Filamentous Ascomycetes

temperatures (Perkins 1986; O'Gorman et al. 2009). This method has been applied for example in *Neurospora, Paecilomyces,* and *Aspergillus* sect. *Fumigati* species.

1. Transfer the fruiting bodies from the agar plate into microfuge tubes containing 0.05 % Tween 80 or 0.01 % Tween 20 (ideally buffered at pH 6 with, e.g., 0.01 M sodium phosphate).
2. Agitate the suspension with glass beads in order to break open the ascomata and asci, or simply crush the ascomata on the side of the tube before thoroughly mixing in the buffered Tween.
3. Check by microscopy whether the majority of ascospores are released into the suspension. If not, repeat step 2.
4. Heat-treat the suspension in a water bath or on a thermal cycler (time and temperature is species-dependent). Note that most conidia are eliminated by a treatment of 10 min at 60 °C.
5. Make a decimal dilution in 0.05 % Tween 80 or 0.01 % Tween 20.
6. Spread plate 0.1 mL of each dilution on agar plates (duplicate).
7. Incubate at a suitable temperature allowing ascospore germination.
8. After incubation, pick individual young germinating colonies from the highest dilutions (preferable) or locate single ascospores using a dissecting microscope as described above.

2.5.6.3 Ejected Ascospores

Many species forcibly eject ascospores from the ascomata. The fact that ascospores are shot out of the ascomata provides a convenient way to collect pure ascospores, free of contaminating hyphae and conidia (Dyer et al. 1993). However, some caution must be exercised as some species (notably lichen-forming fungi) eject ascospores in packets, so it can prove difficult to isolate individual ascospores. There are two related methods available according to how the ejected ascospores are trapped, as follows.

Method 1

1. Pour a thin layer of clear agar (e.g. water agar) into the top (i.e. uppermost) lid of a Petri dish.

2. Place the lid over a Petri dish containing either individual ascoma or sets of ascomata (transferred on agar to keep the cultures moist).
3. Incubate the cultures for a suitable period (e.g., overnight or up to 48 h) during which time ascospores will be ejected onto the overlying agar.
4. Inspect the lids containing the ascospores using a dissecting microscope.
5. Subculture either ascospores or hyphal tips from germinating ascospores as described above.

Method 2

1. Aliquot 3 mL of 0.05 % Tween 80 or 0.01 % Tween 20 into the bottom of a sterile 5 mL Petri dish.
2. Attach either an individual ascoma or sets of ascomata to the lid of the Petri dish using Vaseline (petroleum jelly), ensuring that some of the underlying growth media is included to prevent desiccation of the ascoma.
3. Incubate the cultures for a suitable period (e.g., overnight or up to 48 h) during which time ascospores will be ejected into the underlying liquid.
4. Spread plate 0.5 mL aliquots of the Tween solution onto 9 cm Petri dishes containing a suitable clear media, and leave the dishes left to dry.
5. Incubate further and inspect the plates using a dissecting microscope.
6. Subculture either ascospores or hyphal tips from germinating ascospores as described above.

2.5.6.4 Isolation from Sclerotial Fruiting Bodies

If the fruiting bodies are firm and sclerotial, the following method can be used.

1. Harvest the fruiting bodies by adding 10 mL 0.05 % Tween 80 to the culture (slant or plate).
2. Scrape of the agar surface with a transfer loop.
3. Filter the suspension through a 100 μm filter.
4. Transfer the retained sclerotia/stromata to a vial.
5. Vortex the suspension followed by decanting to remove residual conidia. Repeat this procedure at least five times.

6. Filter the suspension onto Whatman #4 filter paper.
7. Clean the stromata used for obtaining single-ascospore cultures further by vortexing 1–2 min in 10 mL sterile water containing 1 g glass beads (200–350 µm diameter).
8. Remove from the beads.
9. Vortex and decant twice with water.
10. Dissect sclerotia/stromata with a micro-scalpel under the stereomicroscope.
11. Transfer the ascospores onto an agar medium (e.g. water agar) as described above.
12. Incubate overnight at a temperature allowing germination.
13. Locate single ascospores using a (dissecting) microscope and transfer to separate agar plates.

2.5.7 Analysis of Progeny

A selection of the collected single-ascospore isolates should be assessed, to confirm that recombination has occurred in the case of putative outcrossing. Ascospore analysis can also be used to confirm the breeding system of the species in question. Various DNA fingerprinting techniques, such as RAPD and AFLP analysis, together with use of MLST markers and segregation of the *MAT* locus itself have been used to evaluate variation in the parents and progeny of a cross (e.g. Murtagh et al. 2000; Seymour et al. 2005; O'Gorman et al. 2009; Horn et al. 2009c; Swilaiman et al. 2013). It would be expected that self-fertilization via a homothallic breeding system would lead to uniformity in the ascospore progeny, whereas heterothallism and outcrossing would lead to genetic variation among the offspring (Murtagh et al. 2000). Indeed, a 1:1 segregation of mating types among the ascospore progeny is a clear indication of the presence of a heterothallic breeding system. However, it is cautioned that some ascospore progeny will be identical due to the fact that each set of eight ascospores in an ascus is composed of four sets of sister ascospores, derived from a mitotic division post-meiosis and tetrad formation.

2.6 Utilization of the Sexual Cycle as a Tool for Gene Identification and Manipulation

Once a reliable method has been developed to induce the sexual cycle in a given fungal species, it can then be used as a valuable laboratory tool for a range of applications such as classical genetic analysis, strain improvement, and as a complement to modern genetic manipulation experiments. A comprehensive description of the applications and uses of the sexual cycle is beyond the scope of the present chapter. Instead brief mention will now be made of ways in which the sexual cycle can be used in the context of fungal gene identification and transformation relevant to current accompanying chapters.

2.6.1 Genetics of Traits of Interest and Gene Identification

For many studies of gene transformation, candidate genes will have been identified at the onset of studies. However, in some cases the gene(s) responsible for a particular trait (and associated phenotype) of interest might be unknown. When studying such a trait it is very useful to know at the onset of gene manipulation studies whether that trait has a monogenic (i.e., determined by a single gene) or polygenic (i.e., determined by several genes) basis. This can influence the design of subsequent experimental and gene transformation studies. The sexual cycle provides an ideal tool to determine the genetic basis of a trait of interest because crosses can be set up between parents which differ in that trait. The ascospore progeny can then be collected and assessed for the trait of interest, with different patterns predicted in the frequency of offspring according to the genetic basis of the trait. In the case of a monogenic trait determined by a single dominant gene, it would be expected that haploid ascospore progeny will show a 1:1 segregation pattern for that trait. This can ideally be confirmed by backcrossing to the relevant parent. Conversely, if a trait has a polygenic basis with several genes segregating simultaneously then a more complex pattern of inheritance will be evident, with the progeny often failing to show distinct classes but rather a continuous distribution of phenotypes between the two parents (Caten 1979; Dyer et al. 2000).

Once the genetic basis of a trait is known, various techniques exploiting the sexual cycle are then available to locate and try and identify the specific gene or genes of interest, or other genetic causal factor(s). Examples include firstly the use of classical genetic mapping techniques. Genetic maps now available for almost 30 fungal species, although these vary in their coverage (Foulongne-Oriol 2012). By using two- and three-point crossing data it should be possible to locate the position of the gene between two known markers on a genetic map (Perkins 1986) and then chromosome walking and bioinformatic approaches can be used to try and identify specific genes. This topic has recently been reviewed by Foulongne-Oriol (2012) and examples of the use of mapping of Mendelian traits are provided in the review of Hall (2013). Secondly, a method termed 'bulk segregant analysis' (BSA) can be used to identify DNA marker(s) linked to a region of the genome responsible for a particular phenotype based on analysis of sexual progeny (Michelmore et al. 1991). This involves making pooled bulks of DNA from the progeny based on the presence or absence of the phenotype of interest. These pooled samples can either then be screened for the presence of DNA markers (e.g., PCR fingerprints) or subjected to next generation sequencing (NGS). In theory the only differences between the pools should arise from the genetic marker of interest together with regions of the genome linked to that marker. Examples of the application of BSA to filamentous fungi using molecular markers include the work of Chun and Lee (1999), Jurgenson et al. (2002), Jin et al. (2007), Lewis et al. (2007), and Dettman et al. (2010). Meanwhile, BSA has been applied in conjunction with NGS by Lambreghts et al. (2009), Pomraning et al. (2011), and Nowrousian et al. (2012). A third method for gene localization based on the analysis of sexual progeny is the technique of quantitative trait loci (QTL) analysis. The QTL method is especially suitable for providing insights into the genetic basis of more complex polygenic traits and can provide an estimation of the number of genes contributing to a particular trait and the identification of regions of the genomes linked to a particular trait (Miles and Wayne 2008). Hall (2013) has described how QTL mapping can be applied to genetic analysis of *Neurospora* species, including a review of the various mapping methods, and readers are referred there for further details. Recent examples of QTL analysis in filamentous fungi include those of Christians et al. (2011) in *A. nidulans* and Turner et al. (2011) in *Neurospora*. This method has the pre-requirement of parents which differ genetically with respect to the trait of interest, together with the presence of a dense genetic map. Thus, QTL mapping is only applicable to certain studies. It is also cautioned that although both BSA and QTL approaches can involve considerable work, they most often end with the identification of a genome region of interest with various candidate genes, rather than the actual identification of specific genes. Thus, further work is normally required after such studies.

2.6.2 Gene Manipulation by Sexual Reproduction: Strain Improvement and Gene Complementation

Other chapters in this book describe how the genetic composition of fungi can be manipulated by various methods of gene transformation. Although often overlooked, the sexual cycle can be also used as an efficient method for gene transformation.

The sexual cycle can be used to combine together genes of interest by crossing parents with the individual gene(s) and then selecting for ascospore offspring showing the desirable mixture of genes. For example, it might be desirable in certain gene function studies to produce mutant strains with multiple gene deletions. This can be achieved by lengthy rounds of transformation and marker recycling (Yoon et al. 2011). However, the sexual cycle can provide an efficient alternative method because strains with complementary gene deletions can simply be crossed together. The sexual progeny are then screened for the presence or absence of the genes (e.g. by PCR assay) and isolates containing the desired combinations of multiple gene deletion selected for further study. In a parallel fashion, Böhm et al. (2013) illustrated how such an approach could be used for industrial strain improvement, with sexual progeny of *P. chrysogenum* screened for isolates that exhibited high penicillin titre but lack of a contaminating secondary metabolite chrysogenin. More broadly, sexual crosses can be set up to generate novel genetic diversity, allowing the offspring to be screened for isolates exhibiting, for example, either enhanced or decreased production of a particular metabolite, a phenomenon known as 'transgressive segregation' (Rieseberg et al. 1999).

Finally, the sexual cycle can be used for gene complementation purposes. When gene deletion has been shown to lead to a particular loss (or gain) of function, it is often required to then return the original gene back to the mutant strain to show that the original phenotype can be restored. This can be problematic if a limited number of selectable markers are available. However, by crossing the gene deletion strain with a strain containing the wild-type allele, it is possible to reintroduce the wild-type gene (Paoletti et al. 2007). If a consistent correlation between the presence of the wild-type gene (this can be screened for by PCR) and the presence of the restored phenotype can be demonstrated, or progeny which recapitulate the original genotype can be shown to exhibit the wild-type phenotype, then the role of the gene has been proven, i.e., proof of gene function through gene restoration by sexual crossing.

2.7 Conclusions and Outlook

Many industrially and clinically important fungal species were once thought to reproduce only asexually. Using a selection of the methods listed above, it has been shown that certain of these species may also reproduce sexually. For some other species a sexual state has still never been observed, although analyses of molecular markers indicate recombination. The term heterothallic is used for outcrossing species where a sexual state has been observed, and we suggest the term "proto-heterothallic" for such asexual species where genetic evidence, such as the presence of complementary *MAT* loci, indicates the presence of a sexual cycle. Similarly, for species with a homothallic mating type organization lacking a sexual state we propose the term "proto-homothallic".

In this manuscript, we describe tools for the induction of a sexual state in heterothallic fungi. Based on the review, we show that no single protocol can secure induction of a sexual state in heterothallic fungi, not even in species belonging to the same genus, e.g., *Aspergillus*. There are likely to be more surprises waiting as we observe a 'sexual revolution' in fungi (Dyer and O'Gorman 2011). Recently, sclerotia production was induced in the proto-heterothallic species *Aspergillus niger*. The formation of sclerotia, thought to be sterile fruiting bodies in certain *Aspergillus* species, can be induced by inoculating *A. niger* onto fruits such as raisins, blueberries, cranberries, mulberries, apricot, prune, and mango on a CYA agar (Frisvad et al. 2014). By following (one of the) methods mentioned above and applying these specific growth conditions, a sexual state might be discovered in the near future for this biotechnologically important species. Finally, experimental crossings in vitro do not strictly reflect what happens in natura. For example, heat-resistant ascospores of the heterothallic *P. variotii* frequently spoil pasteurised fruit drinks and other food products, indicating a common occurrence of these ascospores and therefore the sexual state in nature. However, incubation times up to 6 weeks and specific agar media (PDA) are needed to induce sexual recombination in the lab (Houbraken et al. 2008).

Acknowledgments J.H. thanks Tineke van Doorn and Richard Summerbell for the discussions on the term proto-heterothallic.

References

Aanen DK, Hoekstra RF (2007) Why sex is good: on fungi and beyond. In: Heitman J, Kronstad JW, Taylor JW, Casselton LA (eds) Sex in fungi: molecular determination and evolutionary principles. ASM, Washington, pp 527–534

Arabatzis M, Velegraki A (2013) Sexual reproduction in the opportunistic human pathogen *Aspergillus terreus*. Mycologia 105:71–79

Arie T, Christiansen SK, Yoder OC, Turgeon BG (1997) Efficient cloning of ascomycete mating type genes by PCR amplification of the conserved *MAT* HMG box. Fungal Genet Biol 32:118–130

Astell CR, Ahlstrom-Jonasson L, Smith M, Tatchell K, Nasmyth KA, Hall BD (1981) The sequence of the DNAs coding for the mating-type loci of *Saccharomyces cerevisiae*. Cell 27:15–23

Barrs VR, van Doorn TM, Houbraken J, Kidd SE, Martin P, Pinheiro DM, Richardson M, Varga J, Samson RA (2013) *Aspergillus felis* sp. nov., an emerging agent of invasive aspergillosis in humans, cats and dogs. PLoS One 8:e64871

Bistis GN (1998) Physiological heterothallism and sexuality in euascomycetes: a partial history. Fungal Genet Biol 23:213–222

Blakeslee AF (1904) Sexual reproduction in the Mucorineae. Proc Natl Acad Sci U S A 40:205–319

Böhm J, Hoff B, O'Gorman CM, Wolfers S, Klix V, Binger D, Zadra I, Kürnsteiner H, Pöggeler S, Dyer PS, Kück U (2013) Sexual reproduction and mating-type-mediated strain development in the penicillin-producing fungus *Penicillium chrysogenum*. Proc Natl Acad Sci U S A 110:1476–1481

Burnett J (2003) Fungal populations and species. Oxford University Press, New York

Caten CE (1979) Quantitative genetic variation in fungi. In: Thompson JN, Thoday JM (eds) Quantitative genetic variation. Academic, New York, pp 35–59

Cavindera B, Trail F (2012) Role of *Fig1*, a component of the low-affinity calcium uptake system, in growth and sexual development of filamentous fungi. Eukaryot Cell 11:978–988

Choi H-W, Kim J-M, Hong S-K, Kim WG, Chun S-C, S-H Y (2009) Mating types and optimum culture conditions for sexual state formation of *Fusarium fujikuroi* isolates. Mycobiology 37:247–250

Christians JK, Cheema MS, Vergara IA, Watt CA, Pinto LJ, Chen N, Moore MM (2011) Quantitative trait locus (QTL) mapping reveals a role for unstudied genes in *Aspergillus* virulence. PLoS One 6:e19325

Chun SJ, Lee Y-H (1999) Genetic analysis of a mutation on appressorium formation in *Magnaporthe grisea*. FEMS Microbiol Lett 173:133–137

Covert SF, Aoki T, O'Donnell K, Starkey D, Holliday A, Geiser DM, Cheung F, Town C, Strom A, Juba J, Scandiani M, Yang XB (2007) Sexual reproduction in the soybean sudden death syndrome pathogen *Fusarium tucumaniae*. Fungal Genet Biol 44:799–807

Covert SF, Briley A, Wallace MM, McKinney VT (1999) Partial MAT-2 gene structure and the influence of temperature on mating success in *Gibberella circinata*. Fungal Genet Biol 28:43–54

Crous PW, Verkleij GJM, Groenewald JZ, Samson RA (eds) (2009) Fungal biodiversity. CBS laboratory manual series 1. Centraalbureau voor Schimmelcultures, Utrecht, The Netherlands

Darbyshir HL, van de Vondervoort PJI, Dyer PS (2013) Discovery of sexual reproduction in the black aspergill. Fungal Genet REp 60(Suppl):687

Debuchy R, Berteaux-Lecellier V, Silar P (2010) Mating systems and sexual morphogenesis in ascomycetes. In: Borkovich KA, Ebbole DJ (eds) Cellular and molecular biology of filamentous fungi. ASM, Washington, pp 501–535

Debuchy R, Turgeon BG (2006) Mating-type structure, evolution, and function in euascomycetes. In: Kües U, Fischer R (eds) The mycota I: growth, differentiation and sexuality. Springer, Berlin, Germany, pp 293–323

Dettman JR, Anderson JB, Kohn LM (2010) Genome-wide investigation of reproductive isolation in experimental lineages and natural species of *Neurospora*: identifying candidate regions by microarray-based genotyping and mapping. Evolution 64:694–709

Dodge BO (1957) Rib formation in ascospores of *Neurospora* and questions of terminology. Bull Torrey Bot Club 84:182–188

Dyer PS (2007) Sexual reproduction and significance of *MAT* in the aspergilli. In: Heitman J, Kronstad JW, Taylor JW, Casselton LA (eds) Sex in fungi: molecular determination and evolutionary principles. ASM, Washington, pp 123–142

Dyer PS, Bateman GL, Wood HW (2001a) Development of apothecia of the eyespot pathogen *Tapesia* on cereal crop stubble residue in England. Plant Pathol 50:356–362

Dyer PS, Furneaux PA, Douhan G, Murray TD (2001b) A multiplex PCR test for determination of mating type applied to the plant pathogens *Tapesia yallundae* and *Tapesia acuformis*. Fungal Genet Biol 33:173–180

Dyer PS, Hansen J, Delaney A, Lucas JA (2000) Genetic control of resistance to the DMI fungicide prochloraz in the cereal eyespot pathogen *Tapesia yallundae*. Appl Environ Microbiol 66:4599–4604

Dyer PS, Ingram DS, Johnstone K (1992) The control of sexual morphogenesis in the Ascomycotina. Biol Rev 67:421–458

Dyer PS, Nicholson P, Lucas JA, Peberdy JF (1995) Genetic control of sexual compatibility in *Tapesia yallundae*, 79. Abstracts, eighteenth fungal genetics conference, Asilomar. University of California, USA

Dyer PS, Nicholson P, Lucas JA, Peberdy JF (1996) *Tapesia acuformis* as a causal agent of eyespot disease of cereals and evidence for a heterothallic mating system using molecular markers. Mycol Res 100:1219–1226

Dyer PS, Nicholson P, Rezanoor HN, Lucas JA, Peberdy JF (1993) Two-allele heterothallism in *Tapesia yallundae*, the teleomorph of the cereal eyespot pathogen *Pseudocercosporella herpotrichoides*. Physiol Mol Plant Pathol 43:403–414

Dyer PS, O'Gorman CM (2011) A fungal sexual revolution: *Aspergillus* and *Penicillium* show the way. Curr Opin Microbiol 14:649–654

Dyer PS, O'Gorman CM (2012) Sexual development and cryptic sexuality in fungi: insights from *Aspergillus* species. FEMS Microbiol Rev 36:165–192

Dyer PS, Paoletti M (2005) Reproduction in *Aspergillus fumigatus*: sexuality in a supposedly asexual species? Med Mycol 43(Suppl 1):S7–S14

Eagle CE (2009) Mating-type genes and sexual potential in the Ascomycete genera *Aspergillus* and *Penicillium*. PhD thesis, University of Nottingham, UK

Ehrenberg CG (1820) *Syzygites*, eine neue Schimmelgattung, nebst Beobachtungen über sichtbare Bewegung in Schimmeln. Verhandl. Gesamte Naturf., Freunde, Berlin, pp 98–109

Faretra F, Antonacci E, Pollastro S (1988) Sexual behaviour and mating system of *Botryotinia fuckeliana*, teleomorph of *Botrytis cinerea*. J Gen Microbiol 134:2543–2550

Foulongne-Oriol M (2012) Genetic linkage mapping in fungi: current state, applications, and future trends. Appl Microbiol Biotechnol 95:891–904

Fraser JA, Stajich JE, Tarcha EJ, Cole GT, Inglis DO, Sil A, Heitman J (2007) Evolution of the mating type locus: insights gained from the dimorphic primary fungal pathogens *Histoplasma capsulatum*, *Coccidioides immitis*, and *Coccidioides posadasii*. Eukaryot Cell 6:622–629

Fresenius G (1863) Beiträge zur Mykologie. H.L. Brönner, Frankfurt, Germany

Frisvad JC, Petersen LM, Lyhne K, Larsen TO (2014) Formation of sclerotia and production of indoloterpenes by *Aspergillus niger* and other species in section *Nigri*. PLoS One 9:e94857

Galagan JE, Calvo SE, Cuomo C, Ma LJ, Wortman J, Batzoglou S, Lee SL, Batürkmen M, Spevak CC, Clutterbuck J, Kapitonov V, Jurka J, Scazzocchio C, Farman M, Butler J, Purcell S, Harris S, Braus GH, Draht O, Busch S, D'Enfert C, Bouchier C, Goldman GH, Bell-Pedersen D, Griffiths-Jones S, Doonan JH, Yu J, Vienken K, Pain A, Freitag M, Selker EU, Archer DB, Peñalva MA, Oakley BR, Momany M, Tanaka T, Kumagai T, Asai K, Machida M, Nierman WC, Denning DW, Caddick M, Hynes M, Paoletti M, Fischer R, Miller B, Dyer P, Sachs MS, Osmani SA, Birren B (2005) Sequencing of *Aspergillus nidulans* and comparative analysis with *A. fumigatus* and *A. oryzae*. Nature 438:1105–1115

Gordon WL (1961) Sex and mating type in relation to the production of perithecia by certain species of *Fusarium*. Proc Can Phytopathol Soc 28:11

Hall C (2013) Quantitative genetics in *Neurospora*. In: Kasbekar DP, McClusky K (eds) *Neurospora* genomics and molecular biology. Caister Academic, Norfolk, UK, pp 65–84

Han KH, Lee DB, Kim JH, Kim MS, Han KY, Kim WS, Park YS, Kim HB, Han DM (2003) Environmental factors affecting development of *Aspergillus nidulans*. J Microbiol 41:34–40

Heitman J, Carter DA, Dyer PS, Soll DR (2014) Sexual reproduction of human fungal pathogens. In: Casadevall A, Mitchell AP, Berman J, Kwon-Chung KJ, Perfect JR, Heitman J (eds) Fungal pathogens. Cold Spring Harbour Laboratory Press, New York

Henk DA, Eagle CE, Brown K, van den Berg MA, Dyer PS, Peterson SW, Fisher MC (2011) Speciation despite globally overlapping distributions in *Penicillium chrysogenum*: the population genetics of Alexander Fleming's lucky fungus. Mol Ecol 20:4288–4301

Hoff B, Pöggeler S, Kück U (2008) Eighty years after its discovery, Fleming's *Penicillium* strain discloses the secret of its sex. Eukaryot Cell 7:465–470

Horn BW, Moore GG, Carbone I (2009a) Sexual reproduction in *Aspergillus flavus*. Mycologia 101:423–429

Horn BW, Moore GG, Carbone I (2011) Sexual reproduction in aflatoxin-producing *Aspergillus nomius*. Mycologia 103:174–183

Horn BW, Olarte RA, Peterson SW, Carbone I (2013) Sexual reproduction in *Aspergillus tubingensis* from section *Nigri*. Mycologia 105:1153–1163

Horn BW, Ramirez-Prado J, Carbone I (2009b) The sexual state of *Aspergillus parasiticus*. Mycologia 101:275–280

Horn BW, Ramirez-Prado JH, Carbone I (2009c) Sexual reproduction and recombination in the aflatoxin-producing fungus *Aspergillus parasiticus*. Fungal Genet Biol 46:169–175

Houbraken J, Frisvad JC, Seifert KA, Overy DP, Tuthill DM, Valdez JG, Samson RA (2012) New penicillin-producing *Penicillium* species and an overview of section *Chrysogena*. Persoonia 29:78–100

Houbraken J, Varga J, Rico-Munoz E, Johnson S, Samson RA (2008) Sexual reproduction as the cause of heat resistance in the food spoilage fungus *Byssochlamys spectabilis* (anamorph *Paecilomyces variotii*). Appl Environ Microbiol 74:1613–1619

Houbraken J, de Vries RP, Samson RA (2014) Modern taxonomy of biotechnologically important *Aspergillus* and *Penicillium* species. Adv Appl Microbiol 86:199–249

Idnurm A (2011) Sex determination in the first-described sexual fungus. Eukaryot Cell 10:1485–1491

Jin Y, Allen S, Baber L, Bhattarai EK, Lamb TM, Versaw WK (2007) Rapid genetic mapping in *Neurospora crassa*. Fungal Genet Biol 44:455–465

Jurgenson JE, Zeller KA, Leslie JF (2002) Expanded genetic map of *Gibberella moniliformis* (*Fusarium verticillioides*). Appl Environ Microbiol 68:1972–1979

Kawasaki M, Anzawa K, Wakasa A, Takeda K, Mochizuki T, Ishizaki H, Hemashettar B (2010) Matings among three teleomorphs of *Trichophyton mentagrophytes*. Jpn J Med Mycol 51:143–452

Kent CR, Ortiz-Bermúdez P, Gilies SS, Hull CM (2008) Formulation of a defined V8 medium for induction of sexual development of *Cryptococcus neoformans*. Appl Environ Biol 74:6248–6253

Kerényi Z, Moretti A, Waalwijk C, Olah B, Hornok L (2004) Mating type sequences in asexually reproducing *Fusarium* species. Appl Environ Microbiol 70:4419–4423

Klittich CJR, Leslie JF (1988) Nitrate reduction mutants of *Fusarium moniliforme* (*Gibberella fujikuroi*). Genetics 118:417–423

Kwon-Chung KJ, Sugui JA (2009) Sexual reproduction in *Aspergillus* species of medical or economic importance: why so fastidious? Trends Microbiol 17:481–487

Kwon-Chung KJ, Weeks RJ, Larsh HW (1974) Studies on *Emmonsiella capsulata* (*Histoplasma capsulatum*). II. Distribution of the two mating types in 13 endemic states of the United States. Am J Epidemiol 99:44–49

Lambreghts R, Shi M, Belden WJ, deCaprio D, Park D, Henn MR, Galagan JE, Baştürkmen M, Birren BW, Sachs MS, Dunlap JC, Loros JL (2009) A high-density single nucleotide polymorphism map for *Neurospora crassa*. Genetics 181:467–781

Lee SC, Ni M, Li W, Shertz C, Heitman J (2010) The evolution of sex: a perspective from the fungal kingdom. Microbiol Mol Biol Rev 74:298–340

Lewis ZA, Shiver AL, Stiffler N, Miller MR, Johnson EA, Selker EU (2007) High-density detection of restriction-site-associated DNA markers for rapid mapping of mutated loci in *Neurospora*. Genetics 177:1163–1171

López-Villavicencio M, Aguileta G, Giraud T, de Vienne DM, Lacoste S, Couloux A, Dupont J (2010) Sex in *Penicillium*: combined phylogenetic and experimental approaches. Fungal Genet Biol 47:693–706

McAlpin CE, Wicklow DT (2005) Culture media and sources of nitrogen promoting the formation of stromata and ascocarps in *Petromyces alliaceus* (*Aspergillus* section *Flavi*). Can J Microbiol 51:765–771

Metzenberg RL, Glass NL (1990) Mating type and mating strategies in *Neurospora*. Bioessays 12:53–59

Michelmore RW, Paran I, Kesseli RV (1991) Identification of markers linked to disease-resistance genes by bulked segregant analysis: a rapid method to detect markers in specific genomic regions by using segregating populations. Proc Natl Acad Sci U S A 88:9828–9832

Miles CM, Wayne M (2008) Quantitative trait locus (QTL) analysis. Nat Educ 1:208

Milgroom MG (1996) Recombination and the multilocus structure of fungal population. Annu Rev Phytopathol 34:457–477

Mooney JL, Yager LN (1990) Light is required for conidiation in *Aspergillus nidulans*. Genes Dev 4:1473–1482

Moore GG (2014) Sex and recombination in aflatoxigenic Aspergilli: global implications. Front Microbiol 5:32

Murtagh GJ, Dyer PS, Crittenden PD (2000) Sex and the single lichen. Nature 404:564

Nielsen KA, De Obaldia AL, Heitman J (2007) *Cryptococcus neoformans* mates on pigeon guano: implications for the realized ecological niche and globalization. Eukaryot Cell 6:949959

Nowrousian M, Teichert I, Masloff S, Kück U (2012) Whole-genome sequencing of *Sordaria macrospora* mutants identifies developmental genes. G3. Genes Genomes Genetics 2:261–270

Nováková A, Hubka V, Dudová Z, Matsuzawa T, Kubátová A, Yaguchi T, Kolařík M (2014) New species in *Aspergillus* section *Fumigati* from reclamation sites in Wyoming (U.S.A.) and revision of *A. viridinutans* complex. Fung Div 64(1):253–274. doi:10.1007/s13225-013-0262-5

O'Gorman CM, Fuller HT, Dyer PS (2009) Discovery of a sexual cycle in the opportunistic fungal pathogen *Aspergillus fumigatus*. Nature 457:471–474

Paoletti M, Rydholm C, Schwier EU, Anderson MJ, Szakacs G, Lutzoni F, Debeaupuis JP, Latgé JP, Denning DW, Dyer PS (2005) Evidence for sexuality in the opportunistic human pathogen *Aspergillus fumigatus*. Curr Biol 15:1242–1248

Paoletti M, Seymour FA, Alcocer MJC, Kaur N, Calvo AM, Archer DB, Dyer PS (2007) Mating type and the genetic basis of self-fertility in the model fungus *Aspergillus nidulans*. Curr Biol 17:1384–1389

Perkins DD (1986) Hints and precautions for the care, feeding and breeding of *Neurospora*. Fungal Genet Newsl 33:36–41

Pitt JI (1979) The genus *Penicillium* and its teleomorph states *Eupenicillium* and *Talaromyces*. Academic, London, UK

Pöggeler S (2001) Mating-type genes for classical strain improvements of ascomycetes. Appl Microbiol Biotechnol 56:589–601

Pomraning KR, Smith KM, Freitag M (2011) Bulk segregant analysis followed by high-throughput sequencing reveals the *Neurospora* cell cycle gene, *ndc-1*, to be allelic with the gene for ornithine decarboxylase, *spe-1*. Eukaryot Cell 10:724–733

Pringle A, Baker DM, Platt JL, Wares JP, Latgé JP, Taylor JW (2005) Cryptic speciation in the cosmopolitan and clonal human pathogenic fungus *Aspergillus fumigatus*. Evolution 59:1886–1899

Raju NB, Perkins DD (1994) Diverse programs of ascus development in pseudohomothallic species of *Neurospora*, *Gelasinospora* and *Podospora*. Dev Genet 15:104–118

Ramirez-Prado JH, Moore GG, Horn BW, Carbone I (2008) Characterization and population analysis of the mating-type genes in *Aspergillus flavus* and *Aspergillus parasiticus*. Fungal Genet Biol 45:1292–1299

Raper KB, Fennell DI (1965) The genus *Aspergillus*. Williams & Wilkins Co, Baltimore

Rieseberg LH, Archer MA, Wayne RK (1999) Transgressive segregation, adaptation and speciation. Heredity 83:363–372

Robellet X, Oestreicher N, Guitton A, Vélot C (2010) Gene silencing of transgenes inserted in the *Aspergillus nidulans alcM* and/or *alcS* loci. Curr Genet 56:341–348

Ropars J, López-Villavicencio M, Dupont J, Snirc A, Gillot G, Coton M, Jany J-L, Coton E, Giraud T (2014) Induction of sexual reproduction and genetic diversity in the cheese fungus *Penicillium roqueforti*. Evol Appl 7:433–441

Rydholm C, Dyer PS, Lutzoni F (2007) DNA sequence characterization and molecular evolution of MAT1 and MAT2 mating-type loci of the self-compatible ascomycete mold *Neosartorya fischeri*. Eukaryot Cell 6:868–874

Samson RA, Houbraken J, Thrane U, Frisvad JC, Andersen B (2010) Food and indoor fungi. CBS laboratory manual Series 2. CBS KNAW Fungal Biodiversity Centre, Utrecht, The Netherlands

Samson RA, Varga J, Dyer PS (2009) Morphology and reproductive mode of *Aspergillus fumigatus*. In: Latgé JP, Steinbach WJ (eds) *Aspergillus fumigatus* and Aspergillosis. ASM, Washington, pp 7–13

Schoch CL, Seifert KA, Huhndorf S, Robert V, Spouge JL, Levesque CA, Chen W, Fungal Barcoding Consortium (2012) Nuclear ribosomal internal transcribed spacer (ITS) region as a universal DNA barcode marker for Fungi. Proc Natl Acad Sci U S A 109:6241–6246

Seidl V, Seibel C, Kubicek CP, Schmoll M (2009) Sexual development in the industrial workhorse *Trichoderma reesei*. Proc Natl Acad Sci U S A 106:13909–13914

Seymour FA, Crittenden PD, Dickinson MJ, Paoletti M, Montiel D, Cho L, Dyer PS (2005) Breeding systems in the lichen-forming fungal genus *Cladonia*. Fungal Genet Biol 42:554–563

Short DP, O'Donnell K, Thrane U, Nielsen KF, Zhang N, Juba JH, Geiser DM (2013) Phylogenetic relationships among members of the *Fusarium solani* species complex in human infections and the descriptions of *F. keratoplasticum* sp. nov. and *F. petroliphilum* stat. nov. Fungal Genet. Biol 53:59–70

Singh G, Dyer PS, Ashby AM (1999) Intra-specific and inter-specific conservation of mating-type genes from the discomycete plant-pathogenic fungi *Pyrenopeziza brassicae* and *Tapesia yallundae*. Curr Genet 36:290–300

Steenkamp ET, Wingfield BD, Coutinho TA, Zeller KA, Wingfield MJ, Marasas WF, Leslie JF (2000) PCR-based identification of MAT-1 and MAT-2 in the *Gibberella fujikuroi* species complex. Appl Environ Microbiol 66:4378–4382

Sugui JA, Losada L, Wang W, Varga J, Ngamskulrungroj P, Abu-Asab M, Chang YC, O'Gorman CM, Wickes BL, Nierman WC, Dyer PS, Kwon-Chung KJ (2011) Identification and characterization of an *Aspergillus fumigatus* "supermater" pair. MBio 2:e00234–11

Sugui JA, Vinh DC, Nardone G, Shea YR, Chang YC, Zelazny AM, Marr KA, Holland SM, Kwon-Chung KJ (2010) *Neosartorya udagawae* (*Aspergillus udagawae*), an emerging agent of aspergillosis: how different is it from *Aspergillus fumigatus*? J Clin Microbiol 48:220–228

Swilaiman SS, O'Gorman CM, Balajee SA, Dyer PS (2013) Discovery of a sexual cycle in *Aspergillus lentulus*, a close relative of A. fumigatus. Eukaryot Cell 12:962–969

Takada M, Udagawa S (1988) A new species of heterothallic *Talaromyces*. Mycotaxon 31:417–425

Takan JP, Chipili J, Muthumeenakshi S, Talbot NJ, Manyasa EO, Bandyopadhyay R, Sere Y, Nutsugah SK, Talhinhas P, Hossain M, Brown AE, Sreenivasaprasad S (2012) *Magnaporthe oryzae* populations adapted to finger millet and rice exhibit distinctive patterns of genetic diversity, sexuality and host interaction. Mol Biotechnol 50:145–158

Todd RB, Davis MA, Hynes MJ (2007) Genetic manipulation of *Aspergillus nidulans*: meiotic progeny for genetic analysis and strain construction. Nat Protoc 2:811–821

Turner E, Jacobson DJ, Taylor JW (2011) Genetic architecture of a reinforced, postmating reproductive isolation barrier between *Neurospora* spceies indicates evolution via natural selection. PLoS Genet 7:e1002204

Tschanz AT, Horst RK, Nelson PE (1976) The effect of environment on sexual reproduction of *Gibberella zeae*. Mycologia 68:327–340

Turgeon BG, Yoder OC (2000) Proposed nomenclature for mating type genes of filamentous ascomycetes. Fungal Genet Biol 31:1–5

Yoon J, Maruyama J, Kitamoto K (2011) Disruption of ten protease genes in the filamentous fungus *Aspergillus oryzae* highly improves production of heterologous proteins. Appl Microbiol Biotechnol 89:747–759

What Have We Learned by Doing Transformations in *Neurospora tetrasperma*?

3

Durgadas P. Kasbekar

3.1 Introduction to *Neurospora tetrasperma* and Comparison with *N. crassa*

The germination of an ascospore marks the "birth" of a *Neurospora* strain. Ascospores are the end products of a sexual cross between two strains, one of mating type *mat A*, and the other *mat a*. Heat, or chemicals produced by heating the substrate, induces ascospores to germinate and to send out a germ tube that becomes the first hypha. Ascospores also serve as units of dispersal, although dispersal can additionally occur during vegetative growth, via powdery wind-borne vegetative spores (called conidia), that bud off from the tips of aerial hyphae (Pandit and Maheshwari 1996).

Conidia are spheroid cells containing 2–10 nuclei that can live for up to several weeks, whereas ascospores are more resistant to stress and longer lived (months to years). Under favorable conditions a conidium sends out a germ tube to produce a new hypha. Conidia also function as the paternal fertilizing element during a sexual cross. The maternal element is the protoperithe-

cium, a specialized knot of hyphae that is produced from the vegetative hyphae following nutrient deprivation. Specialized hyphae called trichogynes emanate from the protoperithecia and in response to mating-type-specific sex hormone from conidia of the opposite mating type show chemotropic growth towards the conidia (Bistis 1996). Fertilization of protoperithecia by conidia of the opposite mating type is the prelude to their differentiation into perithecia. Within the perithecia, the dikaryotic ascogenous hyphae undergo several rounds of karyogamy (nuclear fusion) between nuclei of opposite mating types, and the diploid zygote nucleus produced by each nuclear fusion immediately undergoes meiosis in a cell, called the penultimate cell, that then differentiates to become an ascus. The four haploid nuclei from meiosis then undergo a post-meiotic mitosis. In *Neurospora crassa*, the resultant eight nuclei (4 *mat A* + 4 *mat a*) are then partitioned into the eight ascospores that develop within an ascus. Additional mitotic divisions occur within each ascospore and produce more nuclei and the ascospore then matures and becomes dormant. Finally, octets of ascospores produced in each ascus are shot out through the ostiole, an aperture at the top of the perithecium. All the nuclei in a strain (mycelium) produced by germination of an individual ascospore are of the same genotype (i.e., the mycelium is homokaryotic), therefore to complete the sexual cycle mycelium from an ascospore of the opposite mating type is needed.

D.P. Kasbekar, Ph.D. (✉)
Centre for DNA Fingerprinting and Diagnostics,
Tuljaguda Complex Nampally, Hyderabad 500001,
India
e-mail: kas@cdfd.org.in

Since the products of two ascospores are required for completion of the sexual cycle, *N. crassa* is designated as a heterothallic species.

In contrast, in *N. tetrasperma* the eight haploid nuclei produced by meiosis and post-meiotic mitosis are packaged as four non-sister pairs (1 *mat A* + 1 *mat a*) into each of four ascospores that form per ascus. Each ascospore thus contains nuclei of both mating types, and the resulting mycelium is heterokaryotic (i.e., having nuclei of more than genotype) and it is competent to complete the sexual cycle without the need for a mycelium from another ascospore. Since the sexual cycle can be completed by the mycelium from a single heterokaryotic ascospore, *N. tetrasperma* is a pseudohomothallic species.

A subset of the vegetative conidia produced by a heterokaryotic *N. tetrasperma* mycelium can by chance be homokaryotic, and germination of such conidia can give rise to vegetatively derived single-mating-type strains that are self-sterile. Homokaryotic strains can cross with like strains of opposite mating type. *N. tetrasperma* asci occasionally produce five or more (up to eight) ascospores instead of the normal four by replacement of a dikaryotic ascospore by a pair of homokaryotic ascospores, each homokaryotic ascospore being slightly smaller in size than a dikaryotic ascospore. The mycelium generated from the small ascospores is self-sterile, but it can cross with a homokaryon of the opposite mating type. The dominant *Eight-spore* (*E*) mutation can substantially increase the frequency of replacement of dikaryotic ascospores by pairs of smaller homokaryotic ascospores and a majority of asci from crosses heterozygous for the *E* mutation contain five to eight ascospores. Therefore, *N. tetrasperma* is actually a facultatively heterothallic species.

The webpage http://www.fgsc.net/Neurospora/sectionB2.htm provides excellent figures to explain the differences between the heterothallic, pseudohomothallic, and homothallic lifecycles.

In *N. tetrasperma*, a marker that has undergone second-division segregation can become segregated into both the *mat A* and *mat a* nuclei of a subset of progeny ascospores. A self-cross of

the resulting culture would thus be homozygous for the marker. In *N. crassa*, a novel mutation arising in one of the haploid parents of a cross can be made homozygous only in a subsequent cross between f1 segregants of the opposite mating type, whereas in *N. tetrasperma* homozygosity for a newly arisen mutation that recombines with the centromere can be automatically achieved.

3.2 An *ERG-3* Mutant Enables Transformation of *N. tetrasperma*

Transformation experiments in *N. crassa* for the most part use the bacterial hygromycin (*hph*) gene as a selectable marker and select for transformants on medium supplemented with the antibiotic hygromycin. However, this protocol did not work in *N. tetrasperma* because it was found to be naturally resistant to hygromycin. We discovered that *ergosterol-3* (*erg-3*) mutations increased hygromycin-sensitivity in *N. crassa* (Bhat et al. 2004), which led us to surmise that *erg-3* mutations might also increase hygromycin-sensitivity in *N. tetrasperma*. The *erg-3* gene encodes the enzyme sterol C-14 reductase, which is essential for ergosterol biosynthesis. We crossed a strain bearing an *erg-3* mutation in the *N. crassa* gene with the *N. crassa / N. tetrasperma* hybrid strain C4, T4 (Metzenberg and Ahlgren 1969; Perkins 1991), and then used an *erg-3* mutant segregant to initiate a series of backcrosses with the *N. tetrasperma* reference strains *85 A* or *a*. We anticipated that an *erg-3* strain of *N. tetrasperma* would possess dual mating specificities; that is, it would be capable of crossing with both *85 A* and *85 a*; however, it would be self-sterile, since *erg-3* strains of *N. crassa* are female-sterile. We began recovering *erg-3* strains with dual mating specificity in the third backcross. A dual mating specificity *erg-3* mutant strain from the fourth backcross was designated Te-4 and adopted as the reference *N. tetrasperma erg-3* strain. UV spectroscopy confirmed the absence of ergosterol from Te-4, and we confirmed that Te-4 was self-sterile but that it could cross with both *85 A* and *a*. The wild-type and Te-4 differed strikingly in their sensitivity to hygromycin. Strain *85* conidia could grow in the presence of as much as 220 µg/mL of hygromycin whereas Te-4 conidia were sensitive, and the sensitivity phenotype segregated with the *erg-3* mutation in crosses. The hygromycin-sensitive phenotype made it feasible to use the Te-4 strain to select for transformants on hygromycin medium by complementation with the *hph* gene (Bhat et al. 2004).

3.3 Screening for RIP-Defective Mutants in *N. tetrasperma*

The genome defense process called repeat-induced point mutation (RIP) occurs during a sexual cross, in the haploid nuclei of the premeiotic dikaryon, and subjects duplicated DNA sequences to G:C to A:T mutations and cytosine methylation (Selker 1990). Only a few recessive RIP-defective mutants have been reported and they were identified by a "candidate gene" approach in *N. crassa* (Freitag et al. 2002). Isolation of additional mutants would define additional genes required for RIP. The difficulty of achieving homozygosity for unknown mutations affecting a diplophase-specific process such as RIP makes it impractical to use *N. crassa* to screen for such mutations, but such screens are feasible, at least in principle, in *N. tetrasperma* wherein a novel mutation can automatically become homozygous via second-division segregation. Our approach was to create a tagged duplication of the *erg-3* gene by transformation. In a sexual cross the duplication would target RIP to the endogenous *erg-3* gene. Ascospores bearing *erg-3* mutations produce colonies with a distinct morphology on Vogel's-sorbose agar medium, thereby allowing RIP efficiency to be determined by simply counting the number of wild-type and mutant progeny colonies under a dissection microscope. RIP in self-crosses of a *N. tetrasperma* strain duplicated for *erg-3* sequences would produce mutations in *erg-3*, and following this, crossing over between the centromere and the *erg-3* mutation results in a fraction of the ascospores becoming homoallelic for the mutation. Alternatively, some of the small ascospores might be homokaryotic for the mutation. In either case, if a self-cross failed to produce any *erg-3* mutant progeny, it signaled a potential homoallelism for a novel mutation conferring a RIP defect. We used transformation of Te-4 to construct self-fertile strains that contained a mutant *erg-3* allele at the endogenous locus and were homoallelic for an ectopic *erg-3⁺* transgene. RIP-induced *erg-3* mutant ascospores were generated in self-crosses of these strains at frequencies in the 2–20 % range, and they produced colonies with the mutant morphology on Vogel's sorbose agar. By screening for self-crosses that failed to produce *erg-3* mutant progeny, presumably due to homozygosity for novel recessive RIP-deficient mutations, we isolated a UV-induced mutant with a putative partial RIP defect (Bhat et al. 2004).

We also performed co-transformations of Te-4 with the *hph* gene together with PCR-amplified DNA of other genes to construct strains duplicated for the amplified DNA. In this way, we isolated RIP-induced mutants in *rid-1* and *sad-1*, which are essential genes, respectively, for RIP and another genome defense mechanism called meiotic silencing by unpaired DNA (Bhat et al. 2004).

3.4 Meiotic Silencing by Unpaired DNA in *N. tetrasperma*

Meiotic silencing by unpaired DNA (also MSUD, or simply, meiotic silencing) is a gene silencing mechanism discovered in *N. crassa* that is presumed to employ RNAi to eliminate the transcripts of any gene that does not pair properly in meiosis with a homolog in the same chromosomal position (Aramayo and Metzenberg 1996; Shiu et al. 2001, 2006). Meiotic silencing can be assayed by using tester strains such as ::*Bml^r* and ::*mei-3* that contain an extra copy of the β−tubulin, or *mei-3* (RAD51 ortholog) gene inserted ectopically into the *his-3* locus on chromosome 1. In the cross of a tester with a standard laboratory OR strain of opposite mating type, the ectopically duplicated gene lacks a homolog to pair with, and therefore it silences itself as well as its endogenous copies, regardless of the latter being paired. Since the gene product is essential for ascus development, the silencing results in an ascus-development defect and reduced ascospore production (Raju et al. 2007; Kasbekar et al. 2011). In a *tester A × tester a* homozygous cross the ectopic gene is paired, and consequently it is not silenced, and the cross shows normal ascus development and ascus production. Semi-dominant *Sad-1*, *Sad-2*, and *Sms-2*, and other mutations suppress meiotic silencing, presumably by disrupting the normal pairing of their wild-type alleles (i.e., *sad-1+*, *sad-2+*, *sms-2+*, etc), and induce the latter to silence themselves. Crosses of the testers with the *Sad-1* and *Sad-2* suppressors also show normal ascus development (Raju et al. 2007; Kasbekar et al. 2011).

When 80 wild-isolated *N. crassa* strains were examined by crossing them with the::*Bml[r]* and ::*mei-3* testers, only eight behaved like OR and showed silencing in crosses with both the testers (Ramakrishnan et al. 2011). These eight strains were designated the "OR" type. Four wild strains showed suppression of meiotic silencing of the *bml* and *mei-3* genes, and typified the "Sad type". Crosses with the 68 other wild strains showed suppression of silencing in *mei-3*, but not *bml*, and additional results suggested that in crosses of these strains, the silencing is greater in perithecia produced early in the cross relative to that in the perithecia produced later in the cross. We designated them an intermediate "Esm type" (for early silencing in meiosis). We hypothesized that the Sad or Esm phenotype arises from heterozygosity for sequence polymorphism in the cross with the OR-derived testers. The polymorphisms might reduce pairing and thereby silence meiotic silencing genes. These results prompted us to ask whether *N. tetrasperma* is of Sad or Esm type, since earlier results of Jacobson et al. (2008) had already suggested that meiotic silencing is greatly reduced or absent in *N. tetrasperma*. Jacobson et al. (2008) had hypothesized that structural differences between the mating-type chromosomes [chromosome 1 which bears the *mat* idiomorphs] might result in a substantial region to remain unsynapsed during normal meiosis and thus cause self-silencing of the *sad-1* gene.

We used the *asm-1* (*ascus maturation-1*) gene to study whether self-crosses in strain 85 were Esm or Sad type. The *asm-1* gene extends from nucleotide 3,977,115 to 3,980,557 of chromosome 5 and encodes a key regulator of sexual development. In *N. crassa*, a cross heterozygous for a deletion mutant (i.e., *asm-1+ × ΔAsm-1*) silences the *asm-1* gene and a large majority of the asci contain ascospores that remain white, immature, and inviable (Aramayo and Metzenberg 1996). We amplified a 1,724 bp segment (nucleotides 3,977,199–3,979,922) of *N. crassa asm-1* by PCR, then purified it by gel electrophoresis and gel elution, and cloned it into the vector pCSN43 which carries the bacterial *hph* gene (Staben et al. 1989). The resulting plasmid was transformed into the Te-4 strain and transformants were selected on Vogel's-sorbose agar medium supplemented with hygromycin. One transformant was crossed with the single-mating-type *85 A* strain and self-fertile progeny (i.e., containing both *mat A* and *mat a* nuclei), were identified. Southern analysis identified six strains that also contained the *Dp(asm)* transgene. If we disregard second-division segregation, self-crosses of these progeny are expected to be heterozygous for the transgene. A large proportion of asci from self-crosses of these progeny were white-spored, whereas white-spored asci were never seen from self-crosses of the 85 strain. By analogy with *N. crassa*, we presume that during meiosis *Dp(asm)* remains unpaired and this triggers *asm-1* silencing causing all the four ascospores of the ascus to remain white, immature, and inviable. The proportion of white-spored asci was greater (range 31–76 %) among the early tetrads and then showed a decline (range 4–33 %) after about a day. These results suggested that self-crosses in the strain 85 background are of the Esm type.

Additionally, we crossed the initial *Dp(asm)* transformant with the *E A* strain. As mentioned above, the *E* mutation increases the proportion of small homokaryotic ascospores. We identified two self-sterile progeny that presumably were of single-mating type, and that by Southern analysis were verified to contain the *Dp(asm)* transgene.

The two homokaryotic $Dp(asm)$ progeny were of opposite mating type therefore we could perform a cross that was homozygous for the $Dp(asm)$ transgene. Only three of the first 15 asci examined from this cross were white-spored, but all the subsequent asci (of >100 examined) were black-spored. These results were consistent with our expectation that the endogenous asm-1 locus is not subject to meiotic silencing in crosses homozygous for the $Dp(asm)$ transgene.

3.5 Conclusions and Future Prospects

The foregoing account documents our ability to successfully transform *N. tetrasperma* and use RIP to induce mutations in any gene. This ability opens up the prospect for quickly bringing the insights developed in *N. crassa* into *N. tetrasperma* to take advantage of the latter's pseudo-homothallic life cycle. For example, it is now possible to generate RIP-induced mutation in the *mus-51*, *mus-52*, and *mus-53* genes and thereby disable integration of transforming DNA via non-homologous end joining (NHEJ). In such mutants integration of transforming DNA would occur only by homologous recombination (Ninomiya et al. 2004). NHEJ-defective strains have been used to systematically knockout individual *N. crassa* genes (Dunlap et al. 2007). While the Fungal Genetics Stock Center (FGSC, USA) collection includes strains carrying knock-out mutations in genes involved in NHEJ for diverse species including *N. crassa*, *Magnaporthe grisea*, *Aspergillus nidulans*, *A. niger*, *A. flavus*, and *A. fumigatus*, it does not include a *N. tetrasperma* strain for targeted transformation. NHEJ-defective *N. tetrasperma* strains could be used to make knock-out mutants in *N. tetrasperma*. Indeed, given the high sequence homology between most genes in the two species (>94 %), it might even be possible to knock-out *N. tetrasperma* genes using the same DNA constructs used to make the *N. crassa* knock-out mutants.

NHEJ-defective *N. tetrasperma* strains can also potentially be employed to replace un-16^+ with un-16^{ts} to obtain un-16^{ts}; mus double mutant strains for selection of targeted transformants using the *Magnaporthe* orthologue of *N. crassa* un-16^+ (ncu01949) as a selective marker. The un-16^+ orthologue complements the "no growth at 37 °C" phenotype of *N. crassa* un-16^{ts} mutants (McCluskey et al. 2007). Insertion of the transforming DNA by homologous recombination will knock-out the target locus, confer temperature-independence to the transformants, and allow their selection at 37 °C. This approach will allow us to use temperature selection in an erg$^+$ background instead of hph selection. Development of a selectable marker based on complementation of a temperature sensitive (ts) lethal mutation in *Neurospora* means that transformation can be accomplished while leaving dominant markers such as hygromycin or phosphinothricin resistance for subsequent manipulations. It also avoids the mutagenicity associated with histidine supplementation required for use of the his-targeting system in *N. crassa*. Replacing un-16^+ with un-16^{ts} is non-trivial. One approach is to replace the un-16^+ allele in a *mus*; *erg-3* double mutant strain with *erg-3*$^+$ and select for *erg-3*$^+$ transformants on pisatin-medium. The transformants will include unwanted integrants into *erg-3*. Since un-16^+ is essential for viability, we expect the transformants to be [(un-16::erg-3^+)+(un-16^+)] heterokaryons. Next, one could use homologous recombination to replace un-16::erg-3^+ with un-16^{ts} and select for homokaryotic transformants on nystatin-medium. This strategy makes use of the pisatin-sensitive and nystatin-resistant phenotype of *erg-3* (Grindle 1973, 1974; Papavinasasundaram and Kasbekar 1993).

Another promising area, though not genomic transformation in the strict sense, is the introgression of *N. crassa* translocations into *N. tetrasperma*. The idea is to produce self-fertile [(*T*)+(*N*)] strains whose self-crosses can generate both [(*T*)+(*N*)] and [(*Dp*)+(*Df*)] progeny (Kasbekar 2014; Dev Ashish Giri and Durgadas P. Kasbekar, unpublished results). If the [(*Dp*)+(*Df*)] progeny turn out to be self-sterile, then it might provide the first evidence for the existence of "nucleus-limited" genes required for fertility. A nucleus-limited gene is one in which a wild-type allele

(*WT*) fails to complement a null allele (Δ) in a [(*WT*)+(Δ)] heterokaryon. Such genes have not yet been found, but their existence is predicted based on the putative nucleus-limited phenotype of the *scon^c* mutant (Burton and Metzenberg 1972), and the more recently discovered MatIS gene silencing in *A. nidulans* (Czaja et al. 2013).

Acknowledgement I thank Kevin McCluskey for many useful suggestions. My research in CDFD is supported by the Haldane Chair.

References

Aramayo R, Metzenberg RL (1996) Meiotic transvection in fungi. Cell 86:103–113

Bhat A, Tamuli R, Kasbekar DP (2004) Genetic transformation of *Neurospora tetrasperma*, demonstration of repeat-induced point mutation (RIP) in self-crosses, and a screen for recessive RIP-defective mutants. Genetics 167:1155–1164

Bistis GN (1996) Trichogynes and fertilization in uni-and bimating type colonies of *Neurospora tetrasperma*. Fungal Genet Biol 20:93–98

Burton EG, Metzenberg RL (1972) Novel mutations causing derepression of several enzymes of sulfur metabolism in *Neurospora crassa*. J Bacteriol 109:140–151

Czaja W, Miller KY, Miller BL (2013) Novel sexual-cycle-specific gene silencing in *Aspergillus nidulans*. Genetics 193:1149–1162

Dunlap JC, Borkovich KA, Henn MR, Turner GE, Sachs MS, Glass NL, McCluskey K, Plamann M, Galagan JE, Birren BW et al (2007) Enabling a community to dissect an organism: overview of the Neurospora functional genomics project. Adv Genet 57:49–96

Freitag M, Williams RL, Kothe GO, Selker EU (2002) A cytosine methyltransferase homologue is essential for repeat- induced point mutation in *Neurospora crassa*. Proc Natl Acad Sci U S A 99:8802–8807

Grindle M (1973) Sterol mutants of *Neurospora crassa*: their isolation, growth characteristics and resistance to polyene antibiotics. Mol Gen Genet 120:283–290

Grindle M (1974) The efficacy of various mutagens and polyene antibiotics for the induction and isolation of sterol mutants of *Neurospora crassa*. Mol Gen Genet 130:81–90

Jacobson DJ, Raju NB, Freitag M (2008) Evidence for the absence of meiotic silencing by unpaired DNA in Neurospora tetrasperma. Fungal Genet Biol 45:351–362

Kasbekar DP (2014) Are any fungal genes nucleus-limited? J Biosci 39(3):341–346

Kasbekar DP, Singh PK, Ramakrishnan M, Kranthi Raj B (2011) Carrefour Mme. Gras: a wild-isolated Neurospora crassa strain that suppresses meiotic silencing by unpaired DNA and uncovers a novel ascospore stability defect. Fungal Genet Biol 48:612–620

McCluskey K, Walker SA, Yedlin RL, Madole D, Plamann M (2007) Complementation of un-16 and the development of a stable marker for transformation of Neurospora crassa. Fungal Genet Newslett 54:9–11

Metzenberg RL, Ahlgren SK (1969) Hybrid strains useful in transferring genes from one species of Neurospora to another. Neurospora Newsl 15:9–10

Ninomiya Y, Suzuki K, Ishii C, Inoue H (2004) Highly efficient gene replacements in Neurospora strains deficient for nonhomologous end joining. Proc Natl Acad Sci U S A 101:12248–12253

Pandit A, Maheshwari R (1996) Life-history of *Neurospora intermedia* in a sugarcane field. J Biosci 21:57–79

Papavinasasundaram KG, Kasbekar DP (1993) Pisatin resistance in *Dictyostelium discoideum* and *Neurospora crassa*: comparison of mutant phenotypes. J Gen Microbiol 139:3035–3041

Perkins DD (1991) Transfer of genes and translocations from *Neurospora crassa* to *N. tetrasperma*. Fungal Genet Newslett 38:84

Raju NB, Metzenberg RL, Shiu PKT (2007) Neurospora spore killers Sk-2 and Sk-3 suppress meiotic silencing by unpaired DNA. Genetics 176:43–52

Ramakrishnan M, Naga Sowjanya T, Raj KB, Kasbekar DP (2011) Meiotic silencing by unpaired DNA is expressed more strongly in the early than the late perithecia of crosses involving most wild-isolated *Neurospora crassa* strains and in self-crosses of *N. tetrasperma*. Fungal Genet Biol 48:1146–1152

Selker EU (1990) Premeiotic instability of repeated sequences in *Neurospora crassa*. Annu Rev Genet 24:579–613

Shiu PK, Raju NB, Zickler D, Metzenberg RL (2001) Meiotic silencing by unpaired DNA. Cell 107:905–916

Shiu PKT, Zickler D, Raju NB, Ruprich-Robert G, Metzenberg RL (2006) SAD-2 is required for meiotic silencing by unpaired DNA and perinuclear localization of SAD-1 RNA-directed RNA polymerase. Proc Natl Acad Sci U S A 103:2243–2248

Staben C, Jensen B, Singer M, Pollock J, Schechtman M, Kinsey JA, Selker EU (1989) Use of a bacterial hygromycin B resistance gene as a dominant selectable marker in Neurospora crassa transformations. Fungal Genet Newslett 36:79

Part II

Endogenous DNA: Repetitive Elements

Repeat-Induced Point Mutation: A Fungal-Specific, Endogenous Mutagenesis Process

4

James K. Hane, Angela H. Williams, Adam P. Taranto, Peter S. Solomon, and Richard P. Oliver

4.1 Introduction

4.1.1 Observations of RIP in *Neurospora crassa*

RIP was first identified in *Neurospora crassa* by Selker and colleagues (Selker et al. 1987a) and subsequently identified in a number of fungal species. The main outcome of RIP is to induce transition mutations (interchanges between pyrimidines or between purine bases, i.e., C↔T or G↔A) in repeated sequences during the sexual stage of the fungal life-cycle. The effects of RIP were first identified in the progeny of a cross of *N. crassa* isolates carrying a transformation vector designed to investigate DNA methylation control in *Neurospora* (Selker et al. 1987b). This study found that the introduced DNA sequences that were homologous to those in the host genome were rapidly mutated

J.K. Hane, BMolBiol. (Hons.), Ph. D. Bioinformatics (✉)
R.P. Oliver, B.Sc., Ph.D. Biochem.
Centre for Crop and Disease Management, Curtin University, Perth, WA, Australia
e-mail: James.Hane@curtin.edu.au

A.H. Williams, B. Appl. Sci. (Hons.)
The Institute of Agriculture, The University of Western Australia, Crawley, WA, Australia

A.P. Taranto, B.Sc. (Hons.) • P.S. Solomon, B. Appl. Sci. (Hons.), Ph.D.
Plant Sciences Division, Research School of Biology, The Australian National University, Canberra, ACT, Australia

in ascogenous hyphae, with both copies of the duplicated sequences displaying the mutations. The phenomenon of RIP was noted to occur after fertilization but before meiosis, during a brief premeiotic phase in which a dikaryon is formed prior to nuclear duplication and karyogamy (nuclear fusion) (Selker et al. 1987a; Selker 1990). For this reason, the acronym "RIP" for "rearrangement induced premeiotically" was initially used to describe the process but was retroactively changed to "repeat-induced point (mutation)".

These early studies of RIP identified numerous transition mutations from C:G to T:A base pairs within the regions of repetitive DNA (Selker et al. 1987a; Selker 1990). Furthermore, the frequency of cytosine transitions was found to be strongly biased towards cytosine bases adjacent to a downstream adenine base (CpA) (Cambareri et al. 1989). The CpA dinucleotide is thus changed to TpA, while the TpG sequence on the opposite strand is also converted to TpA. This dinucleotide mutation bias has since been widely observed across the Pezizomycotina (filamentous Ascomycota) (Clutterbuck 2011).

In *N. crassa* RIP was observed to operate in successive sexual cycles until the sequence identity between pairs of repeated sequences was reduced below the minimum similarity threshold of ~80 % (Cambareri et al. 1989). Repetitive sequences were observed to be mutated by RIP for up to six generations (Cambareri et al. 1991), until they became too dissimilar. In a single

M.A. van den Berg and K. Maruthachalam (eds.), *Genetic Transformation Systems in Fungi, Volume 2*, Fungal Biology, DOI 10.1007/978-3-319-10503-1_4,
© Springer International Publishing Switzerland 2015

sexual cycle, up to 30 % of the G:C pairs in a duplicated sequence can be mutated (Cambareri et al. 1989), although the efficiency can vary across strains (Noubissi et al. 2000). Additionally, RIP has been observed to "leak" beyond the bounds of duplicated sequences, creating mutational effects on unduplicated neighboring sequences within at least 4 kbp of the nearest repeated sequence (Irelan et al. 1994).

The major consequences of RIP in *N. crassa* are that repeated DNA segments, such as would result from the transposition of a retrotransposon, or the duplication of a gene, are mutated and potentially inactivated (Galagan et al. 2003). Transposable elements (TEs) were first detected in maize in the late 1940s (McClintock 1950) and have since been observed in the sequenced genomes of all prokaryotic and eukaryotic organisms (Kempken and Kuck 1998). Unlike normal genes they are able to change their position within the host genome. The subsequent transposition into coding sequences and their initiation of chromosomal rearrangements can have a profound effect on gene expression and genome evolution (Kempken and Kuck 1998). Their presence in filamentous fungi was first identified in 1989 in the species *Passalora fulva* (syn. *Fulvia fulva*, *Cladosporium fulvum*) (McHale et al. 1989) and *Neurospora crassa* (Kinsey and Helbe 1989). TEs are divided into two major classes. Class II transposons transpose at the DNA level, usually by excision and reinsertion at a new position in the genome (Kempken and Kuck 1998), thus the deleterious effects on fungal genomes of class II transposons are comparatively few. However, class I transposons transpose via RNA intermediates which involves replication of new copies while leaving behind the original copy (Kempken and Kuck 1998). The uncontrolled replication of transposons has the potential to significantly alter genome sequence organization and total size. The reference genome assembly for *N. crassa* (strain N150, 74-OR23-1VA, Fungal Genetics Stock Center 2489) has a moderately low repetitive content of ~10 % and all gene duplication and divergence is proposed to have occurred prior to the evolution of RIP (Galagan et al. 2003). All sequences derived from transposons show the hallmarks of RIP and active transposons

are absent from virtually all *Neurospora* strains (Selker 2002; Galagan et al. 2003). A rare exception came from reports of an African strain FGSC430, called Adiopodoumé, which was found to contain over 40 functional copies of Tad—a LINE1-like non-LTR class 1 transposon (Kinsey et al. 1994). While it was initially speculated that Adiopodoumé was RIP-deficient, it was subsequently determined to be capable of RIP (Kinsey et al. 1994) but was able to suppress RIP in crosses with other strains (Noubissi et al. 2000). Another report of an active transposon in *N. crassa* was the mini-transposon "guest" (Yeadon and Catcheside 1995), which at 98 bp in length is one of the shortest transposable elements ever reported and is thus too short to be recognized by RIP (Watters et al. 1999).

4.1.2 The Taxonomic Range of RIP

RIP has been fully documented in extant isolates of just five species but its hallmarks have been found in many others. To demonstrate active RIP, the ability to perform sexual crossing and genetic transformation is required, yet one or both capabilities are lacking for many species under current study. Despite few fungal species being amenable to such analysis, RIP has been experimentally demonstrated for *Magnaporthe oryzae* (Ikeda et al. 2002; Dean et al. 2005), *Podospora anserina* (Graia et al. 2001), *Leptosphaeria maculans* (Idnurm and Howlett 2003), and *Fusarium graminearum* (Cuomo et al. 2007).

For a larger number of fungal species, the prior action of RIP has been inferred from the detection of RIP-like polymorphism between repetitive DNA in whole or partial genomic sequence datasets (Table 4.1). Because RIP converts cytosine to thymine bases, it leaves behind a tell-tale signature within the genome characterized by a depletion of G:C content and, in most species in which RIP has been reported, an abundance of TpA dinucleotides. The occurrence of RIP can also be identified when alignments of repeat family members are examined for an increase in the number of directional transitions (interchanges between pyrimidines or between purine bases, i.e. C↔T or G↔A) over transversions

Table 4.1 Experimental and computational evidence for RIP across various fungal species

Phylum	Sub-phylum	Class	Species	Level of support for RIP	RIP-target dinucleotide bias	References
Ascomycota	Pezizomycotina	Dothideomycetes	*Leptosphaeria maculans*	Experimental	(T/C)pCp(A/G)	Idnurm and Howlett (2003); Fudal et al. (2009); Van de Wouw et al. (2010); Rouxel et al. (2011)
		Dothideomycetes	*Parastagonospora nodorum*	In silico	CpA	Hane and Oliver (2008, 2010)
		Dothideomycetes	*Zymoseptoria tritici*	In silico	CpA	Goodwin et al. (2011)
		Eurotiomycetes	*Aspergillus* spp.	In silico	Cp(A/G)	Neuveglise et al. (1996); Nielsen et al. (2001); Montiel et al. (2006); Braumann et al. (2008)
		Eurotiomycetes	*Penicillium chysogenum*	In silico	CpA	Braumann et al. (2008)
		Eurotiomycetes	*Penicillium roqueforti*	In silico	CpA	Ropars et al. (2012)
		Orbiliomycetes	*Arthobotryus oligospora*	In silico	CpA	Meerupati et al. (2013)
		Orbiliomycetes	*Monacrosporium haptotylum*	In silico	CpA	Meerupati et al. (2013)
		Sordariomycetes	*Colletotrichum cereale*	In silico	CpA	Crouch et al. (2008)
		Sordariomycetes	*Fusarium graminearum*	Experimental	CpA	Cuomo et al. (2007)
		Sordariomycetes	*Fusarium oxysporum*	In silico	Cp(A/G)	Julien et al. (1992); Hua-Van et al. (1998, 2001)
		Sordariomycetes	*Magnaporthe oryzae*	Experimental	(A/T)pCp(A/T)	Nakayashiki et al. (1999); Ikeda et al. (2002); Dean et al. (2005); Farman (2007)
		Sordariomycetes	*Metarrhizium* spp.	In silico	CpA	Gao et al. (2011); Pattemore et al. (2014)
		Sordariomycetes	*Neurospora crassa*	Experimental	CpA	Selker et al. (1987a); Selker and Garrett (1988); Cambareri et al. (1989, 1991); Selker (1990); Galagan et al. (2003)
		Sordariomycetes	*Neurospora tetrasperma*	In silico	CpA	Bhat et al. (2004)
		Sordariomycetes	*Podospora anserina*	Experimental	CpA	Graia et al. (2001)

(continued)

Table 4.1 (continued)

Phylum	Sub-phylum	Class	Species	Level of support for RIP	RIP-target dinucleotide bias	References
Basidiomycota	Agaricomycotina	Agaricomycetes	*Rhizoctonia solani*	In silico	CpG	Hane et al. (2014)
	Pucciniomycotina	Microbotyomycetes	*Microbotryum lychnidis-dioicae*	In silico	TpCpG	Horns et al. (2012)
		Microbotyomycetes	*Microbotryum violaceum*	In silico	TpCpG	Hood et al. (2005)
		Microbotyomycetes	*Rhodotorula graminis*	In silico	TpCpG	Horns et al. (2012)
		Pucciniomycetes	*Melampsora laricis-populina*	In silico	TpCpG	Horns et al. (2012)
		Pucciniomycetes	*Puccinia graminis*	In silico	TpCpG	Horns et al. (2012)
	Ustilaginomycotina	Ustilaginomycetes	*Ustilago hordei*	In silico	CpG	Laurie et al. (2013) but contradicted by Lefebvre et al. (2013)

(interchanges of pyrimidine (C or T) for purine (A or G) base or vice versa) and RIP-like single nucleotide polymorphisms (Hane and Oliver 2008) (For more information about bioinformatics inference of RIP, please refer to Chap. 5 of this volume). Using these techniques, RIP has been inferred in silico for several species, detailed in Table 4.1.

A strong bias for cytosine transition at CpA dinucleotides was first observed for RIP in *N. crassa* and subsequent reports of RIP within the sub-phylum Pezizomyotina remain consistent with this (Hane and Oliver 2008, 2010; Clutterbuck 2011) (Table 4.1). These reports of RIP within the Pezizomycotina include early diverging lineages (Meerupati et al. 2013) suggesting phylum-wide conservation and early evolution of this process (Meerupati et al. 2013). However the mechanism by which RIP occurs may have later been lost in specific lineages, such as those including the ancestors of some members of the genus *Metarhizium*, which exhibit varying levels of RIP (Gao et al. 2011; Meerupati et al. 2013; Pattemore et al. 2014). In contrast, yeast species of the non-filamentous Ascomycota (Saccharyomycotina and Taphrinomycotina) do not appear to exhibit signs of RIP (Clutterbuck 2011).

There have been some reports of RIP-like cytosine transition mutations in species of the phylum Basidiomycota based on in silico evidence only. The RIP-like mutations have a bias towards mutation of CpG rather than CpA (Table 4.1). It is important to note that transition of cytosine to thymine biased towards CpG dinucleotides is also characteristic of the process of cytosine deamination, a far more widely conserved process than RIP, which involves the methylation of cytosine to 5-methylcytosine (5mC) followed by deamination converting 5mC to thymine (Nabel et al. 2012). Reports of RIP biased towards mutation of CpG should therefore be treated with a degree of caution.

In the case of the sub-phylum Pucciniomycotina, a more specific bias for mutation of the trinucleotide TpCpG has been observed (Hood et al. 2005; Horns et al. 2012). A recent survey of RIP-like mutations across a nine basidiomycete species reported CpG-biased mutation for species of the class Pucciniomycetes but not for the classes Agaricomycetes or Ustilaginomycetes (Horns et al. 2012). However, there have also been reports of RIP-like CpG mutations in *Ustilago hordeii* of the Ustilaginomycetes (Laurie et al. 2013) and *Rhizoctonia solani* AG8 of the Agaricomycetes (Hane et al. 2014). Cells of *R. solani* AG8 are multi-nucleated and are commonly observed to contain between 6 and 15 nuclei (Sneh et al. 1991). Unusually, RIP-like SNP mutations with a moderate CpG bias were reported in *R. solani* within both single-copy (per nucleus) genes and repetitive retrotransposon sequences alike, with RIP-like SNP diversity occurring primarily between heterokaryons (Hane et al. 2014). However in *R. solani*, RIP-like CpG mutations were also more strongly associated with repetitive DNA than genes, as their frequency was observed to increase with proximity to transposons (Hane et al. 2014).

4.1.3 The Molecular Machinery of RIP

The only gene currently known to be essential for RIP is the *rid* (RIP-defective) gene, a DNA methyltransferase first identified in *N. crassa* (Freitag et al. 2002). Mutations in *rid* [GenBank: AAM27408.1] resulted in a loss of RIP activity in a homozygous mutant cross (Freitag et al. 2002). The protein product encoded by the *rid* gene contains a conserved domain for C-5-cytosine-specific DNA methylation [Pfam: PF00145]. A homologous protein appears to be present in most species of the Ascomycota in which RIP is known to occur; however, it has not yet been reported in species of the Basidiomycota that are purported to exhibit RIP. Consensus alignments of these proteins show high amino acid conservation within the domain region, but much lower conservation across other regions of the protein sequences (Freitag et al. 2002; Braumann et al. 2008; Williams, Hane, Lichtenzveig, Singh and Oliver unpublished).

In a recent survey of the Ascomycota (Clutterbuck 2011), 48 Pezizomycotina species showing evidence of RIP also contained a *rid* homolog, with the single exception of the early-diverging black truffle species *Tuber melanosporum* in which RIP is not observed (Martin et al. 2010; Clutterbuck 2011). However, the genome of *Sordaria macrospora* contains a *rid* homolog but is not reported to exhibit active RIP (Nowrousian et al. 2010), indicating that the presence of a *rid* homolog alone does not confer RIP-activity.

4.1.4 Regional Variability of RIP-Activity Across the Fungal Genome

RIP has been proposed to act by a processive mechanism as it affects both repeats independently (Selker 1990). However, the molecular machinery of RIP also appears to be limited by DNA sequence homology and possibly by certain DNA-binding proteins. Studies in *N. crassa* have determined that at least for that species, RIP mutations only accumulate within repeated sequences that are at least 400 bp in length (Watters et al. 1999) and share at least 80 % sequence identity (Cambareri et al. 1991). Furthermore, RIP is absent within regions of the genome containing large tandem arrays of ribosomal DNA (rDNA) repeats called nucleolus organizer regions (NORs). In *Parastagonospora* (*syn. Stagonospora*) *nodorum*, rDNA repeats within the NOR were largely unaffected by RIP; however, rDNA repeats outside the NOR were heavily RIP-affected. Additionally the highly repetitive 5S ribosomal repeat which is present both within the NOR array and elsewhere in the genome as individual repeats avoided RIP mutation altogether due to its short length which was presumably below the threshold for RIP for this species (Hane and Oliver 2008, 2010).

In addition to low-activity in the NOR array, RIP also appears to affect certain regions of the genome at higher than normal levels. RIP-degraded copies of the Pholy class 1 transposon of *Leptosphaeria maculans* have been reported to be more frequently located within pericentromeric regions than elsewhere in the genome (Attard et al. 2005). Also in *L. maculans*, conditionally dispensable chromosomes (which are not required for cell viability) were observed to contain genes with higher levels of RIP than on highly conserved "core" chromosomes (Balesdent et al. 2013). What is not yet clear is whether regional variability in RIP intensity is primarily determined by the physical properties of these DNA regions or if observations of this variability emerge as a by-product of either purifying or diversifying selection pressures.

4.1.5 Variability of RIP-Activity Between Fungal Species and Isolates

While not extensively surveyed, there appears to be a broad spectrum of RIP activity across various fungal species. For example, the proportion of cytosines mutated across various species and sequence regions has been reported (Montiel et al. 2006) to be between 0.06 and 0.72 % in MAGGY transposons of various isolates of *Magnaporthe oryzae* (Ikeda et al. 2002), 2 % in Pot elements of *Podospora anserina* (Graia et al. 2001), 28 % in Fot1 elements of *Fusarium oxysporum* (Daboussi et al. 2002), and 32.8 % in Pyret elements of *M. oryzae* (Ikeda et al. 2002). This variation may be dependent on several factors including: activity of the rid (and potentially other) proteins, frequency of meiotic cycling, and length of exposure to RIP. The effects of RIP are particularly prominent in *N. crassa*, where gene family expansion is negligible due to the inactivating effects of RIP (Galagan et al. 2003). In contrast, species within the genus *Metarhizium* (Gao et al. 2011) exhibit lower degrees of RIP. One species within this genus, *M. robertsii*, exhibits little or no RIP resulting in the presence of a large number of transposase genes (Gao et al. 2011).

Variability in the degree of RIP across species or isolates may be caused by different enzymatic activities of their rid proteins. There is significant variation in relative RIP activity between RIP-competent species. In *M. oryzae* the RIP process is described as being less efficient at recognizing repeated sequences as well as inducing a lower number of mutations compared to *N. crassa* (Dean et al. 2005). In species with lower RIP efficiencies some repeats may potentially "escape" RIP mutation, leading to a homogenous mixture of RIP-affected and unaffected repeats within their genomes. Even in the highly RIP-affected *N. crassa*, it is possible for some repeats to escape RIP, at least for a brief time, i.e. during a single round of meiosis. It has been reported that when three copies of a repeat are present, RIP occurs within two of these but the third is unaffected (Fincham et al. 1989). Transposon repeats surviving RIP in this manner may still be mutated in subsequent rounds, however if RIP-activity is disrupted, can go on to replicate freely (Braumann et al. 2008). It has also been proposed that RIP can be lost in different isolates of a species or genus (Dean et al. 2005; Gao et al. 2011) although the mechanism by which this may occur is unclear.

RIP-activity may also be affected by the frequency at which a species undergoes meiosis (Arnaise et al. 2008). For example, *Fusarium graminearum* undergoes frequent meiotic cycling and is heavily RIP-affected (Cuomo et al. 2007; Brown et al. 2012), whereas species such as *P. fulva* and *M. oryzae* that undergo less frequent meiotic cycling are less RIP-affected and consequently exhibit significant transposon expansion (Dean et al. 2005; de Wit et al. 2012). Furthermore, complexities of mating-type and heterokaryon incompatibilities may also limit the frequency of RIP across species (Arnaise et al. 2008). It appears that RIP mutation even in a given strain is not consistent across the course of its evolution. In genomes with relatively high repeat contents such as *L. maculans* and *Ustilago hordei* (34 % and 30 % respectively) (Rouxel et al. 2011; Laurie et al. 2013), it has been proposed that TEs expanded during a phase when RIP was not active, perhaps during an extended period of asexual reproduction (Stuckenbrock and Croll 2014). RIP would not be expected to occur in asexual species, yet RIP-like mutations have also been observed in species with no reported sexual life-cycle stage, including *Metarhizium* spp., *Aspergillus niger*, *Penicillium chrysogenum* (Braumann et al. 2008), *Verticillium* spp. (Gao et al. 2011; Amyotte et al. 2012), and *Phoma medicaginis* var. *medicaginis* (Williams et al., unpublished data). There are several possible explanations for this; firstly, that RIP may have occurred during meiosis in a RIP-competent ancestor that subsequently became asexual as

may have occurred following geographical isolation of mating types of a heterothallic RIP-competent population; secondly, that "RIP-like" mutations may occur under non-pre-meiotic conditions, such as where two copies of a similar sequence are present on separate nuclei during anastomosis; thirdly, that two or more RIP-polymorphic repeats may have been laterally transferred from a RIP-competent donor. However, the most likely explanation may be simply that these species undergo sexual recombination but have hitherto been unobserved to do so or alternatively may have cryptic sexual lifestyles.

The extent of accumulation of RIP mutations also appears to be dependent on the duration of the pre-meiotic dikaryon phase, which occurs between fertilization and ascospore production. The gene *amil1* of *Podospora anserina* regulates positioning and distribution of nuclei. Loss of function of this gene was found to significantly enhance RIP in *P. anserina*, due to increased length of exposure to the RIP machinery (Bouhouche et al. 2004). Additionally, a longer period of time spent in the pre-meiotic phase has been observed to lead to higher levels of RIP mutation in *N. crassa* (Singer and Selker 1995).

4.2 Rip and Methylation

4.2.1 5-Methylcytosine DNA-Methylation

In addition to being a model for the study of RIP, *N. crassa* is also a model system for the study of epigenetic modifications in fungi as it is competent in DNA-methylation, RNA interference, and a number of histone modifications common to higher eukaryotes (Aramayo and Selker 2013). RIP appears to be an independent process from epigenetic modifications, but it has been well established in *N. crassa* that RIP-degraded transposon "relics" frequently exhibit 5-methylcytosine (5mC) DNA methylation following RIP processing (Lewis et al. 2009). It should be noted, however, that in the genomes of some RIP-competent species, including *P. anserina* and *L. maculans*, a clear association between RIP and methylation has not been reported (Graia et al. 2001; Idnurm and Howlett 2003; Arnaise et al. 2008). It is also worth noting that 5mC methylation in fungi can result in increased levels of C→T transitions (Mishra et al. 2011), via the deamination of 5mC (Nabel et al. 2012), which are not attributable to RIP.

The determining factor for the targeted de novo methylation of "RIP-relics" appears to be the depletion of G:C content that results from RIP. This has been demonstrated through the introduction of various AT-rich oligomers of 25–100 bp in length into *N. crassa*, which were observed to be 5mC-methylated in a manner directly proportional to their TpA content and most prominently for the motifs TTAA and TAAA (Tamaru and Selker 2003). Other studies also report that weakly RIP-affected sequences, with few CpG to TpA transitions, were unable to induce de novo DNA-methylation (Singer and Selker 1995; Selker 2002). Tamaru and Selker have speculated that de novo 5mC methylation may be mediated by an as yet unidentified AT-rich DNA-binding protein. This hypothesis is supported by their observation that distamycin, a compound which competes for binding of the minor-groove of AT-rich DNA, blocked de novo methylation of AT-rich sequences (Tamaru and Selker 2003).

Just as RIP may leak beyond the boundaries of repeats into neighboring single-copy sequences, RIP-relics and other AT-rich sequences have also been shown to promote the spread of 5mC into adjacent sequences that would normally lack methylation (Miao et al. 2000; Tamaru and Selker 2003). In *N. crassa*, "methylation-leakage" is regulated by the DNA methylation modulator (DMM) proteins. DMM-2 binds HP1-associated DNA and recruits DMM-1, which inhibits the uncontrolled spread of 5mC and H3K9 methylation beyond AT-rich RIP-relics (Honda et al. 2010). Furthermore, RIP has been found to direct the methylation of histones and is strongly associated with H3K9me3 modifications (Lewis et al. 2009). Histone H3K9me3 modifications lead to the binding of heterochromatin protein 1 (HP1) which in turn leads to constitutive repression of

local gene expression via interaction with the DNA-methyltransferase Dim-2 (James and Elgin 1986; Honda and Selker 2008). Unsurprisingly, the centromeres of the *N. crassa* genome which are largely comprised of RIP-relics are also heavily enriched with 5mC and H3K9me3 methylation (Smith et al. 2011).

4.2.2 RNA-Directed DNA-Methylation

RNA-directed DNA-methylation (RdDm) involves the complementary binding of small interfering (si) RNA which triggers RNA-induced transcriptional silencing (RITS), the end result of which is H3K9 histone modification and the subsequent repression of local gene expression (Volpe and Martienssen 2011). *N. crassa* possesses the genes required for RdDm and produces siRNAs from hemi-methylated RIP-relic loci during the early S-phase of meiosis, consistent with RITS-directed DNA-methylation (Chicas et al. 2004). However, as outlined in the previous section, *N. crassa* and other species competent in both RIP and RIP-directed 5mC methylation are capable of H3K9-mediated gene silencing without requiring RdDm and RITS (Chicas et al. 2004; Freitag et al. 2004). Given that all repeats of *N. crassa* are RIP-degraded, and therefore become targets of AT-directed de novo methylation, RdDm would appear to be redundant for the purpose of repeat methylation in *N. crassa*.

4.3 Rip Is Both a Driver and Antagonist of Genome Diversity

Although the purpose of RIP has been previously proposed as a defence against uncontrolled replication of invading transposable elements (Selker 1990), RIP can also act as both a barrier to and a catalyst for generating genomic diversity. In certain fungal species, notably pathogens of plants, insects, and animals (Meerupati et al. 2013), it has been proposed that RIP leads to increased

4.3.1 RIP Control of Gene Family Expansion

As RIP acts indiscriminately upon any duplicated sequences, including whole-chromosome, segmental, and gene duplications, it has the potential to alter the sequence of repeated endogenous genes as well as transposon repeats. It has been suggested that genomes can protect themselves from transposon invasion and do so at the expense of beneficial gene variability that transposons can bring (Galagan and Selker 2004). RIP has been reported to have limited the evolution of new genes arising from gene duplication and subsequent mutation in *N. crassa*, as there were no pairs of genes identified in the genome with greater than 80 % similarity (Selker 1990; Kelkar et al. 2001; Galagan et al. 2003), the experimentally determined threshold for RIP activity. This pattern is also observed in *M. haptotylum* and *A. oligospora* (Meerupati et al. 2013), with no genes of *A. oligospora* longer than 400 bp showing greater than 80 % nucleotide identity, suggesting that active RIP reduces the potential for gene family expansion. However in other species, RIP has also been proposed to lead to rapid evolution through accelerating the rate of divergence of duplicated genes (Fudal et al. 2009; Van de Wouw et al. 2010; Goodwin et al. 2011; Hane et al. 2014). In two closely related members of the *Metarhizium* genus with differential RIP activities, the species with higher RIP activity exhibits decreased numbers of expanded gene families (Gao et al. 2011; Meerupati et al. 2013). *M. robertsii*, which appears to have lost the mechanism for RIP compared to *M. acridum*, possesses a large number of lineage-specific gene duplications that are proposed to have contributed to its comparably broader host range (Gao et al. 2011).

4.3.2 RIP Leakage Mutates Single Copy Genes Flanking Repeats

In cases where RIP "leaks" beyond the bounds of duplicated repeat sequences into single-copy DNA regions (Irelan et al. 1994; Fudal et al. 2009; Van de Wouw et al. 2010), RIP is capable of introducing mutations and subsequent epigenetic modifications into neighboring unique genes (Galagan et al. 2003). The extent that "RIP-leakage" can influence gene mutation is dependent on the frequency and spread of repetitive DNA in a fungal genome. The *L. maculans* genome is largely compartmentalized into two types of DNA region typified by variation in G:C content, repetitive DNA content, and gene density. G:C equilibrated regions make up the majority of the genome, which is littered with numerous G:C depleted (AT-rich), repeat-rich, and gene-sparse "AT-isochores". Numerous RIP polymorphisms were detected between single-copy avirulence genes within AT-isochores of *L. maculans*, demonstrating that the frequency of "leaked" RIP mutations was higher closer to neighboring repetitive DNA, e.g., *AvrLm4* and *AvrLm6* in *L. maculans* (Van de Wouw et al. 2010). In addition to *L. maculans*, isochore-like genome organization has been suggested for the plant pathogens *P. fulva* and *Colletotrichum orbiculare*, in which there appears to be an enrichment of predicted "effector-like" genes within the repeat-rich regions of their respective genomes (Rouxel et al. 2011; de Wit et al. 2012; Gan et al. 2013). The lineage-specific (non-homologous) genes of *Verticillium dahliae* are also reported to neighbor RIP-degraded repeats (Klosterman et al. 2011), possibly indicating that these novel sequences arose due to RIP-leakage.

4.3.3 RIP-Induced Nonsense Mutation May Accelerate Evolution of Small Secreted Proteins

In *N. crassa*, G:C pairs are mutated by RIP at different frequencies, the majority of C to T mutations occur in cytosines that are immediately 5′

of adenines, followed by CpT, CpG, and rarely CpC (Selker 1990). Most species of the Pezizomycotina studied thus far share this same bias for CpA mutations. Significantly, the end-product of RIP for both CpA and its reverse complement (TpG) is TpA, making RIP highly efficient at randomly introducing amber (TpApG) and ochre (TpApA) nonsense mutations into repeated open-reading frames. Reports of RIP in *N. crassa* suggest that after a gene duplication event, each paralog has an 80 % probability of acquiring an in-frame stop codon after only a single round of RIP (Galagan et al. 2003). Presumably, the probability of RIP introducing nonsense mutations is likely to be higher than reported observations, as observed RIP in post-meiotic progeny may be influenced by selection pressures to retain at least one functional paralog. Furthermore, in *M. graminicola*, all transposons with more than ten copies contained stop codons within their coding regions, indicating they had been effectively inactivated (Goodwin et al. 2011). The introduction of nonsense mutations (stop codons) by RIP has been proposed to contribute to the evolution of pathogenicity in many fungi, by driving the evolution of lineage-specific small secreted proteins (SSPs) that frequently have important roles as pathogenicity effectors (de Jonge et al. 2011; Oliver 2012; Vleeshouwers and Oliver 2014). SSPs have been proposed to have been converted from long secreted proteins via RIP-induced nonsense mutations, as a preliminary step in the evolution of effector proteins (Meerupati et al. 2013). Genes encoding SSPs proposed to have been created via RIP mutation have been identified as undergoing rapid divergence and lacking conserved domains or homologs in other species (Van de Wouw et al. 2010; Meerupati et al. 2013). Lineage and species-specific genes in pathogenic species of the Orbiliomycetes were significantly shorter than conserved "core" genes shared with other fungi and the percentage of genes affected by RIP in *M. haplotypum* was higher (76.7 %) for genes encoding SSPs than for all genes (38.2 %) (Meerupati et al. 2013).

While the majority of reported RIP-affected fungal species are predominantly haploid, RIP has also been reported for a handful of predominantly heterokaryotic species, particularly within the Basidiomycota. Pathogens within this group include the bi-nucleated rust fungi of the Pucciniomycetes (Hood et al. 2005; Horns et al. 2012) and the multi-nucleated *R. solani* (Hane et al. 2014), which exhibit RIP-like mutations biased towards CpG dinucleotides. In *R. solani*, predicted "effector-like" genes exhibited higher rates of non-synonymous mutation relative to non-"effector-like" genes, presumably due to widespread RIP-like mutations. Opal stop codons (TGA) were also more abundant than ambre or ochre stop codons (TAG and TAA), suggesting that some genes may be undergoing RIP-induced shortening of their open-reading frames. This suggests that as in *L. maculans* (Van de Wouw et al. 2010), albeit by different means, RIP may be contributing to the adaptation of pathogenicity genes under diversifying selection in *R. solani* (Hane et al. 2014).

4.3.4 RIP, Transformation and Reverse Genetics

Reverse genetic methods—inferring the phenotype resulting from a genotypic change—were in their infancy when RIP was first discovered, even in the model *N. crassa* (Paietta and Marzluf 1985). High efficiency methods for homologous recombination of transformation vectors did not emerge until strains lacking the Ku70 and Ku80 orthologs became available (Ninomiya et al. 2004). Therefore, it was attractive to use RIP to mutate single-copy genes without the need for site-specific integration. Insertion of a copy of a gene of interest, followed by a meiotic cycle, would mutate both copies and allow the determination of the phenotype. Furthermore, the function of RIP-inactivated genes could be restored through a process called "RIP-and-rescue", which involved complementation with an ectopic functional copy residing on a plasmid vector (Ferea and Bowman 1996).

Reverse genetic methods based on RIP are of course restricted to species that exhibit strong RIP and can both be transformed and crossed efficiently. In practice the method was only tried in *N. crassa* and even in this species, because of RIP-leakage, there was the potential to introduce off-target effects in neighboring genes depending on their distance from the gene targeted for knock-out. As such, RIP-based transformation in fungi was soon superseded by advances in homologous recombination and other techniques.

4.4 Summary

RIP is a process of repeat-targeted point mutation that has broad consequences for the evolution of genes and biological processes such as pathogenicity in certain branches of the Fungi. With the exception of a few specialized regions of the genome, it acts upon all repeated sequences indiscriminately, mutating cytosine to thymine bases within and nearby repeated DNA. Because of this, RIP can also be exploited as an endogenous mutagen for the purpose of gene disruption and/or complementation. The widespread conservation of RIP (or RIP-like phenomena) across the Pezizomycotina and certain species of the Basidiomycota implies that it is generally beneficial to fungal fitness and survival. RIP is widely considered to act primarily as a genome defence mechanism against degradation of the genome through uncontrolled replication of retrotransposons. However, emerging data suggests that RIP is that and more, with additional roles in genome evolution, promoting, or constraining gene diversity and the innovation of novel genes.

References

Amyotte SG, Tan X, Pennerman K, Jimenez-Gasco Mdel M, Klosterman SJ, Ma LJ, Dobinson KF, Veronese P (2012) Transposable elements in phytopathogenic *Verticillium* spp.: insights into genome evolution and inter- and intra-specific diversification. BMC Genomics 13:314

Aramayo R, Selker EU (2013) *Neurospora crassa*, a model system for epigenetics research. Cold Spring Harb Perspect Biol 5(10):a017921

Arnaise S, Zickler D, Bourdais A, Dequard-Chablat M, Debuchy R (2008) Mutations in mating-type genes greatly decrease repeat-induced point mutation process in the fungus *Podospora anserina*. Fungal Genet Biol 45(3):207–220

Attard A, Gout L, Ross S, Parlange F, Cattolico L, Balesdent MH, Rouxel T (2005) Truncated and RIP-degenerated copies of the LTR retrotransposon Pholy are clustered in a pericentromeric region of the *Leptosphaeria maculans* genome. Fungal Genet Biol 42(1):30–41

Balesdent MH, Fudal I, Ollivier B, Bally P, Grandaubert J, Eber F, Chevre AM, Leflon M, Rouxel T (2013) The dispensable chromosome of *Leptosphaeria maculans* shelters an effector gene conferring avirulence towards *Brassica rapa*. New Phytol 198(3):887–898

Bhat A, Tamuli R, Kasbekar DP (2004) Genetic transformation of *Neurospora tetrasperma*, demonstration of repeat-induced point mutation (RIP) in self-crosses and a screen for recessive RIP-defective mutants. Genetics 167(3):1155–1164

Bouhouche K, Zickler D, Debuchy R, Arnaise S (2004) Altering a gene involved in nuclear distribution increases the repeat-induced point mutation process in the fungus *Podospora anserina*. Genetics 167(1):151–159

Braumann I, van den Berg M, Kempken F (2008) Repeat induced point mutation in two asexual fungi, *Aspergillus niger* and *Penicillium chrysogenum*. Curr Genet 53:287

Brown NA, Antoniw J, Hammond-Kosack KE (2012) The predicted secretome of the plant pathogenic fungus *Fusarium graminearum*: a refined comparative analysis. PLoS One 7(4):e33731

Cambareri EB, Jensen BC, Schabtach E, Selker EU (1989) Repeat-induced G-C to A-T mutations in *Neurospora*. Science 244(4912):1571–1575

Cambareri EB, Singer MJ, Selker EU (1991) Recurrence of repeat-induced point mutation (RIP) in *Neurospora crassa*. Genetics 127(4):699–710

Chicas A, Cogoni C, Macino G (2004) RNAi-dependent and RNAi-independent mechanisms contribute to the silencing of RIPed sequences in *Neurospora crassa*. Nucleic Acids Res 32(14):4237–4243

Clutterbuck AJ (2011) Genomic evidence of repeat-induced point mutation (RIP) in filamentous ascomycetes. Fungal Genet Biol 48(3):306–326

Crouch JA, Glasheen BM, Giunta MA, Clarke BB, Hillman BI (2008) The evolution of transposon repeat-induced point mutation in the genome of *Colletotrichum cereale*: Reconciling sex, recombination and homoplasy in an "asexual" pathogen. Fungal Genet Biol 45(3):190–206

Cuomo CA, Guldener U, Xu JR, Trail F, Turgeon BG, Di Pietro A, Walton JD, Ma LJ, Baker SE, Rep M, Adam G, Antoniw J, Baldwin T, Calvo S, Chang YL, Decaprio D, Gale LR, Gnerre S, Goswami RS, Hammond-Kosack K, Harris LJ, Hilburn K, Kennell JC, Kroken S, Magnuson JK, Mannhaupt G, Mauceli E, Mewes HW, Mitterbauer R, Muehlbauer G, Munsterkotter M, Nelson D, O'Donnell K, Ouellet T, Qi W, Quesneville H, Roncero MI, Seong KY, Tetko

IV, Urban M, Waalwijk C, Ward TJ, Yao J, Birren BW, Kistler HC (2007) The *Fusarium graminearum* genome reveals a link between localized polymorphism and pathogen specialization. Science 317(5843):1400–1402

Daboussi MJ, Daviere JM, Graziani S, Langin T (2002) Evolution of the Fot1 transposons in the genus *Fusarium*: discontinuous distribution and epigenetic inactivation. Mol Biol Evol 19(4):510–520

de Jonge R, Bolton MD, Thomma BP (2011) How filamentous pathogens co-opt plants: the ins and outs of fungal effectors. Curr Opin Plant Biol 14(4):400–406

de Wit PJGM, van der Burgt A, Okmen B, Stergiopoulos I, Abd-Elsalam KA, Aerts AL, Bahkali AH, Beenen HG, Chettri P, Cox MP, Datema E, de Vries RP, Dhillon B, Ganley AR, Griffiths SA, Guo Y, Hamelin RC, Henrissat B, Kabir MS, Jashni MK, Kema G, Klaubauf S, Lapidus A, Levasseur A, Lindquist E, Mehrabi R, Ohm RA, Owen TJ, Salamov A, Schwelm A, Schijlen E, Sun H, van den Burg HA, van Ham RCHJ, Zhang S, Goodwin SB, Grigoriev IV, Collemare J, Bradshaw RE (2012) The genomes of the fungal plant pathogens *Cladosporium fulvum* and *Dothistroma septosporum* reveal adaptation to different hosts and lifestyles but also signatures of common ancestry. PLoS Genet 8(11):e1003088

Dean RA, Talbot NJ, Ebbole DJ, Farman ML, Mitchell TK, Orbach MJ, Thon M, Kulkarni R, Xu JR, Pan H, Read ND, Lee YH, Carbone I, Brown D, Oh YY, Donofrio N, Jeong JS, Soanes DM, Djonovic S, Kolomiets E, Rehmeyer C, Li W, Harding M, Kim S, Lebrun MH, Bohnert H, Coughlan S, Butler J, Calvo S, Ma LJ, Nicol R, Purcell S, Nusbaum C, Galagan JE, Birren BW (2005) The genome sequence of the rice blast fungus *Magnaporthe grisea*. Nature 434(7036):980–986

Farman ML (2007) Telomeres in the rice blast fungus *Magnaporthe oryzae*: the world of the end as we know it. FEMS Microbiol Lett 273(2):125–132

Ferea TL, Bowman BJ (1996) The vacuolar ATPase of *Neurospora crassa* is indispensable: inactivation of the *vma-1* gene by repeat-induced point mutation. Genetics 143(1):147–154

Fincham JR, Connerton IF, Notarianni E, Harrington K (1989) Premeiotic disruption of duplicated and triplicated copies of the *Neurospora crassa am* (glutamate dehydrogenase) gene. Curr Genet 15(5):327–334

Freitag M, Williams RL, Kothe GO, Selker EU (2002) A cytosine methyltransferase homologue is essential for repeat-induced point mutation in *Neurospora crassa*. Proc Natl Acad Sci 99(13):8802–8807

Freitag M, Lee DW, Kothe GO, Pratt RJ, Aramayo R, Selker EU (2004) DNA methylation is independent of RNA interference in *Neurospora*. Science 304(5679):1939

Fudal I, Ross S, Brun H, Besnard AL, Ermel M, Kuhn ML, Balesdent MH, Rouxel T (2009) Repeat-induced point mutation (RIP) as an alternative mechanism of evolution toward virulence in *Leptosphaeria maculans*. Mol Plant Microbe Interact 22(8):932–941

Galagan JE, Selker EU (2004) RIP: the evolutionary cost of genome defense. Trends Genet 20(9):417–423

Galagan JE, Calvo SE, Borkovich KA, Selker EU, Read ND, Jaffe D, FitzHugh W, Ma LJ, Smirnov S, Purcell S, Rehman B, Elkins T, Engels R, Wang S, Nielsen CB, Butler J, Endrizzi M, Qui D, Ianakiev P, Bell-Pedersen D, Nelson MA, Werner-Washburne M, Selitrennikoff CP, Kinsey JA, Braun EL, Zelter A, Schulte U, Kothe GO, Jedd G, Mewes W, Staben C, Marcotte E, Greenberg D, Roy A, Foley K, Naylor J, Stange-Thomann N, Barrett R, Gnerre S, Kamal M, Kamvysselis M, Mauceli E, Bielke C, Rudd S, Frishman D, Krystofova S, Rasmussen C, Metzenberg RL, Perkins DD, Kroken S, Cogoni C, Macino G, Catcheside D, Li W, Pratt RJ, Osmani SA, DeSouza CP, Glass L, Orbach MJ, Berglund JA, Voelker R, Yarden O, Plamann M, Seiler S, Dunlap J, Radford A, Aramayo R, Natvig DO, Alex LA, Mannhaupt G, Ebbole DJ, Freitag M, Paulsen I, Sachs MS, Lander ES, Nusbaum C, Birren B (2003) The genome sequence of the filamentous fungus *Neurospora crassa*. Nature 422(6934):859–868

Gan P, Ikeda K, Irieda H, Narusaka M, O'Connell RJ, Narusaka Y, Takano Y, Kubo Y, Shirasu K (2013) Comparative genomic and transcriptomic analyses reveal the hemibiotrophic stage shift of *Colletotrichum* fungi. New Phytol 197(4):1236–1249

Gao Q, Jin K, Ying SH, Zhang Y, Xiao G, Shang Y, Duan Z, Hu X, Xie XQ, Zhou G, Peng G, Luo Z, Huang W, Wang B, Fang W, Wang S, Zhong Y, Ma LJ, St Leger RJ, Zhao GP, Pei Y, Feng MG, Xia Y, Wang C (2011) Genome sequencing and comparative transcriptomics of the model entomopathogenic fungi Metarhizium anisopliae and M. acridum. PLoS Genet 7(1):e1001264

Goodwin SB, M'Barek SB, Dhillon B, Wittenberg AH, Crane CF, Hane JK, Foster AJ, Van der Lee TA, Grimwood J, Aerts A, Antoniw J, Bailey A, Bluhm B, Bowler J, Bristow J, van der Burgt A, Canto-Canche B, Churchill AC, Conde-Ferraez L, Cools HJ, Coutinho PM, Csukai M, Dehal P, De Wit P, Donzelli B, van de Geest HC, van Ham RC, Hammond-Kosack KE, Henrissat B, Kilian A, Kobayashi AK, Koopmann E, Kourmpetis Y, Kuzniar A, Lindquist E, Lombard V, Maliepaard C, Martins N, Mehrabi R, Nap JP, Ponomarenko A, Rudd JJ, Salamov A, Schmutz J, Schouten HJ, Shapiro H, Stergiopoulos I, Torriani SF, Tu H, de Vries RP, Waalwijk C, Ware SB, Wiebenga A, Zwiers LH, Oliver RP, Grigoriev IV, Kema GH (2011) Finished genome of the fungal wheat pathogen *Mycosphaerella graminicola* reveals dispensome structure, chromosome plasticity, and stealth pathogenesis. PLoS Genet 7(6):e1002070

Graia F, Lespinet O, Rimbault B, Dequard-Chablat M, Coppin E, Picard M (2001) Genome quality control: RIP (repeat-induced point mutation) comes to *Podospora*. Mol Microbiol 40(3):586–595

Hane JK, Oliver RP (2008) RIPCAL: a tool for alignment-based analysis of repeat-induced point mutations in fungal genomic sequences. BMC Bioinform 9:478

Hane JK, Oliver RP (2010) *In silico* reversal of repeat-induced point mutation (RIP) identifies the origins of repeat families and uncovers obscured duplicated genes. BMC Genomics 11:655

Hane JK, Anderson JP, Williams AH, Sperschneider J, Singh KB (2014) Genome sequencing and comparative genomics of the broad host-range pathogen *Rhizoctonia solani* AG8. PLoS Genet 10(5):e1004281

Honda S, Selker EU (2008) Direct interaction between DNA methyltransferase DIM-2 and HP1 is required for DNA methylation in *Neurospora crassa*. Mol Cell Biol 28(19):6044–6055

Honda S, Lewis ZA, Huarte M, Cho LY, David LL, Shi Y, Selker EU (2010) The DMM complex prevents spreading of DNA methylation from transposons to nearby genes in Neurospora crassa. Genes Dev 24(5):443–454

Hood ME, Katawczik M, Giraud T (2005) Repeat-induced point mutation and the population structure of transposable elements in *Microbotryum violaceum*. Genetics 170(3):1081–1089

Horns F, Petit E, Yockteng R, Hood ME (2012) Patterns of repeat-induced point mutation in transposable elements of basidiomycete fungi. Genome Biol Evol 4(3):240–247

Hua-Van A, Hericourt F, Capy P, Daboussi MJ, Langin T (1998) Three highly divergent subfamilies of the impala transposable element coexist in the genome of the fungus *Fusarium oxysporum*. Mol Gen Genet 259(4):354–362

Hua-Van A, Langin T, Daboussi MJ (2001) Evolutionary history of the impala transposon in *Fusarium oxysporum*. Mol Biol Evol 18(10):1959–1969

Idnurm A, Howlett BJ (2003) Analysis of loss of pathogenicity mutants reveals that repeat-induced point mutations can occur in the Dothideomycete *Leptosphaeria maculans*. Fungal Genet Biol 39(1):31–37

Ikeda K, Nakayashiki H, Kataoka T, Tamba H, Hashimoto Y, Tosa Y, Mayama S (2002) Repeat-induced point mutation (RIP) in *Magnaporthe grisea*: implications for its sexual cycle in the natural field context. Mol Microbiol 45(5):1355–1364

Irelan JT, Hagemann AT, Selker EU (1994) High frequency repeat-induced point mutation (RIP) is not associated with efficient recombination in *Neurospora*. Genetics 138(4):1093–1103

James TC, Elgin SC (1986) Identification of a nonhistone chromosomal protein associated with heterochromatin in *Drosophila melanogaster* and its gene. Mol Cell Biol 6(11):3862–3872

Julien J, Poirier-Hamon S, Brygoo Y (1992) Foret1, a reverse transcriptase-like sequence in the filamentous fungus *Fusarium oxysporum*. Nucleic Acids Res 20(15):3933–3937

Kelkar HS, Griffith J, Case ME, Covert SF, Hall RD, Keith CH, Oliver JS, Orbach MJ, Sachs MS, Wagner JR, Weise MJ, Wunderlich JK, Arnold J (2001) The *Neurospora crassa* genome: cosmid libraries sorted by chromosome. Genetics 157(3):979–990

Kempken F, Kuck U (1998) Transposons in filamentous fungi - facts and perspectives. Bioessays 20(8):652–659

Kinsey J, Helbe J (1989) Isolation of a transposable element from *Neurospora crassa*. Proc Natl Acad Sci U S A 86:1929–1933

Kinsey JA, Garrett-Engele PW, Cambareri EB, Selker EU (1994) The *Neurospora* transposon Tad is sensitive to repeat-induced point mutation (RIP). Genetics 138(3):657–664

Klosterman SJ, Subbarao KV, Kang S, Veronese P, Gold SE, Thomma BPHJ, Chen Z, Henrissat B, Lee Y-H, Park J, Garcia-Pedrajas MD, Barbara DJ, Anchieta A, de Jonge R, Santhanam P, Maruthachalam K, Atallah Z, Amyotte SG, Paz Z, Inderbitzin P, Hayes RJ, Heiman DI, Young S, Zeng Q, Engels R, Galagan J, Cuomo CA, Dobinson KF, Ma L-J (2011) Comparative genomics yields insights into niche adaptation of plant vascular wilt pathogens. PLoS Pathog 7(7):e1002137

Laurie JD, Linning R, Wong P, Bakkeren G (2013) Do TE activity and counteracting genome defenses, RNAi and methylation, shape the sex lives of smut fungi? Plant Signal Behav 8(4):e23853

Lefebvre F, Joly DL, Labbe C, Teichmann B, Linning R, Belzile F, Bakkeren G, Belanger RR (2013) The transition from a phytopathogenic smut ancestor to an anamorphic biocontrol agent deciphered by comparative whole-genome analysis. Plant Cell 25(6):1946–1959

Lewis ZA, Honda S, Khlafallah TK, Jeffress JK, Freitag M, Mohn F, Schubeler D, Selker EU (2009) Relics of repeat-induced point mutation direct heterochromatin formation in *Neurospora crassa*. Genome Res 19(3):427–437

Martin F, Kohler A, Murat C, Balestrini R, Coutinho PM, Jaillon O, Montanini B, Morin E, Noel B, Percudani R, Porcel B, Rubini A, Amicucci A, Amselem J, Anthouard V, Arcioni S, Artiguenave F, Aury JM, Ballario P, Bolchi A, Brenna A, Brun A, Buee M, Cantarel B, Chevalier G, Couloux A, Da Silva C, Denoeud F, Duplessis S, Ghignone S, Hilselberger B, Iotti M, Marcais B, Mello A, Miranda M, Pacioni G, Quesneville H, Riccioni C, Ruotolo R, Splivallo R, Stocchi V, Tisserant E, Viscomi AR, Zambonelli A, Zampieri E, Henrissat B, Lebrun MH, Paolocci F, Bonfante P, Ottonello S, Wincker P (2010) Perigord black truffle genome uncovers evolutionary origins and mechanisms of symbiosis. Nature 464(7291):1033–1038

McClintock B (1950) The origin and behavior of mutable loci in maize. Proc Natl Acad Sci U S A 36(6):344–355

McHale MT, Roberts IN, Talbot NJ, Oliver RP (1989) Expression of reverse transcriptase genes in *Fulvia fulva*. Mol Plant Microbe Interact 2(4):165–168

Meerupati T, Andersson KM, Friman E, Kumar D, Tunlid A, Ahren D (2013) Genomic mechanisms accounting for the adaptation to parasitism in nematode-trapping fungi. PLoS Genet 9(11):e1003909

Miao VP, Freitag M, Selker EU (2000) Short TpA-rich segments of the zeta-eta region induce DNA methylation in *Neurospora crassa*. J Mol Biol 300(2):249–273

Mishra PK, Baum M, Carbon J (2011) DNA methylation regulates phenotype-dependent transcriptional activity in *Candida albicans*. Proc Natl Acad Sci U S A 108(29):11965–11970

Montiel MD, Lee HA, Archer DB (2006) Evidence of RIP (repeat-induced point mutation) in transposase sequences of *Aspergillus oryzae*. Fungal Genet Biol 43(6):439–445

Nabel CS, Manning SA, Kohli RM (2012) The curious chemical biology of cytosine: deamination, methylation, and oxidation as modulators of genomic potential. ACS Chem Biol 7(1):20–30

Nakayashiki H, Nishimoto N, Ikeda K, Tosa Y, Mayama S (1999) Degenerate MAGGY elements in a subgroup of *Pyricularia grisea*: a possible example of successful capture of a genetic invader by a fungal genome. Mol Gen Genet 261(6):958–966

Neuveglise C, Sarfati J, Latge JP, Paris S (1996) Afut1, a retrotransposon-like element from *Aspergillus fumigatus*. Nucleic Acids Res 24(8):1428–1434

Nielsen ML, Hermansen TD, Aleksenko A (2001) A family of DNA repeats in *Aspergillus nidulans* has assimilated degenerated retrotransposons. Mol Genet Genomics 265(5):883–887

Ninomiya Y, Suzuki K, Ishii C, Inoue H (2004) Highly efficient gene replacements in *Neurospora* strains deficient for nonhomologous end-joining. Proc Natl Acad Sci U S A 101(33):12248–12253

Noubissi FK, McCluskey K, Kasbekar DP (2000) Repeat-induced point mutation (RIP) in crosses with wild-isolated strains of *Neurospora crassa*: evidence for dominant reduction of RIP. Fungal Genet Biol 31(2):91–97

Nowrousian M, Stajich JE, Chu M, Engh I, Espagne E, Halliday K, Kamerewerd J, Kempken F, Knab B, Kuo HC, Osiewacz HD, Poggeler S, Read ND, Seiler S, Smith KM, Zickler D, Kuck U, Freitag M (2010) *De novo* assembly of a 40 Mb eukaryotic genome from short sequence reads: *Sordaria macrospora*, a model organism for fungal morphogenesis. PLoS Genet 6(4):e1000891

Oliver R (2012) Genomic tillage and the harvest of fungal phytopathogens. New Phytol 196(4):1015–1023

Paietta JV, Marzluf GA (1985) Gene disruption by transformation in *Neurospora crassa*. Mol Cell Biol 5(7):1554–1559

Pattemore JAH, Hane JK, Williams AH, Wilson BAL, Stodart BJ, Ash GJ (2014) The genome sequence of the biocontrol fungus Metarhizium anisopliae and comparative genomics of Metarhizium species. BMC Genomics 15:660

Ropars J, Dupont J, Fontanillas E, Rodriguez de la Vega RC, Malagnac F, Coton M, Giraud T, Lopez-Villavicencio M (2012) Sex in cheese: evidence for sexuality in the fungus *Penicillium roqueforti*. PLoS One 7(11):e49665

Rouxel T, Grandaubert J, Hane JK, Hoede C, van de Wouw AP, Couloux A, Dominguez V, Anthouard V, Bally P, Bourras S, Cozijnsen AJ, Ciuffetti LM, Degrave A, Dilmaghani A, Duret L, Fudal I, Goodwin SB, Gout L, Glaser N, Linglin J, Kema GHJ, Lapalu N, Lawrence CB, May K, Meyer M, Ollivier B, Poulain J, Schoch CL, Simon A,

Spatafora JW, Stachowiak A, Turgeon BG, Tyler BM, Vincent D, Weissenbach J, Amselem J, Quesneville H, Oliver RP, Wincker P, Balesdent M-H, Howlett BJ (2011) Effector diversification within compartments of the *Leptosphaeria maculans* genome affected by Repeat-Induced Point mutations. Nat Commun 2:202

Selker EU (1990) Premeiotic instability of repeated sequences in *Neurospora crassa*. Annu Rev Genet 24(1):579–613

Selker EU (2002) Repeat-induced gene silencing in fungi. Adv Genet 46:439–450

Selker E, Garrett P (1988) DNA sequence duplications trigger gene inactivation in *Neurospora crassa*. Proc Natl Acad Sci U S A 85:6870–6874

Selker EU, Cambareri EB, Jensen BC, Haack KR (1987a) Rearrangement of duplicated DNA in specialized cells of *Neurospora*. Cell 51(5):741–752

Selker EU, Jensen BC, Richardson GA (1987b) A portable signal causing faithful DNA methylation *de novo* in *Neurospora crassa*. Science 238(4823):48–53

Singer MJ, Selker EU (1995) Genetic and epigenetic inactivation of repetitive sequences in *Neurospora crassa*: RIP, DNA methylation, and quelling. Curr Top Microbiol Immunol 197:165–177

Smith KM, Phatale PA, Sullivan CM, Pomraning KR, Freitag M (2011) Heterochromatin is required for normal distribution of *Neurospora crassa* CenH3. Mol Cell Biol 31(12):2528–2542

Sneh B, Burpee L, Ogoshi A (1991) Identification of *Rhizoctonia* species. APS Press, St. Paul, MN

Stuckenbrock EH, Croll D (2014) The evolving fungal genome. Fung Biol Rev 28(1):1–12

Tamaru H, Selker EU (2003) Synthesis of signals for *de novo* DNA methylation in *Neurospora crassa*. Mol Cell Biol 23(7):2379–2394

Van de Wouw AP, Cozijnsen AJ, Hane JK, Brunner PC, McDonald BA, Oliver RP, Howlett BJ (2010) Evolution of linked avirulence effectors in *Leptosphaeria maculans* is affected by genomic environment and exposure to resistance genes in host plants. PLoS Pathog 6(11):e1001180

Vleeshouwers VG, Oliver RP (2014) Effectors as tools in disease resistance breeding against biotrophic, hemibiotrophic, and necrotrophic plant pathogens. Mol Plant Microbe Interact 27(3):196–206

Volpe T, Martienssen RA (2011) RNA interference and heterochromatin assembly. Cold Spring Harb Perspect Biol 3(9):a003731

Watters MK, Randall TA, Margolin BS, Selker EU, Stadler DR (1999) Action of repeat-induced point mutation on both strands of a duplex and on tandem duplications of various sizes in *Neurospora*. Genetics 153(2):705–714

Yeadon PJ, Catcheside DE (1995) Guest: a 98 bp inverted repeat transposable element in *Neurospora crassa*. Mol Gen Genet 247(1):105–109

Calculating RIP Mutation in Fungal Genomes Using RIPCAL

5

James K. Hane

5.1 Introduction

Repeat-induced point mutation (RIP) is targeted to repetitive DNA sequences and appears only to occur within certain fungal taxons (Galagan and Selker 2004; Clutterbuck 2011; Horns et al. 2012). RIP was first observed in the saprophyte *Neurospora crassa* (Selker et al. 1987; Selker and Stevens 1987) to occur prior to meiosis, randomly mutating cytosine to thymine bases of repetitive DNA. In *N. crassa* and in broader studies across the Pezizomycotina (filamentous Ascomycetes), a consistent bias towards preferential mutation of CpA dinucleotides (Table 5.1) has been observed (Selker et al. 1987; Selker and Stevens 1987; Hane and Oliver 2008; Clutterbuck 2011). An alternative bias towards CpG dinucleotides has been reported in Basidiomycete species (Hood et al. 2005; Hane and Oliver 2008; Clutterbuck 2011; Horns et al. 2012); however care should be taken to distinguish this from cytosine deamination which targets 5-methylcytosine and also mutates $C:G \rightarrow T:A$ with a CpG bias (Walsh and Xu 2006).

Ostensibly, RIP appears to deactivate transposons through the introduction of nonsense muta-tions into their protein-coding open-reading frames. Indeed, the commonly observed CpA bias is the most efficient possible dinucleotide target for randomly generating stop codons, resulting in a TpA dinucleotide in both DNA strands which can potentially make up a TAG or TAA stop codon. The accelerated rate of muta-tion that RIP confers may also influence the evo-lution of pathogenicity (Fudal et al. 2009; Van de Wouw et al. 2010) via leakage of RIP from repeats into neighbouring gene sequences.

Early studies of RIP were limited by the avail-ability of sequence data, but since RIP depletes G:C content and in Pezizomycotina also leads to the accumulation of TpA dinucleotides, these studies primarily employed ratio-based methods (or "indices") of dinucleotide frequencies indi-cating increased TpA content, reviewed in Hane and Oliver (2008). Dinucleotide ratios are rough indicators which can be affected by factors other than RIP, but remain useful when whole-genome sequences are unavailable or for scanning a genome for regions containing "ancient" RIP—which due to RIP and other mutations may no longer be similar enough to sister repeats to be recognised as a coherent repeat family. Detection of RIP-like polymorphisms within multiple alignments of repeat sequences is recommended if whole-genome data is available. The number of whole-genome sequences of fungal species has increased exponentially in the past decade, enabling multiple alignment-based RIP analysis

J.K. Hane, B. MolBiol. (Hons.), Ph.D. Bioinformatics (✉)
Centre for Crop and Disease Management,
Curtin University, Kent Street, Bentley,
Perth, WA 6102, Australia
e-mail: James.Hane@curtin.edu.au

M.A. van den Berg and K. Maruthachalam (eds.), *Genetic Transformation Systems in Fungi, Volume 2*, Fungal Biology, DOI 10.1007/978-3-319-10503-1_5,
© Springer International Publishing Switzerland 2015

Table 5.1 Potential RIP-like CpN dinucleotide mutations and their reverse complements

RIP-like mutation	Reverse-complementary RIP-like mutation
CpA → TpA	TpG → TpA
CpC → TpC	GpG → GpA
CpG → TpG	CpG → CpA
CpT → TpT	ApG → ApA

Table 5.2 Application of RIPCAL to various fungal species

Species	References
Arthrobotrys oligospora	Yang et al. (2011)
Colletotrichum orbiculare	Gan et al. (2013)
Colletotrichum gloeosporioides	Gan et al. (2013)
Grosmannia clavigera	DiGuistini et al. (2011)
Leptosphaeria maculans	Van de Wouw et al. (2010), Rouxel et al. (2011), Daverdin et al. (2012)
Macrophomina phaseolina	Islam et al. (2012)
Metarhizium acridum	Gao et al. (2011)
Metarhizium robertsii	Gao et al. (2011)
Mycosphaerella fijiensis	Santana et al. (2012)
Penicillium roqueforti	Ropars et al. (2012)
Phaeosphaeria nodorum	Hane and Oliver (2008), Hane and Oliver (2010)
Pseudozyma flocculosa	Lefebvre et al. (2013)
Pyrenophora teres f. maculata	Ellwood et al. (2012)
Pyrenophora teres f. teres	Ellwood et al. (2012)
Pyrenophora tritici-repentis	Manning et al. (2013)
Verticillium albo-atrum	Klosterman et al. (2011), Amyotte et al. (2012)
Verticillium dahliae	Klosterman et al. (2011), Amyotte et al. (2012)
Unknown aphid symbiont	Vogel and Moran (2013)

and leading to the development of RIPCAL. To date, RIPCAL has been applied to several fungal species (Table 5.2).

RIPCAL is capable of predicting whether a whole genome, repeat family or single sequence has been mutated by RIP, the relative extent of RIP between repeat families, presence of any biases for the mutation of certain dinucleotide and locational biases in RIP-frequency or type within the genome or at certain positions of a repeat. RIPCAL has various analysis "modes", which can be applied to a series of repeat families within a whole genome, a single repeat family or a single sequence. The first and primary mode measures RIP-like polymorphisms within a multiple alignment of repeats. RIPCAL's second mode determines dinucleotide frequencies in a set of sequences. The third mode reports genome regions containing RIP-index values above or below specified thresholds.

RIPCAL also includes deRIP, which predicts a consensus sequence representing an ancestral repeat family prior to the effects of RIP. Accumulated RIP can alter sequences to the point where they are unrecognisable by sequence similarity searches; however in silico reversal of RIP can provide insight into the nature and origin of some repeats (Hane and Oliver 2010).

5.2 General Methods

5.2.1 Installation

RIPCAL is available for Windows and Linux operating systems (OS); however not all features will be available through the graphical-user interface (GUI) and will require commands to be run through a command-line interface (CLI). Windows OS use of CLI features will also require installation of ActivePerl (http://www.activestate.com/activeperl). In both OS, locally performed multiple alignment in RIPCAL will require ClustalW (http://www.clustal.org/clustal2/) (Larkin et al. 2007).

1. Download latest RIPCAL package from www.sourceforge.net/projects/ripcal
2. Extract zip contents to new folder
3. To open RIPCAL GUI:
   - Windows OS: double click the exe.
   - Linux OS: set ripcal_x_x_x.pl to executable and run
4. To run RIPCAL CLI, open terminal shell client and run ripcal executable with the argument "-c". See below for additional command-line arguments relevant to different analysis modes

5.2.2 Input Formats

RIPCAL requires FASTA (http://www.ncbi.nlm.nih.gov/BLAST/fasta.shtml), multiple-aligned FASTA (http://www.bioperl.org/wiki/FASTA_multiple_alignment_format) or GFF (http://www.sequenceontology.org/gff3.shtml) format input files. Input requirements are context sensitive, depending on the chosen analysis mode. Alignment-based and dinucleotide analysis modes both require FASTA and GFF inputs for whole-genome analysis and FASTA only for single repeat family analysis. Index mode requires FASTA input only. Note: RIPCAL will attempt to distinguish nonaligned from aligned FASTA inputs by looking for gap ("-") characters.

Both FASTA and GFF input files can be of any extension as long as they conform to their respective formats; however file-browsing in GUI mode looks for *.fa, *.fas and *.fasta and *.gff and *.gff3 file extensions by default.

Repeat families are groups of related repetitive sequences, which are defined in a GFF file using the "target" attribute (i.e. "Target = repeat-familyID X Y", note: coordinates X and Y are ignored by RIPCAL). If a particular member of a repeat family is intended to be the "model" sequence (which all other sequences in the family are compared to) then this can also be indicated with a "note = model" GFF attribute; however there should only be one model sequence defined per repeat family. If a GFF input is not provided, all sequences in a fasta file are treated as a single repeat family. In this case, if a "user-defined" model is specified then the first sequence in the multi-FASTA file is assumed to be the model.

5.2.3 Analysis Modes

5.2.3.1 Alignment-Based Mode
Pre-analysis Considerations
De novo repeat finders such as RepeatScout (Price et al. 2005) and tools for mapping repeat family consensus sequences to a whole-genome assembly such as RepeatMasker (Smit et al. 1996–2010) have idiosyncrasies which may affect the reliability of a subsequent alignment-based RIPCAL analysis. De novo repeat finders are likely to report some redundant repeat families. Repeat family consensus sequences should be compared, i.e. by BLASTN (Altschul et al. 1990) and if necessary merged manually. Merging of redundant repeat families is recommended to be performed by generating a consensus sequence from a multiple alignment between repeats from one or more combined families. RepeatMasker and other repeat-mappers attempt to predict how multiple short repeat matches to the same repeat family within a small region of the genome may join up into a larger match. If this is predicted incorrectly it leads to repeat sequences not representing their full repeats and inflation of the number of repeats in a repeat family.

If multiple alignments are performed prior to RIPCAL analysis, the following considerations are useful when setting the alignment parameters. Some repeats can have large insertions/deletions, so gap extension penalties should be lowered accordingly. Some repeats also have short internal repeats and low-complexity sequences; therefore window size (or equivalent parameters) should be set as large as possible to reduce misalignments. As RIP involves $C \rightarrow T$ mutations (or $G \rightarrow A$ if reverse complemented), if your alignment tool of choice allows the use of custom matrices, then an adjusted matrix allowing higher scores for matches between C and T and between G and A will also improve the accuracy of fungal repeat alignments.

Manual inspection of multiple alignments is also highly recommended, which can be performed with Jalview (http://www.jalview.org/) (Waterhouse et al. 2009). Some repeat families, particularly larger ones with large insertion/deletions or low-complexity sequence regions, fail to align properly. In these cases misalignments can be manually corrected with Jalview or CINEMA (http://utopia.cs.man.ac.uk/utopia/cinema) (Lord et al. 2002).

Selection of Comparative "Model" Sequences
Alignment-based analyses quantify RIP-like mutations in a multiple alignment by comparing each aligned sequence to a "model" sequence.

Table 5.3 Degenerate base letters and their corresponding nucleotides

Degenerate base letter	Corresponds to
W	A/T
S	G/C
M	A/C
K	G/T
R	G/A
Y	T/C
B	G/T/C OR not A
D	A/G/T OR not C
H	A/C/T OR not G
V	G/C/A OR not T
N	A/C/G/T

Table 5.4 Probability values used in place of nucleotide integer counts by RIPCAL where a multiple alignment consensus sequence is ambiguous

1/1	1/2	1/3
A	M/R/W	D/H/V
C	M/S/Y	B/H/V
G	K/R/S	B/D/V
T	K/W/Y	B/D/H

There are three methods used for the selection of model (comparison) sequence for alignment-based analyses. Each of these has their own merits depending on the type of input data.

Majority Consensus

This method is recommended for most purposes and determines the most common base at each position of the alignment (where repeat sequence copy number >2). Degenerate base letters are used for positions in the alignment where there is no clear majority, i.e. if 2 or more base counts are equal. The degenerate consensus method assigns degenerate bases W, S, M K, R, Y, B, D, H, V or N (Table 5.3). "N" is used in the degenerate consensus to refer to any base pair combination but is not assigned a probability of RIP mutation when calculating RIP mutation from a degenerate consensus.

Because each sequence in the alignment is now compared to an ambiguous consensus, in this mode RIPCAL converts absolute mutation counts to "probabilities of mutation". Table 5.4 outlines the probability of nucleotide identity for each degenerate base letter. RIP probability at a particular position along an alignment is determined by Table 5.4, i.e. for consensus dinucleotide MpD mutating to TpA in aligned sequences, there is a ($1/2*1/3 = 1/6$) chance that this is a CpA → TpA mutation.

Highest G:C Content

The sequence with the highest total G:C content is selected as the model on the basis that RIP muta-

tion depletes G:C content; therefore highest G:C content should indicate the least RIP affected sequence. This may not be an appropriate method of model selection if there is great variation in length between aligned repeat sequences, as a longer sequence may be chosen over a shorter one.

User Defined

The choice of model sequence is left to the user (see Sect. 1.2.2 describing Input formats for details on how to specify model sequences). If a GFF input is provided but no model is defined for a repeat family and user-defined mode is chosen, RIPCAL will switch to highest GC mode to select the model for that repeat family.

RIP Dominance Metrics

Dominance metrics measure the predominance of a particular type of RIP-like CpN → TpN dinucleotide mutation over another, as outlined in (Hane and Oliver 2008). If there is strong evidence for a RIP dinucleotide bias, the use of RIP dominance metrics is particularly useful for distinguishing RIP-like mutations from other types of mutation.

CpA ↔ TpA dominance

$$\left(\frac{(CpA \leftrightarrow TpA)}{(CpC \leftrightarrow TpC) + (CpG \leftrightarrow TpG) + (CpT \leftrightarrow TpT)} \right)$$

CpC ↔ TpC dominance

$$\left(\frac{(CpC \leftrightarrow TpC)}{(CpA \leftrightarrow TpA) + (CpG \leftrightarrow TpG) + (CpT \leftrightarrow TpT)} \right)$$

CpG ↔ TpG dominance

$$\left(\frac{(CpG \leftrightarrow TpG)}{(CpA \leftrightarrow TpA) + (CpC \leftrightarrow TpC) + (CpT \leftrightarrow TpT)} \right)$$

CpT ↔ TpT dominance

$$\left(\frac{\left(CpT \leftrightarrow TpT \right)}{\left(CpA \leftrightarrow TpA \right) + \left(CpC \leftrightarrow TpC \right) + \left(CpG \leftrightarrow TpG \right)} \right)$$

Direction of RIP Mutations

The RIP dominance formulae shown above contain bi-directional arrows, as in most cases the original sequence that existed prior to being mutated RIP will be unknown. Therefore CpN mutations in both directions are counted. However, counting RIP mutations in a single direction would be appropriate if a known precursor repeat was selected as the comparison model. Although the default dominance metrics and graphical outputs of RIPCAL use bi-directional data, uni-directional mutation frequencies and dominance scores can still be calculated using data from tabular RIPCAL outputs.

5.2.3.2 Dinucleotide Frequency Mode

This mode calculates the relative frequencies of dinucleotides (pairs of adjacent nucleotides), the main purpose of which is to determine if there is a bias among repeats towards certain dinucleotides that indicate RIP, relative to non-repetitive sequences. For example, RIP in species of the Pezizomycotina is biased to CpA nucleotides; therefore dinucleotide analysis of repeats would be expected to indicate increased frequencies of TpA dinucleotides that are the result of RIP. This method can be applied to either whole-genome sequences and multiple repeat families or a single repeat family.

5.2.3.3 RIP-Index Scan Mode

This mode predicts RIP-mutated regions within input sequence based on high or low scoring RIP-indices. In GUI mode, RIPCAL uses default index thresholds only, described below and based on values determined experimentally in *Neurospora crassa* (Margolin et al. 1998). Thresholds can be adjusted in CLI mode.

In index scan mode, RIPCAL breaks down long sequences into smaller chunks (default length 200 bp). If these subregions are above the threshold for RIP for the selected RIP-indices, they are stored in memory and overlapping chunks are merged into longer regions. By

default, regions are reported in GFF format if they are above 300 bp in length.

5.2.4 Output Formats

5.2.4.1 GIF

This output (Fig. 5.1) is created if alignment-based analysis is selected. The alignment diagram is a visual representation of the alignment file received from ClustalW/prealigned input. Sequences appear in identical order to that of the alignment input. Usually this means that sequences are grouped according to similarity (default ClustalW alignment ordering) and this overall sequence similarity usually corresponds to similar RIP profiles. The *y*-axis of the plot at the bottom of the frequency plot represents the overall frequency of various RIP-like mutations (type indicated by colour) along a sliding window. This can show the localised effects of RIP changes in discrete sequence regions.

5.2.4.2 *_RIPALIGN.TXT

This output is created if alignment-based analysis is selected. This is a tabular summary of the data presented in the *.GIF graphical alignment-based output. The tabular data provides more in-depth information than can be presented in graphical form, including the polarity of RIP mutation.

5.2.4.3 *_DINUC.TXT

This output is created if dinucleotide frequency analysis is selected. This creates a tabular dinucleotide frequency table for individual sequences (fasta only) or repeat families (fasta + gff).

5.2.4.4 *_SCAN.TXT

This output is created in RIP-index scan mode. The contents of this file will be GFF format. This groups high and low scoring (by RIP-index) regions as GFF features.

5.3 Detailed Procedure Description

RIPCAL may be run either as a graphical-user interface (GUI) or from a command-line interface (CLI).

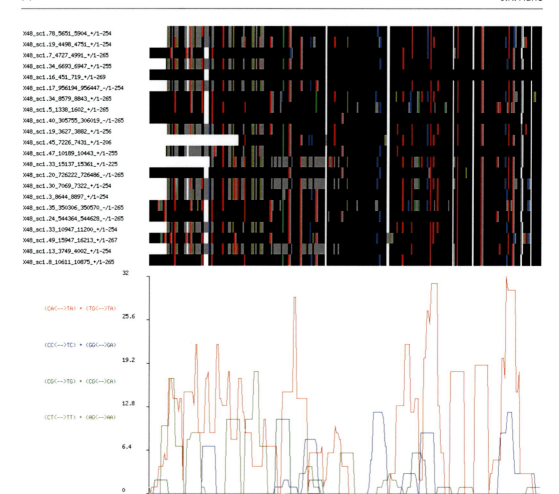

Fig. 5.1 Example RIPCAL alignment-mode graphical output depicting multiple alignment of repeats (*upper* in *black*) and a corresponding RIP mutation frequency plot (*lower*). The four possible CpN mutations are colour-coded in both sections, conserved sequences in the alignment are depicted in *black*, gaps in *white* and non-RIP-like polymorphism in *grey*

5.3.1 General Options

5.3.1.1 GUI

1. Run the exe (windows) or "perl ripcal_x_x_x.pl" through the command-line shell with no arguments, which should open the GUI as shown in Fig. 5.2
2. Select one of the 3 available analysis types (alignment-based, dinucleotide or index scan)
3. Select FASTA+GFF input or FASTA only (GFF input field will become disabled in index scan mode)
4. If alignment-mode selected, select consensus calculation option (highest G:C content, degenerate consensus or user defined)

5.3.1.2 CLI

usage: **perl ripcal_x_x_x.pl --command <arguments>**

Argument	Description	Default
--help OR –h	RIPCAL options help (lists these command-line arguments)	
--command OR –c	Use command-line interface	
--type OR –t	RIP analysis type [align OR index OR scan]	align

- If -fasta and -gff are selected, the input is assumed to contain multiple repeat families
- If -fasta only the input is assumed to contain only a single repeat family

5 Calculating RIP Mutation in Fungal Genomes Using RIPCAL

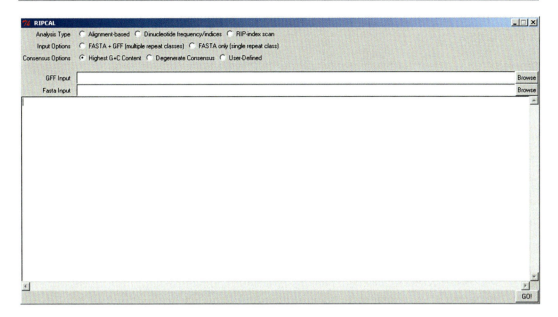

Fig. 5.2 The RIPCAL graphical-user interface (GUI)

- If the value for –m is "user", models are interpreted as the first sequence in the alignment for single family inputs
- For a single repeat family, aligned RIP analysis also accepts prealigned input in FASTA OR CLUSTALW format

--model OR –m	Alignment model [gc OR consensus OR user]	gc
--windowsize OR –z	RIP-frequency graph window	alignment length/50, minimum of 10 bp

5.3.2 Alignment-Based Mode

5.3.2.1 GUI
1. Select "Alignment-based"
2. Select "FASTA+GFF" for whole-genome analysis or "FASTA only" for single repeat family analysis
3. Select Consensus options
4. Browse for FASTA and/or GFF input files
5. Click "GO!"

5.3.2.2 CLI
Usage: **perl ripcal_x_x_x.pl --command --type align <arguments>**

Argument	Description	Default
--seq OR –s	Input sequence file [fasta or clustalw format] (REQUIRED)	
--gff OR –g	Input gff file [gff3 format: http://www.sequenceontology.org/gff3.shtml]	

(continued)

5.3.2.3 Summarising Alignment-Mode Outputs

RIPCAL also comes with a perl script for summarising the results of the *_RIPALIGN.TXT files, that contain detailed comparisons of RIP-like mutations between all sequences individually relative to the model sequence, into a simplified tabular format which summarises results for the whole repeat family. Each repeat family is contained in a separate *_RIPALIGN.TXT output file. The script needs to be run multiple times, but can be directed to a common summary output file.

```
Usage (1st family): perl ripcal_
summarise inputfile 1>outputfile

Usage (subsequent families): perl 
ripcal_summarise   inputfile   >> 
outputfile
```

5.3.3 Dinucleotide Mode

5.3.3.1 GUI

1. Select "Dinucleotide frequency/indices"
2. Select "FASTA+GFF" for whole-genome analysis or "FASTA only" for single repeat family analysis
3. Browse for FASTA and/or GFF input files
4. Click "GO!"

5.3.3.2 CLI

Usage: perl ripcal_x_x_x.pl --command --type index <arguments>

Argument	Description	Default
--seq OR –s	Input sequence file [fasta or clustalw format] (REQUIRED)	
--gff OR –g	Input gff file [gff3 format]	

5.3.4 Index Scan Mode

5.3.4.1 GUI

1. Select "RIP-index scan"
2. Browse for FASTA input file
3. Click "GO!"

5.3.4.2 CLI

Usage: perl ripcal_x_x_x.pl --command --type scan <arguments>

Note: index thresholds can be disabled (ignored) by setting their value to 0.

Argument	Description	Default*
--seq OR –s	Input sequence file [fasta or clustalw format] (REQUIRED)	
-l	Length of scanning subsequence (window size) (bp)	300
-i	Scanning subsequence increment (bp)	150
-q	TpA/ApT threshold (>=)	0.89
-w	$(CpA+TpG)/(ApC+GpT)$ threshold (<=)	1.03
-e	$(CpA+TpG)/TpA$ threshold (<=)	-
-r	$(CpC+GpG)/(TpC+GpA)$ threshold (<=)	-
-y	$CpG/(TpG+CpA)$ threshold (<=)	-

(continued)

-u	$(CpT+ApG)/(TpT+ApA)$ threshold (<=)	-

*As per Margolin et al. (1998)

5.3.5 deRIP

This technique reverses the effects of RIP in silico. The deRIP process is similar to alignment-based mode in scanning a multiple alignment of a repeat family for RIP-like polymorphism, which is followed by reverting the alignment consensus to the putative pre-RIP-mutated sequence. The resultant "deRIPped" sequence is a prediction of what a RIP-mutated repeat DNA may have looked like prior to RIP mutation. deRIP is only available in CLI mode.

Usage: perl deripcal <format> <inputfile> <outputfile>

Argument	Description	Default
<format>	"fasta" or "clustalw"	-
<inputfile>	Multiple alignment file in fasta (i.e. with "-"characters) or clustalw format	-
<outputfile>	The multiple alignment in aligned fasta format with the addition of a new first sequence which is the deRIP consensus	-
	Output file name also used as a prefix for the following 2 outputs	
	<outputfile>.consensus	
	<outputfile>.deripcons	
	Which are the sequences of the majority and derip consensus, respectively. Note: These will still contain gap characters from multiple alignments, which can be simply removed with a replace text command in a text-editor of choice	

5.4 Concluding Remarks

RIPCAL is a useful tool for determining whether a whole genome, repeat family or single sequence is likely to be mutated by RIP and to facilitate the comparison of metrics measuring RIP between repeats or whole genomes. To date, RIPCAL has been applied to the predicted repetitive DNA content of whole genomes in several fungal species (Table 5.2). RIPCAL also includes a tool called deRIP, which detects RIP within a multiple alignment of a repeat family in a similar way to RIPCAL, then predicts a consensus sequence representing the ancestral repeat family prior to the effects of RIP mutation. This technique was applied to repeats of *P. nodorum* (Hane and Oliver 2010), which determined the origin of five previously uncharacterised repeat families, some of which were revealed to be RIP-degraded versions of endogenous genes or gene clusters. A modified version of deRIP was also used to determine the "deRIP-consensus" of single copy genes across multiple isolates of *L. maculans*, which was used to demonstrate that these genes had been affected by leakage of RIP mutations targeted to adjacent repetitive DNA (Van de Wouw et al. 2010).

It is important to bear in mind that the bioinformatic analysis of repetitive sequences can often be problematic and current tools are not perfect. A number of issues can arise during the generation of data that are used as inputs into RIPCAL and deRIP. For example, genome assembly errors can either conceal or duplicate repetitive sequences in whole-genome assemblies. Prediction of repetitive regions using predefined sequence databases will miss novel repeats. De novo repeat predictors may miss low copy repeats, contain redundancy between predicted repeat families or may over-predict copy number due to reporting full-length repeats as a series of consecutive sub-sequences. Furthermore, multiple alignments of large repeat families are likely to be difficult to generate accurately and may require adjustment of commonly used parameters such as gap extension penalty and window size, followed by manual editing. As an alignment-based method of measuring RIP, RIPCAL is wholly dependent on the accuracy of its alignment inputs.

I would encourage readers intending to use RIPCAL to also refer its publications (Hane and Oliver 2008; Hane and Oliver 2010) and to the latest version of the manual, which is available with the latest version of the software at www.sourceforge.net/projects/ripcal.

References

Altschul SF, Gish W et al (1990) Basic local alignment search tool. J Mol Biol 215(3):403–410

Amyotte SG, Tan X et al (2012) Transposable elements in phytopathogenic *Verticillium* spp.: insights into genome evolution and inter- and intra-specific diversification. BMC Genomics 13:314

Clutterbuck AJ (2011) Genomic evidence of repeat-induced point mutation (RIP) in filamentous ascomycetes. Fungal Genet Biol 48(3):306–326

Daverdin G, Rouxel T et al (2012) Genome structure and reproductive behaviour influence the evolutionary potential of a fungal phytopathogen. PLoS Pathog 8(11):e1003020

DiGuistini S, Wang Y et al (2011) Genome and transcriptome analyses of the mountain pine beetle-fungal symbiont *Grosmannia clavigera*, a lodgepole pine pathogen. Proc Natl Acad Sci USA 108(6):2504–2509

Ellwood SR, Syme RA et al (2012) Evolution of three *Pyrenophora* cereal pathogens: recent divergence, speciation and evolution of non-coding DNA. Fungal Genet Biol 49(10):825–829

Fudal I, Ross S et al (2009) Repeat-induced point mutation (RIP) as an alternative mechanism of evolution toward virulence in *Leptosphaeria maculans*. Mol Plant Microbe Interact 22(8):932–941

Galagan JE, Selker EU (2004) RIP: the evolutionary cost of genome defense. Trends Genet 20(9):417–423

Gan P, Ikeda K et al (2013) Comparative genomic and transcriptomic analyses reveal the hemibiotrophic stage shift of *Colletotrichum* fungi. New Phytol 197(4):1236–1249

Gao Q, Jin K et al (2011) Genome sequencing and comparative transcriptomics of the model entomopathogenic fungi *Metarhizium anisopliae* and *M. acridum*. PLoS Genet 7(1):e1001264

Hane JK, Oliver RP (2008) RIPCAL: a tool for alignment-based analysis of repeat-induced point mutations in fungal genomic sequences. BMC Bioinformatics 9:478

Hane JK, Oliver RP (2010) *In silico* reversal of repeat-induced point mutation (RIP) identifies the origins of repeat families and uncovers obscured duplicated genes. BMC Genomics 11:655

Hood ME, Katawczik M et al (2005) Repeat-induced point mutation and the population structure of transposable elements in *Microbotryum violaceum*. Genetics 170(3):1081–1089

Horns F, Petit E et al (2012) Patterns of repeat-induced point mutation in transposable elements of basidiomycete fungi. Genome Biol Evol 4(3):240–247

Islam MS, Haque MS et al (2012) Tools to kill: genome of one of the most destructive plant pathogenic fungi *Macrophomina phaseolina*. BMC Genomics 13:493

Klosterman SJ, Subbarao KV et al (2011) Comparative genomics yields insights into niche adaptation of plant vascular wilt pathogens. PLoS Pathog 7(7):e1002137

Larkin MA, Blackshields G et al (2007) Clustal W and Clustal X version 2.0. Bioinformatics 23(21):2947–2948

Lefebvre F, Joly DL et al (2013) The transition from a phytopathogenic smut ancestor to an anamorphic biocontrol agent deciphered by comparative whole-genome analysis. Plant Cell 25(6):1946–1959

Lord PW, Selley JN et al (2002) CINEMA-MX: a modular multiple alignment editor. Bioinformatics 18(10):1402–1403

Manning VA, Pandelova I et al (2013) Comparative genomics of a plant-pathogenic fungus, *Pyrenophora tritici-repentis*, reveals transduplication and the impact of repeat elements on pathogenicity and population divergence. G3 3(1):41–63

Margolin BS, Garrett-Engele PW et al (1998) A methylated *Neurospora* 5S rRNA pseudogene contains a transposable element inactivated by repeat-induced point mutation. Genetics 149(4):1787–1797

Price AL, Jones NC et al (2005) *De novo* identification of repeat families in large genomes. Bioinformatics 21(Suppl 1):i351–i358

Ropars J, Dupont J et al (2012) Sex in cheese: evidence for sexuality in the fungus *Penicillium roqueforti*. PLoS One 7(11):e49665

Rouxel T, Grandaubert J et al (2011) Effector diversification within compartments of the *Leptosphaeria macu-lans* genome affected by Repeat-Induced Point mutations. Nat Commun 2:202

Santana MF, Silva JC et al (2012) Abundance, distribution and potential impact of transposable elements in the genome of *Mycosphaerella fijiensis*. BMC Genomics 13:720

Selker EU, Stevens JN (1987) Signal for DNA methylation associated with tandem duplication in *Neurospora crassa*. Mol Cell Biol 7(3):1032–1038

Selker EU, Cambareri EB et al (1987) Rearrangement of duplicated DNA in specialized cells of *Neurospora*. Cell 51(5):741–752

Smit AFA, Hubley R et al (1996–2010) RepeatMasker Open-3.0. http://www.repeatmasker.org

Van de Wouw AP, Cozijnsen AJ et al (2010) Evolution of linked avirulence effectors in *Leptosphaeria maculans* is affected by genomic environment and exposure to resistance genes in host plants. PLoS Pathog 6(11):e1001180

Vogel KJ, Moran NA (2013) Functional and evolutionary analysis of the genome of an obligate fungal symbiont. Genome Biol Evol 5(5):891–904

Walsh CP, Xu GL (2006) Cytosine methylation and DNA repair. Curr Top Microbiol Immunol 301:283–315

Waterhouse AM, Procter JB et al (2009) Jalview Version 2—a multiple sequence alignment editor and analysis workbench. Bioinformatics 25(9):1189–1191

Yang J, Wang L et al (2011) Genomic and proteomic analyses of the fungus *Arthrobotrys oligospora* provide insights into nematode-trap formation. PLoS Pathog 7(9):e1002179

Fungal Transposable Elements

6

Linda Paun and Frank Kempken

6.1 Introduction

Transposable elements (TEs) are mobile genetic elements, which can translocate or change their position within the genome. These mobile elements or transposons are ubiquitously distributed throughout all kingdoms and have a huge impact on genome diversity (Daboussi and Capy 2003; Feschotte and Pritham 2007). While in previous decades individual transposons have been studied in great detail, in recent years genome sequencing projects have provided a fast increasing amount of data on TEs. New TEs are being characterized almost constantly. The genomic distribution of TEs is highly diverse (Janicki et al. 2011). Whereas the occurrence of TEs in plants (e.g. maize: 80 % (SanMiguel et al. 1998)) and animals (e.g. human: 46 % (International Human Genome Sequencing Consortium 2001)) is rather high, fungal genomes in general show a low percentage of TEs: 1–4 % (Biémont and Vieira 2006). However, a few fungal genome sequences consist of up to 40 % of TEs, like the Dothiodeomycetes *Mycosphaerella fijiensis* (Ohm et al. 2012). It is not known yet why in some fungi the number of TEs has increased considerably.

First thought to be harmful for the host, nowadays, TEs are believed to be molecular forces which help modifying the genome as they have a huge impact on the genome evolution and the diversity of species (Biémont 2009, 2010). They can have influence on the gene expression by e.g. disrupting open reading frames or integrating very close to protein-encoding genes (Janicki et al. 2011). In addition there is clear evidence that regulatory units and new gene functions have evolved from TEs (Cohen et al. 2009; Janicki et al. 2011; Shaaban et al. 2010). Using TEs for mutagenesis they even function as molecular tools and are helpful in discovering new genes or gene functions (Kempken 1999).

TEs can change their position actively or passively, depending on whether they are autonomous or not, i.e. encode their own transposition enzyme(s). The activity can further be regulated by epigenetical changes, which are introduced by the host genome (Janicki et al. 2011). The human genome e.g. comprises approx. 100 active transposons and 0.3 % of the human diseases are caused by transposons inserting in genes (Belancio et al. 2009).

TEs have been known for almost 100 years now (see Sect. 6.2). They can change their respective position in the genome, by four different mechanisms (Zhang and Saier 2011). This can be either (1) conservative as a DNA transposon, which

---

L. Paun (✉) • F. Kempken
Genetics and Molecular Biology in Botany, Botanical Institute, Christian-Albrechts-University Kiel,
Am Botanischen Garten 5, Kiel 24118, Germany
e-mail: lpaun@bot.uni-kiel.de

M.A. van den Berg and K. Maruthachalam (eds.), *Genetic Transformation Systems in Fungi, Volume 2*, Fungal Biology, DOI 10.1007/978-3-319-10503-1_6,
© Springer International Publishing Switzerland 2015

gets excised from one position and integrates into another, (2) replicative with building a cointegrate between donor and target DNA, (3) excessive by forming a circular molecule or (4) retro as a RNA copy, which is subsequently reverse transcribed into a DNA copy. These processes function in distributing the element throughout the genome and amongst other criteria led to a special classification system (see Sect. 6.3).

This chapter presents basic facts about TEs in eukaryotes and more specific data for fungi. We will present progress in different research fields including bioinformatics and molecular genetics. Emphasis is on the use of TEs as molecular tools. By modifying TEs they may function as a random mutagenesis tool for almost every organism.

6.2 History

The reported history of TEs started 100 years ago with studies on variegation of the maize kernel pericarp (Emerson 1914; Peterson 2013). In the late 40s of the twentieth century, Barbara McClintock observed different color-patterns in corn kernels after crossing different strains. These differences did not derive from genetic inheritance but were reversible. Her observation led to the assumption that some kind of "controlling element" was responsible for these color changes and she called them *Activator* (*Ac*) and *Dissociation* (*Ds*) (McClintock 1951). What she first described as controlling elements later became known as transposons or TEs.

But it was not until the early 1980s that the function and potential of *Ac* and *Ds* was further explored (Döring et al. 1984; Pohlman et al. 1984), and the molecular background, in that case the integration of a transposable sequence into the anthocyanin synthase gene leading to the different colored kernels, was discovered (Wessler 1988). In the following decade many different TEs were found in all kingdoms (Daboussi and Capy 2003). The first detected TEs in fungi were the *Ty1* and δ elements in the yeast *Saccharomyces cerevisiae* (Cameron et al. 1979). First thought to be "junk DNA" or even being harmful for the genome, TEs today are

seen as evolutionary forces, which can positively contribute to diversity (Biémont and Vieira 2006). They can cause genome rearrangements which can lead to deletions or inversions (Schmidt and Anderson 2006; Zhang and Saier 2011). The replicative character of retrotransposons leads to high copy numbers over time, thus increasing genome size (Janicki et al. 2011). However, genome size may also decrease due to homologous recombination between different chromosomes caused by TE repeats (Bennetzen 2005; Grover and Wendel 2010; Vitte and Panaud 2005). Such global genome effects explain the high interest on TEs. One focus is laid on diseases caused by transposons. The human genome consists up to 46 % of TEs with approx. 80–100 active LINE-elements that can lead to disease-causing insertions (Hancks and Kazazian 2012). Additionally, the mutagenic capabilities of TEs stimulated the use as molecular tools for random mutagenesis (Hehl and Baker 1990; Kempken 2003).

6.3 Modern Classification

Finnegan (1989) classified the TEs into two groups on the basis of their mechanism to translocate in the genome. Class I transposons, also called retroelements, transpose via the so-called "copy-and-paste" mechanism (Fig. 6.1a). The genomic copy of the TE is transcribed into RNA and reverse transcribed into cDNA. This cDNA will integrate into the target DNA (Gao and Voytas 2005). Back then this class I consisted of two different types of transposons: LTR (*long terminal repeat*; Retrotransposons) and non-LTR elements, including the LINE-like (*long interspersed nuclear element*) and the SINE-like elements (*short interspersed nuclear element*) (Han 2010). Class II transposons on the other hand translocate directly on DNA-level with a "cut-and-paste" mechanism (Fig. 6.1b) (Finnegan 1989). The transposon gets cut out by an enzyme called transposase and integrates at the target region (Craig et al. 2002; Kazazian 2004). Finding more and more new and different elements led to the extension of the early classification (Kapitonov and Jurka 2008; Wicker et al. 2007).

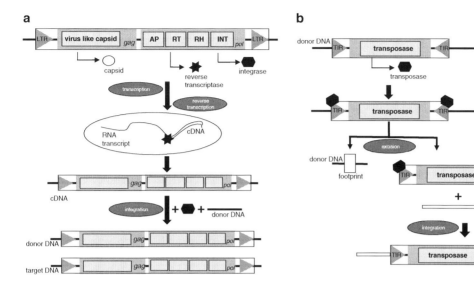

Fig. 6.1 Schematic diagram of two transposons with a different transposition mechanism: (**a**) LTR-element (**b**) TIR-element (Abbreviations *LTR* long terminal repeat, *AP aspartic* proteinase, *RT* reverse transcriptase, *RH* RNaseH, *INT* integrase, *TIR* terminal inverted repeat)

The two classes were divided into various orders, subclasses, and superfamilies based on the mechanism of transposition, similarities in the sequences as well as structural relationships. The current classification is widely accepted and used. Table 6.1 gives an overview over the different classes with some well-studied examples in fungi.

Class I consists of five orders and all of these orders can be found in fungal genomes: LTR and the four non-LTR orders DIRS (*Dictyostelium transposable element*), PLE (*Penelope-like elements*), LINE and SINE with 17 superfamilies all together: *Copia, Gypsy, Bel-Pao, Retrovirus, ERV, DIRS, Ngaro, VIPER, Penelope, R2, RTE, Jockey, L1, I, tRNA, 7SL*, and *5S*. Except for the SINE elements, they all carry a *reverse transcriptase* (RT) gene. LTR elements are flanked by *long terminal repeats*, of variable length in the range of a few 100 bp up to more than 5 kb. LTRs produce a 4–6 bp target site duplication (TSD) upon integration and include two open reading frames (ORFs) *gag* and *pol*. The *gag* gene encodes for a virus-like capsid protein and most of the class I elements include this gene. *pol* includes ORFs for RNaseH (RH), an aspartic proteinase (AP), an integrase (INT), and the RT. Up to this date, the only elements from this order found in fungal genomes are from the *copia* and *gypsy* superfamilies. Both of these elements differ in the order of INT and RT. One example from the LTR order is the *gypsy*-element *Maggy* from *Magnaporthe grisea* which is 5.7 kb in size and the LTRs are 257 bp long (Farman et al. 1996). Figure 6.1a shows a schematic diagram of an LTR-transposon from the *gypsy* superfamily. The second order, DIRS (Cappello et al. 1985), differs from the classical RNA-transposons in not generating TSDs but either *split direct repeats* (SDR) or inverted repeats upon integration (Wicker et al. 2007). This is due to absence of the INT gene. DIRS elements also belong to the so-called YR-retrotransposons, which possess a tyrosine recombinase (Muszewska et al. 2013). PLE (Evgen'ev et al. 1997; Evgen'ev and Arkhipova 2005) encode for a RT which is more closely related to telomerases then to RTs from LTRs and additionally they encode for an endonuclease. Some of them also include an intron. LINEs do not have LTRs and only the superfamilies *L1* and *I* were found in fungi so far. They encode at least for a RT in autonomous elements, but they may also encode for a *nuclease* (Eickbush and Malik 2002;

Table 6.1 Classification of TEs with some well-studied examples in fungi

Class I (retroelement)

Order	Superfamily	Transposon name	Organism	Publication
LTR	Copia	Ty1	Saccharomyces cerevisiae	Cameron et al. (1979)
	Gypsy	Mars2, Mars3	Ascobolus immersus	Goyon et al. (1996)
		CfT-1	Cladosporium fulvum	McHale et al. (1992)
		Maggy	Magnaporthe grisea	Farman et al. (1996)
	Bel-Pao[a]			
	Retrovirus[a]			
	ERV		Laccaria bicolor	Labbé et al. (2012)
DIRS (YR-retrotransposons)	DIRS	Prt1	Phycomyces blakesleeanus	Ruiz-Pérez et al. (1996)
	Ngaro	CcNgaro1	Coprinopsis cinerea and Phanerochaete chrysosporium	Goodwin and Poulter (2004)
		PcNgaro1		
	VIPER[a]			
PLE	Penelope	Coprina_Cc1	Coprinopsis cinerea	Arkhipova (2006), Gladyshev and Arkhipova (2007)
		Coprina_Pc1	Phanerochaete chrysosporium	
LINE	R2[a]			
	RTE[a]			
	Jockey[a]			
	L1	Tad	Neurospora crassa	Kinsey and Helber (1989), Kinsey (1990)
		Zorro	Candida albicans	Goodwin et al. (2001)
		CgT1	Colletotrichum gloeosporioides	He et al. (1996)
	I[b]			
SINE	tRNA	Mg-SINE	Magnaporthe grisea	Kachroo et al. (1995)
	7SL[b]			
	5S[a]			

6 Fungal Transposable Elements

Class II (subclass I—DNA transposons)

Order	Superfamily	Transposon Name	Organism	Publication
TIR	Tc1-Mariner	Fot1	Fusarium oxysporum	Daboussi et al. (1992)
		Vader	Aspergillus niger	Amutan et al. (1996)
		Impala	Fusarium oxysporum	Langin et al. (1995)
	hAT	Restless	Tolypocladium inflatum	Kempken and Kück (1996)
		Hornet1	Fusarium oxysporum	Hua-Van et al. (2000)
	Mutator	Hop	Fusarium oxysporum	Chalvet et al. (2001), Chalvet et al. (2003), Daboussi and Langin (1994)
	Merlin[a]			
	Transib		Phakospora pachyrhizi	Kapitonov and Jurka (2005)
	P[a]			
	PiggyBac		Tuber melanosporum	Martin et al. (2010)
	PIF/Harbinger		Filobasisiella neoformans	Jiang et al. (2003), Zhang et al. (2004)
			Neurospora crassa	
	CACTA[b]			
	Sola		Glomus intraradices	Bao et al. (2009)
	Zator[a]			
Crypton	Crypton	Cn1	Cryptococcus neoformans	Goodwin et al. (2003)

Class II (subclass II—DNA transposons)

Order	Superfamily	Transposon name	Organism	Publication
Helitron	Helitron	Hel_Pc1	Phanerochaete chrysosporium	Poulter et al. (2003)
Maverick	Maverick/Polinton		Glomus intraradices and Phakospora pachyrhizi	Kapitonov and Jurka (2006), Pritham et al. (2007)

[a] Not yet found in fungi
[b] Already found in fungi according to Wicker et al. (2007)

Ostertag and Kazazian 2005). *I* elements in addition encode for a RH gene. SINE elements are nonautonomous and have their origin in polymerases III transcripts rather than in deletion derivatives of other TEs (Kramerov and Vassetzky 2005). They are relatively small (only 80–500 bp) and the type of 5′ end defines the superfamily as this so-called "head" derives either from tRNA, 7SL RNA, or 5S RNA (Kramerov and Vassetzky 2011). They produce a TSD of 5–15 bp and rely on *trans*-activation from other RTs. The attached 3′ polyA tract (Rowold and Herrera 2000) helps with the reverse transcription and is believed to mediate between LINEs and SINEs (Janicki et al. 2011; Jurka 1997; Kajikawa and Okada 2002; Okada et al. 1997). A well-known fungal example is *Mg*-SINE from *Magnaporthe grisea* (Kachroo et al. 1995).

All together class II elements include 14 superfamilies: *Tc1/Mariner*, *hAT*, *Muator*, *Merlin*, *Transib*, *P*, *PiggyBac*, *PIF/Harbinger*, *CACTA*, *Crypton*, *Helitron*, *Maverick*, *Sola*, and *Zator* (Janicki et al. 2011). Of these superfamilies only *Merlin*, *P*, and *Zator* were not yet identified in fungi. The class II elements are separated into two subclasses. Subclass 1 combines two orders, *TIR* (*terminal inverted repeat*) and *Crypton*, which are both known as the classical DNA transposons. Different TIR elements can be specified by the length of the TIR (Fig. 6.1b) and of the differences in the TSD. *Restless*, the first *hAT* transposon found in fungi, e.g. produces an 8 bp TSD upon integration and is enclosed by short TIR (5–27 bp) which is also important for binding the transposase (Kempken and Kück 1996; Kempken and Windhofer 2001). *Tc1-Mariner* elements only form a short TSD of two base pairs, "TA", upon integration and possess a simple structure of two TIRs (Shao and Tu 2001). *Crypton* elements were solely found in fungal genomes so far (Goodwin et al. 2003). They lack TIRs but seem to produce TSDs. They have some similarities to *DIRS* and *IS* elements but do not encode for a RT and the transposition might involve the recombination of the target DNA with a circular molecule (Goodwin et al. 2003; Wicker et al. 2007). Subclass 2 of class II elements comprises *Helitrons* and *Mavericks*. *Helitrons*

transpose via rolling-circle mechanism and do not form TSDs (Kapitonov and Jurka 2001). At both ends they have a TC or CTRR (R for purine) motif with a hairpin structure close to the 3′ end (Wicker et al. 2007). *Mavericks* (or *Polintons*) transpose similar as subclass 1 elements but do not encode for a transposase and are therefore known as self-synthesizing DNA transposons (Kapitonov and Jurka 2006). They are giant TEs (10–20 kb) with long TIRs (Feschotte and Pritham 2005). Class II elements can also be classified regarding their autonomy (Janicki et al. 2011). If they encode a superfamily-specific transposase they are regarded as autonomous (e.g. *Ac-element*). Having no functional transposase they are regarded as non-autonomous (e.g. *Ds-element*). However, autonomy does not necessarily mean that the transposon is active or even functional but it includes all domains that are necessary for transposition even if they yield some kind of deletion or mutation (Wicker et al. 2007). There are four different groups of non-autonomous elements: LARD (*large retrotransposon derivative*), MITE (*miniature inverted-repeat transposable element*), SNAC (*small non-autonomous CACTA transposon*), and TRIM (*terminal repeat retrotransposon in miniature*) and of course the SINE elements from Class I transposons (Wicker et al. 2007).

Each superfamily, regardless of which class, bears different specifications. Newly found transposons can be added to one of the superfamilies by sequence similarities in the transposase gene or by similar TSDs or TIRs (Janicki et al. 2011). The database RepBase is very useful to identify new TEs and updated regularly with the latest research findings (Jurka et al. 2005).

6.4 Origin

The evolutionary origin of transposons remains elusive. As they are present in all prokaryotic and eukaryotic life forms, we must assume an early origin. In case of class I elements there is a clear connection between reverse transcriptase encoding introns and the presumed RNA world (Lambowitz et al. 1999; Toor et al. 2001).

6 Fungal Transposable Elements

DNA transposons on the other hand are much more diverse in structure and mechanisms. *Mavericks* e.g. may have a common origin with linear plasmids originating from linear bacteriophages (Kapitonov and Jurka 2006; Kempken et al. 1992). Transposases from other DNA elements might also be distantly related to intron-encoded recombinases (Bonocora and Shub 2009; Wu and Hao 2014). Horizontal gene transfer (HGT) may have contributed, but the general presence of TEs in all clades argues for vertical transfer in most cases.

6.5 Transposable Elements in Filamentous Fungi

More than 30 years ago, the first TEs were detected in fungi, the *Ty1*, and δ elements in *S. cerevisiae* (Cameron et al. 1979). The first TE in a filamentous fungus was the *Tad1* element from *Neurospora crassa* (Kinsey and Helber 1989). It belongs to the *L1* elements of the non-LTR retrotransposons. Up till now, a great number of TEs were found in different fungi (Table 6.1).

In the past, most TEs were found by coincidence. As eukaryotic TEs usually occur in high copy numbers, it is also possible to analyze repetitive sequences to identify TEs. This approach led to the identification of the *Restless* element from *Tolypocladium inflatum* (Kempken and Kück 1996). In some cases a transposon trap approach proved to be successful as shown for the *Fot1* element from *Fusarium oxysporum* (Daboussi et al. 1992). Nowadays as more and more full genome sequences are available, it is possible to screen the full genome to look for repetitive or transposable elements (Lerat 2010).

Muszewska et al. (2011) give a very good overview of the latest research activities regarding LTR retrotransposons in fungi. Only the two superfamilies *gypsy* and *copia* were found in fungi so far with *gypsy* being the most abundant. They screened 59 fungal genomes for the presence of LTR transposons and found that the variety of the identified TEs is not that strong (Muszewska et al. 2011). The majority of the detected *gypsy* elements belong to the *Chromoviridae*,

elements which additionally comprise a chromointegrase, an integrase with a C-terminal chromodomain (Gao et al. 2008). The different types and functions of non-LTR retrotransposons in fungi were reviewed by Novikova et al. (2009). They did in silico analyses of 57 fungal genomes and nearly all of them harbored non-LTR transposons. Next to the known clades of *Tad*, *CRE* and *L1*, they could also identify two new clades, *Deceiver* and *Inkcap* (Novikova et al. 2009).

6.5.1 Bioinformatical Analyses of Transposable Elements in Fungi

Genome sequencing gets more and more affordable, particularly with next-generation approaches (Nowrousian 2010; Traeger et al. 2013). This led to a flood of sequence information, which needs to be analyzed. In early studies "junk" sequences like repeats were filtered out (Janicki et al. 2011). However, today there are many tools for analyzing repeated sequences. There are several ways to annotate and classify TEs (Lerat 2010). Lerat gives a detailed list of programs which can be used for detecting transposons e.g. REPEATMASKER and REPEATMODELER (Smit et al. 2006; Smit and Hubley 2010). REPEATMASKER is useful for screening genome sequences, as it will find all repeats and annotates them. However, it cannot be used to identify new elements. To achieve this, there are several programs solely to identify LTR (e.g. LTRHARVEST (Ellinghaus et al. 2008)) or non-LTR elements (e.g. RTANALYZER (Lucier et al. 2007)). A new, updated list of bioinformatical tools and databases was recently published by Janicki et al. (2011). The program VisualRepbase, which is based on the public database RepBase, for instance displays TE families and shows the distribution of them on the genome (Tempel et al. 2008).

Until now, more than 100 fungal genomes were sequenced and analyzed. Here we present a compilation of several fungal genome projects (Table 6.2) with a focus on the content of TEs in these fungal genomes. Some fungal genomes

Table 6.2 Examples of genome projects with the TE content

Fungi	Class	Genome size (MB)	Transposable elements (%)	Literature
Agaricus bisporus	Agaricomycetes	31.2	11.2	Morin et al. (2012)
Coprinopsis cinerea	Agaricomycetes	36.2	2.5	Stajich et al. (2010)
Ganoderma lucidum	Agaricomycetes	43.3	7.8	Chen et al. (2012)
Schizophyllum commune	Agaricomycetes	38.5	10.7	Ohm et al. (2010)
Cryptococcus neoformans	Tremellomycetes	19	5	Loftus et al. (2005)
Ustilago maydis	Ustilaginomycetes	20.5	1.1	Kämper et al. (2006)
Alternaria brassicicola	Dothideomycetes	32	5.6	Ohm et al. (2012)
Cladosporim fulvum	Dothideomycetes	61.1	44.2	Ohm et al. (2012)
Cochliobolus sativus	Dothideomycetes	34.4	5.4	Ohm et al. (2012)
Hyserium pulicare	Dothideomycetes	38.4	0.6	Ohm et al. (2012)
Leptoshaerica maculans	Dothideomycetes	44.9	30.9	Ohm et al. (2012)
Mycosphaerella fijiensis	Dothideomycetes	74.1	39	Ohm et al. (2012)
Mycosphaerella graminicola	Dothideomycetes	39.7	11.7	Ohm et al. (2012)
Mycosphaerella populicola	Dothideomycetes	33.2	20.8	Ohm et al. (2012)
Mycosphaerella populorum	Dothideomycetes	29.4	3.6	Ohm et al. (2012)
Pyrenophora teres f. teres	Dothideomycetes	33.6	2	Ohm et al. (2012)
Pyrenophora tritici-repentis	Dothideomycetes	37.9	11.4	Ohm et al. (2012)
Rhytidhysteron rufulum	Dothideomycetes	40.2	0.2	Ohm et al. (2012)
Setoshaeria turcica	Dothideomycetes	43	11.2	Ohm et al. (2012)
Stagnospora nodorum	Dothideomycetes	37.2	2.4	Ohm et al. (2012)
Aspergillus clavatus	Eurotiomycetes	27.9	0.8	Fedorova et al. (2008)
Aspergillus niger	Eurotiomycetes	33.9	105 sequences similar to TEs (e.g. LTR-transposon ANiTa1: 4.9 kb)	Braumann et al. (2007), Pel et al. (2007)
Aspergillus nidulans	Eurotiomycetes	30.1	3	Galagan et al. (2005)
Aspergillus fumigatus	Eurotiomycetes	28	2.9	Galagan et al. (2005)
Aspergillus oryzae	Eurotiomycetes	37	1.3	Galagan et al. (2005)
Coccidioides immitis	Eurotiomycetes	28.9	17[a]	Sharpton et al. (2009)
Coccidioides posadasii	Eurotiomycetes	27	12[a]	Sharpton et al. (2009)
Histoplasma capsulatum	Eurotiomycetes	33	19[a]	Sharpton et al. (2009)
Neosartorya fischeri	Eurotiomycetes	32.6	1	Fedorova et al. (2008)

Penicillium chrysogenum	Eurotiomycetes	32.2	374 sequences similar to TEs (e.g. LTR-transposon PeTra1: 4.9 kb)	Braumann et al. (2007)
Uncinocarpus reesii	Eurotiomycetes	22.3	4[a]	Sharpton et al. (2009)
Sclerotinia sclerotiorum	Leotiomycetes	38.3	7	Amselem et al. (2011)
Tuber melanosporum	Pezizomycetes	125	58	Martin et al. (2010)
Saccharomyces cerevisiae	Saccharomycetes	12.1	3.1	Kim et al. (1998)
Schizosaccharomyces pombe	Schizosaccharomycetes	13.8	1.1	Bowen et al. (2003), Wood et al. (2002)
Beauveria bassiana	Sordariomycetes	33.7	2[a]	Xiao et al. (2012)
Botrytis cinerea T4 B05.10	Sordariomycetes	39.5	0.7	Amselem et al. (2011)
Fusarium circinatum	Sordariomycetes	44.3	1.1	Wiemann et al. (2013)
Fusarium fujikuroi	Sordariomycetes	43.9	2.2	Wiemann et al. (2013)
Fusarium graminearum	Sordariomycetes	36.2	0.3	Ma et al. (2010), Wiemann et al. (2013)
Fusarium mangiferae	Sordariomycetes	45.6	0.5	Wiemann et al. (2013)
Fusarium oxysporum	Sordariomycetes	61.4	4.8	Ma et al. (2010), Wiemann et al. (2013)
Fusarium solani	Sordariomycetes	51.3	1.6	Coleman et al. (2009), Wiemann et al. (2013)
Fusarium verticillioides	Sordariomycetes	41.7	0.5	Ma et al. (2010), Wiemann et al. (2013)
Magnaporthe grisea	Sordariomycetes	37.9	7.3	Dean et al. (2005)
Metarhizium anisopliae	Sordariomycetes	39	1[a]	Gao et al. (2011)
Metarhizium acridum	Sordariomycetes	38	1.5[a]	Bushley et al. (2013), Gao et al. (2011)
Metarhizium robertsii	Sordariomycetes	39	1[a]	Bushley et al. (2013)
Neurospora crassa	Sordariomycetes	38.6	4.6	Galagan et al. (2003)
Podospora anserina	Sordariomycetes	36	3.5	Espagne et al. (2008)
Sordaria macrospora	Sordariomycetes	40	0.14	Nowrousian et al. (2010)
Tolypocladium inflatum	Sordariomycetes	30.4	0.4	Bushley et al. (2013)
Rhizopus oryzae	Phycomycetes	45.3	19.6	Ma et al. (2009)

[a]Repeat density including TE content but also repeats from Satellites, small RNAs or simple repeats

consist of up to 40 % of repetitive sequence whereas the average genome content of TEs is about 1–4 % only. Thus, in general fungal genomes have a low number of transposons compared to plants and animal genomes. However, exceptions like e.g. some Dothideomycetes are noteworthy. It is possible, that in those cases TEs have overcome host copy number control mechanisms. Due to a mechanism called RIP, repetitive elements are being inactivated in many if not most fungi (Cambareri et al. 1989; Clutterbuck 2011; Galagan and Selker 2004; Selker 1997). Clutterbuck (2011) gives a detailed record about RIP in filamentous fungi and its influence on TEs. He analyzed 54 genomes from 49 different fungal species and found hints for RIP in almost every fungi. The TEs suffered from $C \rightarrow T$ and $G \rightarrow A$ single and $CpN \rightarrow TpN$ double mutations. In the *A. niger* strain CBS513.88 e.g. five copies of the *ANiTa* family were found with one differing from the others by one transversion and multiple transitions (Braumann et al. 2008).

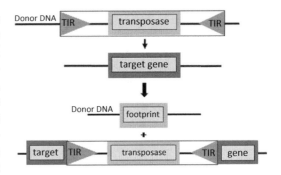

Fig. 6.2 Schematic diagram of the transposon tool mechanism with a TIR-transposable element. The transposon is excised from the donor site by the transposase, which is in this case encoded by the transposon itself. The transposase recognizes the TIRs and cuts out the TE. This process is incorrect and a small part of the transposon remains at the donor site as a so-called footprint. The TE then randomly integrates into a new locus. This might be a gene, which is interrupted, or maybe a promotor region, which then has influence on the transcription level of the controlled gene

6.6 Transposable Elements as Molecular Tools in Fungi

Many fungal genera contain different biotechnological important strains like e.g. *Aspergillus niger*, *Saccharomyces cerevisiae*, and *Tolypocladium inflatum* to name a few only. To better understand gene functions and to identify new genes, transposon-mediated mutagenesis is a powerful tool (Kempken 1999), although homologous recombination with *mus-51* and *mus-52* mutants has become an attractive alternative (Ninomiya et al. 2004). Figure 6.2 shows a schematic diagram of the functionality of a transposon mutagenesis tool. For 20 years now, the potential of TEs as molecular tools has been used with great effect. For plants, the *Ac/Ds* elements are highly explored e.g. in the model organism *Arabidopsis thaliana* (Aarts et al. 1993). Most of the genes in *Drosophila melanogaster* were identified using a TE, the *P Element* (Sentry and Kaiser 1992) and also in bacterial strains like *Plasmodium falciparum*, the human pathogen to cause malaria, random mutagenesis with TEs play an important role. In this organism a TE from the *PiggyBac* superfamily was used for insertional mutagenesis (Ikadai et al. 2013).

The first insertional mutagenesis in a filamentous fungus was reported in the late 1990s (Migheli et al. 1999). There *Fot1* from *Fusarium oxysporum* was shown to be highly active. It belongs to the *Tc1/Mariner* superfamily of DNA elements and the copy number ranges from zero to more than 100 per genome (Daboussi and Langin 1994). Transposon mutagenesis in *T. inflatum* employing *Restless*, an autonomous TE from the *hAT*-superfamily of DNA transposons, led to the identification of the transcription factor gene *tnir1* (*Tolypocladium* nitrogen regulator 1), which is a regulator for the nitrogen metabolism (Kempken and Kück 2000). Another TE from *F. oxysporum* showed activity in heterologous hosts such as *Magnaporthe grisea*, *Aspergillus nidulans*, and *Penicillium griseoroseum* (de Queiroz and Daboussi 2003). It is a useful tool for insertional mutagenesis also in different hosts as it is not dependent on host factors. Another tool, proven to be highly active in its homologous host is *Vader* from *A. niger* (Hihlal et al. 2011).

TEs can also function as diagnostic tools (Fernandez et al. 1998; Kempken et al. 1998; Kempken 1999). They are often strain-specific, and hence exhibit specific patterns in diagnostic polymerase chain reactions. TEs may also be used for population analysis.

6.6.1 Two Case Studies: The *Vader* Element from *Aspergillus niger* and the *Impala* Element from *Fusarium oxysporum*

Vader was first discovered in 1996 in the filamentous fungus *Aspergillus niger var. awamori* (Amutan et al. 1996). It belongs to the *Fot1/Pogo* group within the *Tc1/Mariner* superfamily of DNA TEs and was shown to be a useful tool for random mutagenesis in its homologous host, strain CBS513.88 (Hihlal et al. 2011). It is a non-autonomous transposon which is transactivated by the *Tan1*-transposase (Nyyssönen et al. 1996). A synthetical *Vader* was used with an additional 20 bp anchor sequence to allow the identification of this *Vader* copy after being excised from the donor site (Fig. 6.3). In *A. niger*, a *Vader* excision frequency of 1 in 2.2×10^5 was observed. *Vader* appears to be able to transpose to many different genomic locations on different chromosomes, but based on reintegration sites analyses it prefers integration very close to open reading frames or within introns (Hihlal et al. 2011). Hence, in most cases it has no effect on gene expression, as it gets spliced out during RNA-processing or is outside of the gene. At current, different ways to modify the element are under investigation.

Impala was discovered via transposon trap experiments in *F. oxysporum* (Langin et al. 1995), which is a plant pathogen causing wilt in a wide range of hosts, such as tomato or cotton (Dean et al. 2012). The TE also belongs to the *Tc1/Mariner* superfamily and is an autonomous element, 1.3 kb in size with 37 base pair-long TIRs. *Impala* proved to be a good tool for random mutagenesis in its homologous host (Migheli et al. 2000) as well as in the heterologous host *Penicillium griseoroseum* (de Queiroz and

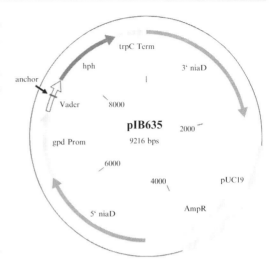

Fig. 6.3 Schematic diagram of the *Vader* Vector pIB635. The vector integrates into the genome via homologous recombination at the niaD-locus (From Hihlal, E., B. Braumann, M. van den Berg, and F. Kempken. 2011. Suitability of *Vader* for Transposon-Mediated Mutagenesis in Aspergillus Niger. © American Society for Microbiology, Applied and Environmental Microbiology, Vol. 77, No. 7, 2011, p. 2332-2336, doi:10.1128/AEM.02688-10 with permission)

Daboussi 2003). It was further used to identify mutants which have a reduced pathogenicity after transposition of *Impala* (Migheli et al. 2000). In that study *F. oxysporum* f. sp. *melonis* was screened for *impala* excision. Two fungal mutants were used, which encode for *niaD::imp160* and *niaD::imp161*, respectively (Langin et al. 1995). These strains have an *impala* element integrated at the *niaD*-locus of *Aspergillus nidulans*. Mutants able to grow on medium containing nitrate were analyzed regarding the excision of *Impala* and its reintegration into the genome. Seventy percent of the analyzed strains showed *Impala* reintegration. The integration sites were distributed randomly in the genome, showing no preferences for a special locus, thus making *Impala* a good tool for mutagenesis. Additionally 3.5 % of the analyzed strains showed a weaker pathogenicity towards melons, their natural host. This study thus proved the potential of transposons to be used for random mutagenesis in fungi (Migheli et al. 2000).

6.7 Conclusions

TEs are a large and diverse group of mobile genetic elements. They can be divided into two main groups, RNA- (Retro-) and DNA transposons, which can transpose in the genome by different mechanisms (Wicker et al. 2007). TEs have a huge influence on genome rearrangements and hence genome evolution and even may cause disease (Janicki et al. 2011). Yet the mechanisms of transposition are not fully understood. As sequencing gets more and more affordable, the screening for new TEs is easier than ever. Several bioinformatics tools are available to analyze repetitive sequences and identify TEs (Lerat 2010). New elements are found every month and their potential still cannot be estimated properly. They may be utilized as tools for random mutagenesis, a method still highly useful to analyze and identify gene functions (Kempken 1999). Besides mutagenesis with chemicals or introduced by UV-radiation, the transposon mutagenesis once established may produce a large pool of mutants.

Acknowledgements L.P. received a Max-Buchner-Fellowship. Part of the laboratory work was funded by an EU grant (MARINE FUNGI EU FP7, 2659269).

References

Aarts MGM, Dirkse WG, Stiekema WJ, Pereira A (1993) Transposon tagging of a male sterility gene in Arabidopsis. Nature 363:715–717

Amselem J, Cuomo CA, van Kan JAL, Viaud M, Benito EP, Couloux A, Coutinho PM, de Vries RP, Dyer PS, Fillinger S, Fournier E, Gout L, Hahn M, Kohn L, Lapalu N, Plummer KM, Pradier J-M, Quévillon E, Sharon A, Simon A, ten Have A, Tudzynski B, Tudzynski P, Wincker P, Andrew M, Anthouard V, Beever RE, Beffa R, Benoit I, Bouzid O, Brault B, Chen Z, Choquer M, Collémare J, Cotton P, Danchin EG, Da Silva C, Gautier A, Giraud C, Giraud T, Gonzalez C, Grossetete S, Güldener U, Henrissat B, Howlett BJ, Kodira C, Kretschmer M, Lappartient A, Leroch M, Levis C, Mauceli E, Neuvéglise C, Oeser B, Pearson M, Poulain J, Poussereau N, Quesneville H, Rascle C, Schumacher J, Ségurens B, Sexton A, Silva E, Sirven C, Soanes DM, Talbot NJ, Templeton M, Yandava C, Yarden O, Zeng Q, Rollins JA, Lebrun M-H, Dickman M (2011) Genomic analysis of the necrotrophic fungal pathogens sclerotinia sclerotiorum and botrytis cinerea. PLoS Genet 7(8):1002230

Amutan M, Nyyssönen E, Stubbs J, Diaz-Torres MR, Dunn-Coleman N (1996) Identification and cloning of a mobile transposon from Aspergillus niger var. Awamori. Curr Genet 29:468–473

Arkhipova I (2006) Distribution and phylogeny of penelope-like elements in eukaryotes. Syst Biol 55(6): 875–885

Bao W, Jurka MG, Kapitonov VV, Jurka J (2009) New superfamilies of eukaryotic DNA transposons and their internal divisions. Mol Biol Evol 26(5):983–993

Belancio VP, Deininger PL, Roy-Engel AM (2009) Review LINE dancing in the human genome: transposable elements and disease. Genome Med 1(10):97

Bennetzen JL (2005) Transposable elements, gene creation and genome rearrangement in flowering plants. Curr Opin Genet Dev 15(6):621–627

Biémont C (2009) Are transposable elements simply silenced or are they under house arrest? Trends Genet 25(8):333–334

Biémont C (2010) A brief history of the status of transposable elements: from junk DNA to major players in evolution. Genetics 186(4):1085–1093

Biémont C, Vieira C (2006) Junk DNA as an evolutionary force. Nature 443:521–524

Bonocora RP, Shub DA (2009) A likely pathway for formation of mobile group I introns. Current Biol CB 19(3):223–28

Bowen NJ, Jordan IK, Epstein JA, Wood V, Levin HL (2003) Retrotransposons and their recognition of Pol II promoters: a comprehensive survey of the transposable elements from the complete genome sequence of Schizosaccharomyces pombe. Genome Res 13(9): 1984–1997

Braumann I, van den Berg M, Kempken F (2007) Transposons in biotechnologically relevant strains of Aspergillus niger and Penicillium chrysogenum. Fungal Genet Biol 44(12):1399–1414

Braumann I, van den Berg M, Kempken F (2008) Repeat induced point mutation in two asexual fungi, Aspergillus niger and Penicillium chrysogenum. Curr Genet 53:287–297

Bushley KE, Raja R, Jaiswal P, Cumbie JS, Nonogaki M, Boyd AE, Owensby CA, Knaus BJ, Elser J, Miller D, Di Y, McPhail KL, Spatafora JW (2013) The genome of tolypocladium inflatum: evolution, organization, and expression of the cyclosporin biosynthetic gene cluster. PLoS Genet 9(6):e1003496

Cambareri EB, Jensen BC, Schabacht E, Selker EU (1989) Repeat-induced G-C to A-T mutations in neurospora. Science 244:1571–1575

Cameron JR, Loh EY, Davis RW (1979) Evidence for transposition of dispersed repetitive DNA families in yeast. Cell 16(4):739–751

Cappello J, Handelsmann K, Lodish HF (1985) Sequence of dictyostelium DIRS-1: an apparent retrotransposon with inverted terminal repeats and an internal circle junction sequence. Cell 43:105–115

Chalvet F, Grimaldi C, Kaper F, Langin T, Daboussi M-J (2003) Hop, an active mutator-like element in the genome of the fungus Fusarium oxysporum. Mol Biol Evol 20(8):1362–1375

Chalvet F, Kaper F, Langin T, Daboussi MJ (2001) Hop, an active MuDR-like element in the filamentous fungus Fusarium oxysporum. Fungal Genet Newsl 48(Suppl):86

Chen S, Xu J, Liu C, Zhu Y, Nelson DR, Zhou S, Li C, Wang L, Guo X, Sun Y, Luo H, Li Y, Song J, Henrissat B, Levasseur A, Qian J, Li J, Luo X, Shi L, He L, Xiang L, Xu X, Niu Y, Li Q, Han MV, Yan H, Zhang J, Chen H, Lv A, Wang Z, Liu M, Schwartz DC, Sun C (2012) Genome sequence of the model medicinal mushroom Ganoderma lucidum. Nat Commun 3(May):913

Clutterbuck AJ (2011) Genomic evidence of repeat-induced point mutation (RIP) in filamentous Ascomycetes. Fungal Genet Biol 48(3):306–326

Cohen CJ, Lock WM, Mager DL (2009) Endogenous retroviral LTRs as promoters for human genes: a critical assessment. Gene 448(2):105–114

Coleman JJ, Rounsley SD, Rodriguez-Carres M, Kuo A, Wasmann CC, Grimwood J, Schmutz J, Taga M, White GJ, Zhou S, Schwartz DC, Freitag M, Ma L-J, Danchin EGJ, Henrissat B, Coutinho PM, Nelson DR, Straney D, Napoli CA, Barker BM, Gribskov M, Rep M, Kroken S, Molnár I, Rensing C, Kennell JC, Zamora J, Farman ML, Selker EU, Salamov A, Shapiro H, Pangilinan J, Lindquist E, Lamers C, Grigoriev IV, Geiser DM, Covert SF, Temporini E, Vanetten HD (2009) The genome of Nectria haematococca: contribution of supernumerary chromosomes to gene expansion. PLoS Genet 5(8):e1000618

Craig NL, Craigie R, Gellert M, Lambowitz A (2002) Mobile DNA II. American Society for Microbiology Press, Washington

Daboussi MJ, Capy P (2003) Transposable elements in filamentous fungi. Annu Rev Microbiol 57:275–299

Daboussi MJ, Langin T (1994) Transposable elements in the fungal plant pathogen Fusarium oxysporum. Genetica 93:49–59

Daboussi MJ, Langin T, Brygoo Y (1992) Fot1, a new family of fungal transposable elements. Mol Gen Genet 232:12–16

Dean R, Van Kan JAL, Pretorius ZA, Hammond Kosack KE, Di Pietro A, Spanu PD, Rudd JJ, Dickman M, Kahmann R, Ellis J, Foster GD (2012) The top 10 fungal pathogens in molecular plant pathology. Mol Plant Pathol 13(4):414–430

Dean RA, Talbot NJ, Ebbole DJ, Farman ML, Mitchell TK, Orbach MJ, Thon M, Kulkarni R, Xu J-R, Pan H, Read ND, Lee Y-H, Carbone I, Brown D, Oh YY, Donofrio N, Jeong JS, Soanes DM, Djonovic S, Kolomiets E, Rehmeyer C, Li W, Harding M, Kim S, Lebrun M-H, Bohnert H, Coughlan S, Butler J, Calvo S, Ma L-J, Nicol R, Purcell S, Nusbaum C, Galagan JE, Birren BW (2005) The genome sequence of the rice blast fungus Magnaporthe grisea. Nature 434(7036):980–986

Döring HP, Tillmann E, Starlinger P (1984) DNA sequence of the maize transposable element dissociation. Nature 307:127–131

Eickbush TH, Malik HS (2002) Origins and evolution of retrotransposons. In: Craig NL, Craigie R, Gellert M, Lambowitz AM (eds) Mobile DNA II. ASM, Washington, pp 1111–1144

Ellinghaus D, Kurtz S, Willhoeft U (2008) LTRharvest, an efficient and flexible software for de novo detection of LTR retrotransposons. BMC Bioinformatics 9:18

Emerson RA (1914) The inheritance of a recurring somatic variation in variegated ears of maize. Am Nat 48(566):87–115

Espagne E, Lespinet O, Malagnac F, Da Silva C, Jaillon O, Porcel BM, Couloux A, Aury J-M, Ségurens B, Poulain J, Anthouard V, Grossetete S, Khalili H, Coppin E, Déquard-Chablat M, Picard M, Contamine V, Arnaise S, Bourdais A, Berteaux-Lecellier V, Gautheret D, de Vries RP, Battaglia E, Coutinho PM, Danchin EGJ, Henrissat B, El Khoury R, Sainsard-Chanet A, Boivin A, Pinan-Lucarré B, Sellem CH, Debuchy R, Wincker P, Weissenbach J, Silar P (2008) The genome sequence of the model ascomycete fungus Podospora anserina. Genome Biol 9(5):R77

Evgen'ev MB, Zelentsova H, Shostak N, Kozistina M, Barskyi V (1997) Penelope, a new family of transposable elements and its possible role in hybrid dysgenesis in Drosophila virilis. Proc Natl Acad Sci U S A 94:196–201

Evgen'ev MB, Arkhipova IR (2005) Penelope-like elements—a new class of retroelements: distribution, function and possible evolutionary significance. Cytogenet Genome Res 110(1–4):510–521

Farman ML, Tosa Y, Nitta N (1996) Maggy, a retrotransposon in the genome of the rice blast fungus Magnaporthe grisea. Mol Gen Genet 251:665–674

Fedorova ND, Khaldi N, Joardar VS, Maiti R, Amedeo P, Anderson MJ, Crabtree J, Silva JC, Badger JH, Albarraq A, Angiuoli S, Bussey H, Bowyer P, Cotty PJ, Dyer PS, Egan A, Galens K, Fraser-Liggett CM, Haas BJ, Inman JM, Kent R, Lemieux S, Malavazi I, Orvis J, Roemer T, Ronning CM, Sundaram JP, Sutton G, Turner G, Venter JC, White OR, Whitty BR, Youngman P, Wolfe KH, Goldman GH, Wortman JR, Jiang B, Denning DW, Nierman WC (2008) Genomic islands in the pathogenic filamentous fungus Aspergillus fumigatus. PLoS Genet 4(4):e1000046

Fernandez D, Quinten M, Tantaoui A, Geiger JP, Daboussi MJ, Langin T (1998) Fot 1 insertions in the Fusarium oxysporum F. sp. albedinis genome provide diagnostic PCR targets for detection of the date palm pathogen. Appl Environ Microbiol 64:633–636

Feschotte C, Pritham EJ (2005) Non-mammalian c-integrases are encoded by giant transposable elements. Trends Genet 21(10):551–552

Feschotte C, Pritham EJ (2007) DNA transposons and the evolution of eukaryotic genomes. Annu Rev Genet 41:331–368

Finnegan DJ (1989) Eukaryotic transposable elements and genome evolution. Trends Genet 5:103–107

Galagan JE, Calvo SE, Borkovich KA, Selker EU, Read ND, Jaffe D, FitzHugh W, Ma L-J, Smirnov S, Purcell S, Rehman B, Elkins T, Engels R, Wang S, Nielsen CB, Butler J, Endrizzi M, Qui D, Ianakiev P, Bell-Petersen D, Nelson MA, Werner Mary W, Selitrennikoff CP, Kinsey JA, Braun EL, Zelter A, Schulte U, Kothe GO, Jedd G, Mewes W, Staben C, Marcotte E, Greenberg D, Roy A, Foley K, Naylor J, Stange-Thomann N, Barret R, Gnerre S, Kamal M, Kamvysselis M, Mauceli E, Bielke C, Rudd S, Frishman D, Krystofova S, Rasmussen C, Metzenberg RL, Perkins DD, Kroken S, Cogoni C, Macino G, Catcheside DE, Li W, Pratt R, Osmani SA, DeSouza CPC, Glass L-N, Orbach MJ, Berglung JA, Voelker R, Yarden O, Plamann M, Seiler S, Dunlap JC, Radford A, Aramayo R, Natvig DO, Alex LA, Mannhaupt G, Ebbole DJ, Freitag M, Paulsen I, Sachs MS, Lander ES, Nusbaum C, Birren B (2003) The genome sequence of the filamentous fungus Neurospora crassa. Nature 422:859–68

Galagan JE, Calvo SE, Cuomo C, Ma L-J, Wortman JR, Batzoglou S, Lee S-I, Basturkmen M, Spevak CC, Clutterbuck J, Kapitonov V, Jurka J, Scazzocchio C, Farman M, Butler J, Purcell S, Harris S, Braus GH, Draht O, Busch S, D'Enfert C, Bouchier C, Goldman GH, Bell-Pedersen D, Griffiths-Jones S, Doonan JH, Yu J, Vienken K, Pain A, Freitag M, Selker EU, Archer DB, Penalva MA, Oakley BR, Momany M, Tanaka T, Kumagai T, Asai K, Machida M, Nierman WC, Denning DW, Caddick M, Hynes M, Paoletti M, Fischer R, Miller B, Dyer P, Sachs MS, Osmani SA, Birren BW (2005) Sequencing of Aspergillus nidulans and comparative analysis with A fumigatus and A oryzae. Nature 438(7071):1105–1115

Galagan JE, Selker EU (2004) RIP: the evolutionary cost of genome defense. Trends Genet 20(9):417–423

Gao Q, Jin K, Ying S-H, Zhang Y, Xiao G, Shang Y, Duan Z, Hu X, Xie X-Q, Zhou G, Peng G, Luo Z, Huang W, Wang B, Fang W, Wang S, Zhong Y, Ma L-J, St Leger RJ, Zhao G-P, Pei Y, Feng M-G, Xia Y, Wang C (2011) Genome sequencing and comparative transcriptomics of the model entomopathogenic fungi Metarhizium anisopliae and M. acridum. PLoS Genet 7(1):e1001264

Gao X, Hou Y, Ebina H, Levin HL, Voytas DF (2008) Chromodomains direct integration of retrotransposons to heterochromatin. Genome Res 18(3):359–369

Gao X, Voytas DF (2005) An eukaryotic gene family related to retroelement integrases. Trends Genet 21(March 2005):133–137

Gladyshev EA, Arkhipova IR (2007) Telomere-associated endonuclease-deficient penelope-like retroelements in diverse eukaryotes. Proc Natl Acad Sci U S A 104(22):9352–9357

Goodwin T, Ormandy J, Poulter R (2001) L1-like non-LTR retrotransposons in the yeast Candida albicans. Curr Genet 39(2):83–91

Goodwin TJ, Butler MI, Poulter RT (2003) Cryptons: a group of tyrosine-recombinase-encoding DNA transposons from pathogenic fungi. Microbiology 149: 3099–3109

Goodwin TJD, Poulter RTM (2004) A new group of tyrosine recombinase-encoding retrotransposons. Mol Biol Evol 21(4):746–759

Goyon C, Rossignol JL, Faugeron G (1996) Native DNA repeats and methylation in ascobolus. Nucleic Acids Res 24:3348–3356

Grover CE, Wendel JF (2010) Recent insights into mechanisms of genome size change in plants. J Bot 2010(4): 1–8

Han JS (2010) Non-long terminal repeat (non-LTR) retrotransposons: mechanisms, recent developments, and unanswered questions. Mob DNA 1(1):15

Hancks DC, Kazazian HH Jr (2012) Active human retrotransposons: variation and disease. Curr Opin Genet Dev 22(3):191–203

He C, Nourse JP, Kelemu S, Irwin JAG, Manners JM (1996) CgT1: a non-LTR retrotransposon with restricted distribution in the fungal phytopathogen Colletotrichum gloeosporioides. Mol Gen Genet 252: 320–331

Hehl R, Baker B (1990) Properties of the maize transposable element activator in transgenic tobacco plants: a versatile inter-species genetic tool. Plant Cell 2(8): 709–721

Hihlal E, Braumann B, van den Berg M, Kempken F (2011) Suitability of vader for transposon-mediated mutagenesis in Aspergillus niger. Appl Enivron Mircobiol 77(7):2332–2336

Hua-Van A, Daviere JM, Kaper F, Langin T, Daboussi MJ (2000) Genome organization in Fusarium oxysporum: clusters of class II transposons. Curr Genet 37: 339–347

Ikadai H, Shaw Saliba K, Kanzok SM, McLean KJ, Tanaka TQ, Cao J, Williamson KC, Jacobs-Lorena M (2013) Transposon mutagenesis identifies genes essential for plasmodium Falciparum gametocytogenesis. Proc Natl Acad Sci U S A 110(18):E1676–E1684

International Human Genome Sequencing Consortium (2001) Initial sequencing and analysis of the human genome. Nature 409:860–921

Janicki M, Rooke R, Yang G (2011) Bioinformatics and genomic analysis of transposable elements in eukaryotic genomes. Chromosome Res 19(6):787–808

Jiang N, Bao Z, Zhang X, Hirochika H, Eddy SR, McCouch SR, Wessler SR (2003) An active DNA transposon family in rice. Nature 421(6919):163–167

Jurka J (1997) Sequence patterns indicate an enzymatic involvement in integration of mammalian retroposons. Proc Natl Acad Sci U S A 94(5):1872–1877

Jurka J, Kapitonov VV, Pavlicek A, Klonowski P, Kohany O, Walichiewicz J (2005) Repbase update, a database of eukaryotic repetitive elements. Cytogenet Genome Res 110(1–4):462–467

Kachroo P, Leong SA, Chattoo BB (1995) Mg-SINE: a short interspersed nuclear element from the rice blast fungus, Magnaphorte grisea. Proc Natl Acad Sci U S A 92:11125–11129

Kajikawa M, Okada N (2002) LINEs mobilize SINEs in the eel through a shared 3′ sequence. Cell 111(3): 433–444

Kämper J, Kahmann R, Bölker M, Ma LJ, Brefort T, Saville BJ, Banuett F, Kronstad JW, Gold SE, Müller O, Perlin MH, Wösten HA, de Vries R, Ruiz-Herrera J, Reynaga-Pena CG, Snetselaar K, McCann M, Pérez-Martin J, Feldbrügge M, Basse CW, Steinberg G, Ibeas JI, Holloman W, Guzman P, Farman M, Stajich JE, Sentandreu R, González-Prieto JM, Kennell JC, Molina L, Schirawski J, Mendoza-Mendoza A, Greilinger D, Münch K, Rössel N, Scherer M, Vranes M, Ladendorf O, Vincon V, Fuchs U, Sandrock B, Meng S, Ho EC, Cahill MJ, Boyce KJ, Klose J, Klosterman SJ, Deelstra HJ, Ortiz-Castellanos L, Li W, Sanchez-Alonso P, Schreier PH, Häuser-Hahn I, Vaupel M, Koopmann E, Friedrich G, Voss H, Schluter T, Margolis J, Platt D, Swimmer C, Gnirke A, Chen F, Vysotskaia V, Mannhaupt G, Güldener U, Münsterkötter M, Haase D, Oesterheld M, Mewes HW, Mauceli EW, DeCaprio D, Wade CM, Butler J, Young S, Jaffe DB, Calvo S, Nusbaum C, Galagan J, Birren BW (2006) Insights from the genome of the biotrophic fungal plant pathogen Ustilago maydis. Nature 444(7115):97–101

Kapitonov VV, Jurka J (2001) Rolling-circle transposons in eukaryotes. Proc Natl Acad Sci U S A 98: 8714–8719

Kapitonov VV, Jurka J (2005) RAG1 core and V(D)J recombination signal sequences were derived from transib transposons. PLoS Biol 3(6):e181

Kapitonov VV, Jurka J (2006) Self-synthesizing DNA transposons in eukaryotes. Proc Natl Acad Sci U S A 103:4540–4545

Kapitonov VV, Jurka J (2008) An universal classification of eukaryotic transposable elements implemented in repbase. Nat Rev Genet 9:411–412

Kazazian HH (2004) Mobile elements: drivers of genome evolution. Science (New York, N Y) 303(5664): 1626–32

Kempken F (1999) Fungal transposons: from mobile elements towards molecular tools. Appl Microbiol Biotechnol 52:756–60

Kempken F (2003) Fungal transposable elements: inducers of mutations and molecular tools. In: Arora DK, Khachatourians GG (eds.) Applied mycology and biotechnology, vol. 3 Fungal genomics. Elsevier Science Annual Review Series. pp. 83–99.

Kempken F, Hermanns J, Osiewacz HD (1992) Evolution of linear plasmids. J Mol Evol 35:502–513

Kempken F, Jacobsen S, Kück U (1998) Distribution of the fungal transposon restless: full-length and truncated copies in closely related strains. Fungal Genet Biol 25:110–118

Kempken F, Kück U (1996) Restless, an active Ac-like transposon from the fungus Tolypocladium inflatum: structure, expression, and alternative RNA splicing. Mol Cell Biol 16:6563–6572

Kempken F, Kück U (2000) Tagging of a nitrogen pathway-specific regulator gene in Tolypocladium inflatum By the transposon restless. Mol Gen Genet 263:302–308

Kempken F, Windhofer F (2001) The hAT family: a versatile transposon group common to plants, fungi, animals, and man. Chromosoma 110:1–9

Kim JM, Vanguri S, Boeke JD, Gabriel A, Voytas DF (1998) Transposable elements and genome organization: a comprehensive survey of retrotransposons revealed by the complete Saccharomyces cerevisiae genome sequence. Genome Res 8(5):464–478

Kinsey JA (1990) Tad, a LINE-like transposable element of neurospora, can transpose between nuclei in heterokaryons. Genetics 126:232–317

Kinsey JA, Helber J (1989) Isolation of a transposable element from Neurospora crassa. Proc Natl Acad Sci U S A 86:1929–1933

Kramerov DA, Vassetzky NS (2011) SINEs. Wiley interdisciplinary reviews. RNA 2(6):772–786

Kramerov DA, Vassetzky NS (2005) Short retroposons in eukaryotic genomes. Int Rev Cytol 247:165–221

Labbé J, Murat C, Morin E, Tuskan GA, Le Tacon F, Martin F (2012) Characterization of transposable elements in the ectomycorrhizal fungus laccaria bicolor. PLoS One 7(8):e40197

Lambowitz AM, Caprara MG, Zimmerly S, Perlman PS (1999) Group I and group II ribozymes as RNPs: clues to the past and guides to the future. Cold Spring Harbor Monograph Archive, Vol. 37 (1999): the RNA WORLD, 2nd ed. The nature of modern RNA suggests a prebiotic RNA world

Langin T, Capy P, Daboussi MJ (1995) The transposable element impala, a fungal member of the Tc1-mariner superfamily. Mol Gen Genet 246:19–28

Lerat E (2010) Identifying repeats and transposable elements in sequenced genomes: how to find your way through the dense forest of programs. Heredity 104(6):520–533

Loftus BJ, Fung E, Roncaglia P, Rowley D, Amedeo P, Bruno D, Vamathevan J, Miranda M, Anderson IJ, Fraser JA, Allen JE, Bosdet IE, Brent MR, Chiu R, Doering TL, Donlin MJ, D'Souza CA, Fox DS, Grinberg V, Fu J, Fukushima M, Haas BJ, Huang JC, Janbon G, Jones SJ, Koo HL, Krzywinski MI, Kwon-Chung JK, Lengeler KB, Maiti R, Marra MA, Marra RE, Mathewson CA, Mitchell TG, Pertea M, Riggs FR, Salzberg SL, Schein JE, Shvartsbeyn A, Shin H, Shumway M, Specht CA, Suh BB, Tenney A, Utterback TR, Wickes BL, Wortman JR, Wye NH, Kronstad JW, Lodge JK, Heitman J, Davis RW, Fraser CM, Hyman RW (2005) The genome of the basidiomycetous yeast and human pathogen Cryptococcus neoformans. Science 307(5713):1321–1324

Lucier J-F, Perreault J, Noël J-F, Boire G, Perreault J-P (2007) RTAnalyzer: a web application for finding new retrotransposons and detecting L1 retrotransposition signatures. Nucleic acids Res 35(Web Server issue): W269–W274

Ma L-J, Van Der Does HC, Borkovich KA, Coleman JJ, Daboussi M, Di Pietro A, Dufresne M, Freitag M, Henrissat B, Houterman PM, Kang S, Shim W, Bluhm BH, Breakspear A, Brown DW, Butchko RAE,

Chapman S, Coulson R, Coutinho PM, Danchin EGJ, Diener A, Gale LR, Gardiner DM, Goff S, Kim E, Hilburn K, Hua-van A, Jonkers W, Li L, Manners JM, Miranda-saavedra D, Mukherjee M, Park G, Park J, Park S, Proctor RH, Ruiz-roldan MC, Sain D, Sakthikumar S, Sykes S, Schwartz DC, Turgeon BG, Wapinski I, Yoder O, Young S (2010) Comparative genomics reveals mobile pathogenicity chromosomes in fusarium. Nature 464(7287):367–373

Ma L-J, Ibrahim AS, Skory C, Grabherr MG, Burger G, Butler M, Elias M, Idnurm A, Lang BF, Sone T, Abe A, Calvo SE, Corrochano LM, Engels R, Fu J, Hansberg W, Kim J-M, Kodira CD, Koehrsen MJ, Liu B, Miranda-Saavedra D, O'Leary S, Ortiz-Castellanos L, Poulter R, Rodriguez-Romero J, Ruiz-Herrera J, Shen Y-Q, Zeng Q, Galagan J, Birren BW, Cuomo CA, Wickes BL (2009) Genomic analysis of the basal lineage fungus rhizopus oryzae reveals a whole-genome duplication. PLoS Genet 5(7):e1000549

Martin F, Kohler A, Murat C, Balestrini R, Coutinho PM, Jaillon O, Montanini B, Morin E, Noel B, Percudani R, Porcel B, Rubini A, Amicucci A, Amselem J, Anthouard V, Arcioni S, Artiguenave F, Aury J-M, Ballario P, Bolchi A, Brenna A, Brun A, Buée M, Cantarel B, Chevalier G, Couloux A, Da Silva C, Denoeud F, Duplessis S, Ghignone S, Hilselberger B, Iotti M, Marçais B, Mello A, Miranda M, Pacioni G, Quesneville H, Riccioni C, Ruotolo R, Splivallo R, Stocchi V, Tisserant E, Viscomi AR, Zambonelli A, Zampieri E, Henrissat B, Lebrun M-H, Paolocci F, Bonfante P, Ottonello S, Wincker P (2010) Périgord black truffle genome uncovers evolutionary origins and mechanisms of symbiosis. Nature 464(7291): 1033–1038

McClintock B (1951) Chromosome organization and genic expression. Cold Spring Harbor Symp Quant Biol 16:13–47

McHale MT, Roberts IN, Noble SM, Beaumont C, Whitehead MP, Seth D, Oliver RP (1992) CfT-1: an LTR-retrotransposon in Cladosporium fulvum, a fungal pathogen of tomato. Mol Gen Genet 233:337–347

Migheli Q, Lauge R, Daviere JM, Gerlinger C, Kaper F, Langin T, Daboussi MJ (1999) Transposition of the autonomous Fot1 element in the filamentous fungus Fusarium oxysporum. Genetics 151:1005–1013

Migheli Q, Steinberg C, Daviere JM, Olivain C, Gerlinger C, Gautheron N, Alabouvette C, Daboussi MJ (2000) Recovery of mutants impaired in pathogenicity after transposition of impala in Fusarium oxysporum f. sp. melonis. Phytopathology 90:1279–1284

Morin E, Kohler A, Baker AR, Foulongne-Oriol M, Lombard V, Nagy LG, Ohm RA, Patyshakuliyeva A, Brun A, Aerts AL, Bailey AM, Billette C, Coutinho PM, Deakin G, Doddapaneni H, Floudas D, Grimwood J, Hildén K, Kües U, Labutti KM, Lapidus A, Lindquist EA, Lucas SM, Murat C, Riley RW, Salamov AA, Schmutz J, Subramanian V, Wösten HAB, Xu J, Eastwood DC, Foster GD, Sonnenberg ASM, Cullen D, de Vries RP, Lundell T, Hibbett DS, Henrissat B, Burton KS, Kerrigan RW, Challen MP, Grigoriev IV,

Martin F (2012) Genome sequence of the button mushroom agaricus bisporus reveals mechanisms governing adaptation to a humic-rich ecological niche. Proc Natl Acad Sci U S A 109(43):17501–17506

Muszewska A, Hoffman-Sommer M, Grynberg M (2011) LTR retrotransposons in fungi. PLoS One 6(12): e29425

Muszewska A, Steczkiewicz K, Ginalski K (2013) DIRS and Ngaro retrotransposons in fungi. PLoS One 8(9): e76319

Ninomiya Y, Suzuki K, Ishii C, Inoue H (2004) Highly efficient gene replacements in neurospora strains deficient for nonhomologous end-joining. Proc Natl Acad Sci U S A 101(33):12248–12253

Novikova O, Fet V, Blinov A (2009) Non-LTR retrotransposons in fungi. Funct Integr Genomics 9(1):27–42

Nowrousian M (2010) Next-generation sequencing techniques for eukaryotic microorganisms: sequencing-based solutions to biological problems. Eukaryot Cell 9(9):1300–1310

Nowrousian M, Stajich JE, Chu M, Engh I, Espagne E, Halliday K, Kamerewerd J, Kempken F, Knab B, Kuo HC, Osiewacz HD, Poggeler S, Read ND, Seiler S, Smith KM, Zickler D, Kuck U, Freitag M (2010) De novo assembly of a 40 Mb eukaryotic genome from short sequence reads: sordaria macrospora, a model organism for fungal morphogenesis. PLoS Genet 6(4):e1000891

Nyyssönen E, Amutan M, Enfield L, Stubbs J, Dunn-Coleman NS (1996) The transposable element Tan1 of Aspergillus niger var. Awamori, a new member of the Fot1 family. Mol Gen Genet 253:50–56

Ohm RA, Feau N, Henrissat B, Schoch CL, Horwitz BA, Barry KW, Condon BJ, Copeland AC, Dhillon B, Glaser F, Hesse CN, Kosti I, LaButti K, Lindquist EA, Lucas S, Salamov AA, Bradshaw RE, Ciuffetti L, Hamelin RC, Kema GHJ, Lawrence C, Scott JA, Spatafora JW, Turgeon BG, de Wit PJGM, Zhong S, Goodwin SB, Grigoriev IV (2012) Diverse lifestyles and strategies of plant pathogenesis encoded in the genomes of eighteen dothideomycetes fungi. PLoS Pathog 8(12):e1003037

Ohm RA, de Jong JF, Lugones LG, Aerts A, Kothe E, Stajich JE, de Vries RP, Record E, Levasseur A, Baker SE, Bartholomew KA, Coutinho PM, Erdmann S, Fowler TJ, Gathman AC, Lombard V, Henrissat B, Knabe N, Kües U, Lilly WW, Lindquist E, Lucas S, Magnuson JK, Piumi F, Raudaskoski M, Salamov A, Schmutz J, Schwarze FWMR, VanKuyk PA, Horton JS, Grigoriev IV, Wösten HAB (2010) Genome sequence of the model mushroom Schizophyllum commune. Nat Biotechnol 28(9):957–963

Okada N, Hamada M, Ogiwara I, Ohshima K (1997) SINEs and LINEs share common 3' sequences: a review. Gene 205:229–243

Ostertag EM, Kazazian HH (2005) Genetics LINEs in mind. Nature 435(7044):890–891

Pel HJ, de Winde JH, Archer DB, Dyer PS, Hofmann G, Schaap PJ, Turner G, de Vries RP, Albang R, Albermann K, Andersen MR, Bendtsen JD, Benen

JAE, van den Berg M, Breestraat S, Caddick MX, Contreras R, Cornell M, Coutinho PM, Danchin EGJ, Debets AJM, Dekker P, van Dijck PWM, van Dijk A, Dijkhuizen L, Driessen AJM, D'Enfert C, Geysens S, Goosen C, Groot GSP, de Groot PWJ, Guillemette T, Henrissat B, Herweijer M, van den Hombergh JPTW, van den Hondel CAMJJ, van der Heijden RTJM, van der Kaaij RM, Klis FM, Kools HJ, Kubicek CP, van Kuyk PA, Lauber J, Lu X, van der Maarel MJEC, Meulenberg R, Menke H, Mortimer MA, Nielsen J, Oliver SG, Olsthoorn M, Pal K, van Peij NNME, Ram AFJ, Rinas U, Roubos JA, Sagt CMJ, Schmoll M, Sun J, Ussery D, Varga J, Vervecken W, van de Vondervoort PJJ, Wedler H, Wosten HAB, Zeng A-P, van Ooyen AJJ, Visser J, Stam H (2007) Genome sequencing and analysis of the versatile cell factory Aspergillus niger CBS 513.88. Nat Biotechol 25(2):221–231

Peterson P (2013) Historical overview of transposable element research. In: Peterson T (ed) Plant transposable elements SE—1, methods in molecular biology. Humana, Totowa, pp 1–9, http://dx.doi.org/10.1007/978-1-62703-568-2_1

Pohlman RF, Fedoroff NV, Messing J (1984) The Nucleotide Sequence of the Maize Controlling Element Activator. Cell 37(2):635–643

Poulter RTM, Goodwin TJD, Butler MI (2003) Vertebrate helentrons and other novel helitrons. Gene 313:201–212

Pritham EJ, Putliwala T, Feschotte C (2007) Mavericks, a novel class of giant transposable elements widespread in eukaryotes and related to DNA viruses. Gene 390:3–17

De Queiroz MV, Daboussi MJ (2003) Impala, a transposon from Fusarium oxysporum, is active in the genome of Penicillium griseoroseum. FEMS Microbiol Lett 218:317–321

Rowold DJ, Herrera RJ (2000) Alu elements and the human genome. Genetica 108(1):57–72

Ruiz-Pérez VL, Murillo FJ, Torres-Martínez S (1996) Prt1, an unusual retrotransposon-like sequence in the fungus Phycomyces blakesleeanus. Mol Gen Genet 253(3):324–333

SanMiguel P, Gaut BS, Tikhonov A, Nakajima Y, Bennetzen JL (1998) The paleontology of intergene retrotransposons of maize. Nat Genet 20(1):43–45

Schmidt AL, Anderson LM (2006) Repetitive DNA elements as mediators of genomic change in response to environmental cues. Biol Rev Camb Philos Soc 81(4):531–543

Selker EU (1997) Epigenetic phenomena in filamentous fungi: useful paradigms or repeat-induced confusion? Trends Genet 13:296–301

Sentry JW, Kaiser K (1992) P element transposition and targeted manipulation of the drosophila genome. Trends Genet 8:329–331

Shaaban M, Palmer JM, El-Naggar WA, El-Sokkary MA, Habib E-SE, Keller NP (2010) Involvement of transposon-like elements in penicillin gene cluster regulation. Fungal Genet Biol 47(5):423–32

Shao H, Tu Z (2001) Expanding the diversity of the IS630-Tc1-Mariner Superfamily : discovery of a unique DD37E transposon and reclassification of the DD37D and DD39D transposons. Genetics 159:1103–1115

Sharpton TJ, Stajich JE, Rounsley SD, Gardner MJ, Wortman JR, Jordar VS, Maiti R, Kodira CD, Neafsey DE, Zeng Q, Hung CY, McMahan C, Muszewska A, Grynberg M, Mandel MA, Kellner EM, Barker BM, Galgiani JN, Orbach MJ, Kirkland TN, Cole GT, Henn MR, Birren BW, Taylor JW (2009) Comparative genomic analyses of the human fungal pathogens coccidioides and their relatives. Genome Res 19(10):1722–1731

Smit A, Hubley R (2010) 2008–2010. RepeatModeler Open-1.0. http://www.repeatmasker.org

Smit A, Hubley R, Green P (2006) 1996–2004. Repeat Masker Open-3.0. http://www.repeatmasker.org

Stajich JE, Wilke SK, Ahrén D, Au CH, Birren BW, Borodovsky M, Burns C, Canbäck B, Casselton LA, Cheng CK, Deng J, Dietrich FS, Fargo DC, Farman ML, Gathman AC, Goldberg J, Guigó R, Hoegger PJ, Hooker JB, Huggins A, James TY, Kamada T, Kilaru S, Kodira C, Kües U, Kupfer D, Kwan HS, Lomsadze A, Li W, Lilly WW, Ma L-J, Mackey AJ, Manning G, Martin F, Muraguchi H, Natvig DO, Palmerini H, Ramesh MA, Rehmeyer CJ, Roe BA, Shenoy N, Stanke M, Ter-Hovhannisyan V, Tunlid A, Velagapudi R, Vision TJ, Zeng Q, Zolan ME, Pukkila PJ (2010) Insights into Evolution of Multicellular Fungi from the assembled chromosomes of the mushroom Coprinopsis cinerea (Coprinus cinereus). Proc Natl Acad Sci U S A 107(26):11889–11894

Tempel S, Jurka M, Jurka J (2008) VisualRepbase: an interface for the study of occurrences of transposable element families. BMC Bioinformatics 9:345

Toor N, Hausner G, Zimmerly S (2001) Coevolution of group II intron RNA structures with their intron-encoded reverse transcriptases. RNA 7:1142–1152

Traeger S, Altegoer F, Freitag M, Gabaldon T, Kempken F, Kumar A, Marcet-Houben M, Pöggeler S, Stajich JE, Nowrousian M (2013) The genome and development-dependent transcriptomes of Pyronema confluens: a window into fungal evolution. PLoS Genet 9(9):e1003820

Vitte C, Panaud O (2005) LTR retrotransposons and flowering plant genome size: emergence of the increase/decrease model. Cytogenet Genome Res 110(1–4):91–107

Wessler SR (1988) Phenotypic diversity mediated by the maize transposable elements Ac and Spm. Science 242(4877):399–405

Wicker T, Sabot F, Hua-Van A, Bennetzen JL, Capy P, Chalhoub B, Flavell A, Leroy P, Morgante M, Panaud O, Paux E, SanMiguel P, Schulman AH (2007) A unified classification system for eukaryotic transposable elements. Nat Rev Genet 8(12):973–982

Wiemann P, Sieber CMK, von Bargen KW, Studt L, Niehaus E-M, Espino JJ, Huß K, Michielse CB, Albermann S, Wagner D, Bergner SV, Connolly LR,

Fischer A, Reuter G, Kleigrewe K, Bald T, Wingfield BD, Ophir R, Freeman S, Hippler M, Smith KM, Brown DW, Proctor RH, Münsterkötter M, Freitag M, Humpf H-U, Güldener U, Tudzynski B (2013) Deciphering the cryptic genome: genome-wide analyses of the rice pathogen fusarium fujikuroi reveal complex regulation of secondary metabolism and novel metabolites. PLoS Pathog 9(6):e1003475

Wood V, Gwilliam R, Rajandream M-A, Lyne M, Lyne R, Stewart A, Sgouros J, Peat N, Hayles J, Baker S, Basham D, Bowman S, Brooks K, Brown D, Brown S, Chillingworth T, Churcher C, Collins M, Connor R, Cronin A, Davis P, Feltwell T, Fraser A, Gentles S, Goble A, Hamlin N, Harris D, Hidalgo J, Hodgson G, Holroyd S, Hornsby T, Howarth S, Huckle EJ, Hunt S, Jagels K, James K, Jones L, Jones M, Leather S, McDonald S, McLean J, Mooney P, Moule S, Mungall K, Murphy L, Niblett D, Odell C, Oliver K, O'Neil S, Pearson D, Quail MA, Rabbinowitsch E, Rutherford K, Rutter S, Saunders D, Seeger K, Sharp S, Skelton J, Simmonds M, Squares R, Squares S, Stevens K, Taylor K, Taylor RG, Tivey A, Walsh S, Warren T, Whitehead S, Woodward J, Volckaert G, Aert R, Robben J, Grymonprez B, Weltjens I, Vanstreels E, Rieger M, Schafer M, Muller-Auer S, Gabel C, Fuchs M, Fritzc C, Holzer E, Moestl D, Hilbert H, Borzym K, Langer I, Beck A, Lehrach H, Reinhardt R, Pohl TM, Eger P, Zimmermann W, Wedler H, Wambutt R, Purnelle B, Goffeau A, Cadieu E, Dreano S, Gloux S, Lelaure V, Mottier S, Galibert F, Aves SJ, Xiang Z, Hunt C, Moore K, Hurst SM, Lucas M, Rochet M, Gaillardin C, Tallada VA, Garzon A, Thode G, Daga RR, Cruzado L, Jimenez J, Sanchez M, del Rey F, Benito J, Dominguez A, Revuelta JL, Moreno S, Armstrong J, Forsburg SL, Cerrutti L, Lowe T, McCombie WR, Paulsen I, Potashkin J, Shpakovski GV, Ussery D, Barrell BG, Nurse P (2002) The genome sequence of Schizosaccharomyces pombe. Nature 415(6874):871–880

Wu B, Hao W (2014) Horizontal transfer and gene conversion as an important driving force in shaping the landscape of mitochondrial introns. G3 (Bethesda, MD) 4(4):605–12

Xiao G, Ying S-H, Zheng P, Wang Z-L, Zhang S, Xie X-Q, Shang Y, St Leger RJ, Zhao G-P, Wang C, Feng M-G (2012) Genomic perspectives on the evolution of fungal entomopathogenicity in Beauveria bassiana. Sci Rep 2:483

Zhang X, Jiang N, Feschotte C, Wessler SR (2004) PIF- and Pong-like transposable elements: distribution, evolution and relationship with tourist-like miniature inverted-repeat transposable elements. Genetics 166(2): 971–986

Zhang Z, Saier MH (2011) Transposon-mediated adaptive and directed mutations and their potential evolutionary benefits. J Mol Microbiol Biotechnol 21(1–2):59–70

In Vivo Targeted Mutagenesis in Yeast Using TaGTEAM

7

Shawn Finney-Manchester and Narendra Maheshri

7.1 Introduction

Evolutionary approaches to engineering proteins or organisms have proven successful in realizing desirable phenotypes (Kim et al. 2012; Goldsmith and Tawfik 2012). This approach requires repeatedly creating genetic diversity in a set of clones, followed by screening or selection for desirable mutants. When applied at the organism level, diversity is often generated in vivo via classical mutagenesis, since multiple, usually unknown genetic loci underpin many phenotypes of biotechnological importance (stress tolerance, production of key metabolites, etc). However, mutational load is limited as most mutations occur in irrelevant and essential genes. As the understanding of key genes and regulatory regions behind complex phenotypes expands with genome-wide approaches, mutagenizing all these targets in vitro is severely limited by the transformation efficiencies that occur in every evolutionary round. To overcome this limitation, we have

developed a method for targeted mutagenesis in vivo in *S. cerevisiae*, dubbed TaGTEAM (TArgeting Gylcosylases To Embedded Arrays for Mutagenesis) (Finney-Manchester and Maheshri 2013). Application of TaGTEAM requires limited genetic manipulation—the genetic diversity is generated *in vivo*. Hence diversity is never limited by transformation efficiencies.

In TaGTEAM, genes to mutagenize are placed adjacent to an array of tetO DNA binding sites on a plasmid or in the chromosome. A chimeric mutator protein—a fusion of a single chain (sc)tetR DNA-binding domain and either the Mag1p DNA glycosylase or a monomeric FokI endonuclease—is also expressed. Both mutators are recruited to the DNA array and their enzymatic activity leads to a double-strand break (DSB) or end (DSE). The use of Mag1p or addition of chemical agents also results in DNA lesions. Most lesions are repaired by host machinery in an error-free way. However, if resolution of the DSB/DSE occurs by homologous recombination (HR), the 5′ ends of the break are resected, exposing long tracts of ssDNA. Any lesion present on ssDNA after resection cannot be repaired; during resynthesis of the opposing strand, a random base pair is placed opposite the lesion, potentially resulting in a point mutation. We observe a $\sim 10^3$-fold elevation in mutation rates in a 20 kb region surrounding the array. Other less desirable repair outcomes are also possible (Fig. 7.1), but can be minimized and are selected against as they result

S. Finney-Manchester, Ph.D.
Zymergen Inc, 6121 Hollis Street, Suite 700,
Emeryville, CA 94608, USA
e-mail: shawn.finneymanchester@gmail.com

N. Maheshri, Ph.D. (✉)
Chemical Engineering Department, Massachusetts
Institute of Technology/Ginkgo Bioworks,
25 Ames St, Cambridge, MA 02139, USA
e-mail: nmaheshri@gmail.com

M.A. van den Berg and K. Maruthachalam (eds.), *Genetic Transformation Systems in Fungi, Volume 2*, Fungal Biology, DOI 10.1007/978-3-319-10503-1_7,
© Springer International Publishing Switzerland 2015

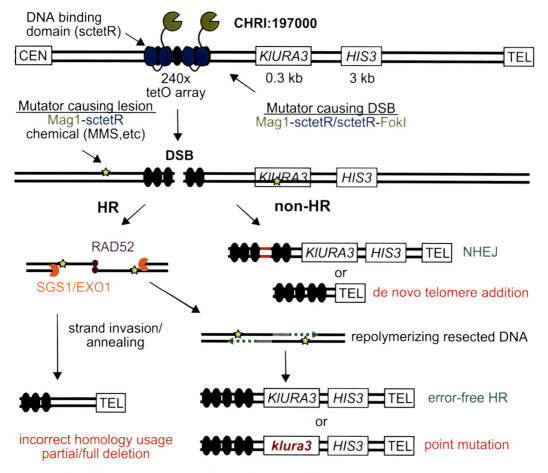

Fig. 7.1 How TaGTEAM works. A 240x non-recombinogenic tetO array is integrated in a subtelomeric region in chromosome I of the yeast genome, along with a *KlURA3* marker to measure mutation rate. Expressing either Mag1-sctetR or FokI-sctetR leads to mutator binding and damage at the array and a DSB or DSE (not depicted). The Mag1-sctetR or chemicals like MMS cause lesions (adducts, abasic sites) in DNA. If the DSB is repaired by HR, The DNA helicase *SGS1* and the 5′–3′ endonuclease *EXO1* chew back or resect 5′ ends. Any lesions present on ssDNA are irreparable as ssDNA is not recognized by host repair machinery. If the *RAD52*-coated ssDNA ends invade a "proper" homologous donor (like the sister chromosome) polymerization of the 3′ ends occurs, allowing reannealing and resynthesis of resected ends. If lesions on the template strand are encountered during resynthesis, translesion polymerases are required to insert some base to continue polymerizing; this may result in a point mutation. HR may also be error-prone if an "improper" donor is used during strand invasion, potentially resulting in deletions of or within the target region. Repair by non-HR processes include NHEJ and de novo telomere addition at the 3′ end of the array. NHEJ may be mutagenic, but only locally at the DSB within the array. By utilizing a *HIS3* marker on the telomere-proximal end, one can select against de novo telomere addition and other undesirable deletions of the target region

in loss of target genes (Finney-Manchester and Maheshri 2013).

While TaGTEAM use has been limited to *S. cerevisiae* strains, the basic mechanism is highly conserved (Yang et al. 2010; Roberts et al. 2012), and we suspect TaGTEAM will be particularly useful in fungal and other species with tractable but low-efficiency genetic transformation. For example, in filamentous fungi, protoplast- or *Agrobacterium tumefaciens*-mediated transformation is possible, albeit with low efficiency (de Groot et al. 1998). Integration of the TaGTEAM array and target sequences for mutagenesis need occur only once. Mutagenesis then proceeds *in vivo* in a controllable fashion. Depending on the species, the balance of repair via HR versus non-homologous end-joining (NHEJ) (Fig. 7.1) may favor NHEJ. If selection against such events is insufficient, the efficiency of error-prone HR can be enhanced by restricting mutagenesis to cell-cycle stages when HR is naturally upregulated (see below) or utilizing an NHEJ mutant.

7.2 Materials and Methods

7.2.1 Strain Design Considerations

The deployment of TaGTEAM into a strain of interest requires transformation of the mutator and of a tetO array flanked with target sequences to be mutagenized. This requires choosing the target size and sequence, the tetO array size and placement, and the mutator used. We have characterized a 240x tetO array integrated at a gene-poor region in chromosome I (ChrI:197000) most extensively. Target sequences for mutagenesis placed adjacent to but within 10 kb of the 240x tetO array experience the highest mutation rates with a drop off further from the array (Fig. 7.2). Because of heightened damage and repair by HR in this region, target sequences containing repetitive regions may lead to undesirable rearrangements or deletions. Many of these can be selected against by including a positive selection marker in the target region (such as the *HIS3* gene in Fig. 7.1). Additional copies of host genes can be used in the targeted region to explore useful "paralogs", though the possibility of recombination with the native locus exists. We envision in the majority of applications target sequences will consist of genes that are heterologous in nature.

Other configurations of the tetO array and target sequences are possible. We find no change in mutation rates for array sizes down to 85x tetO; further reduction decreases the mutation rate (unpublished results). Heretofore unexplored configurations include interspersing target sequences between multiple tetO arrays of varying length. This may lead to elevated mutation rates in potentially larger regions with fewer tetO

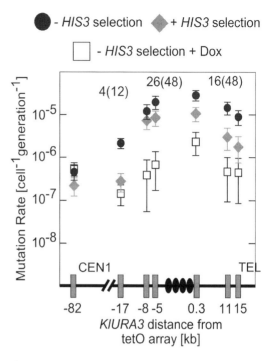

Fig. 7.2 Distance-dependence of Mag1-sctetR-mediated targeted mutagenesis around the tetO array. Fusions of Mag1p to sctetR are expressed from a galactose-inducible promoter on a centromeric plasmid in cells containing a 240x tetO array. The mutation rate marker *KlURA3* is introduced at various positions near the array. Targeted mutation rates at 0.3 kb are elevated 800-fold, while rates at the *CAN1* marker on chromosome V change insignificantly (not shown). This increase persists for at least 10 kb on either side of the array. Addition of dox (*squares*) eliminates targeted mutagenesis completely. Selection for *HIS3* (*diamonds*) decreases mutation rate slightly. Labels on data points represent ability to PCR the *KlURA3* cassette from mutants, PCR+ (total), indicating a significant fraction are point mutations. Error bars represent 95 % confidence limits

sites. However, mutagenesis may be more stable with longer arrays as we have seen a decrease in array size in long-term (>50 generation) mutagenesis (Finney-Manchester 2013).

When integrating the array at ChrI:197000 and expressing mutator, we observe only minimal growth defects. However, when we attempted to mutagenize targets at the *HIS3* locus these strains did not grow upon mutator induction. Because a significant number of mutagenic events lead to partial loss of the target region consistent with de novo telomere addition at the DSB (Fig. 7.1), it is possible that large-scale deletions

Table 7.1 Mutation rates and spectrum of various mutator/drug combinations

Targeted mutator	Mag1-sctetR				sctetR-FokI		
Lesion generator	–	–	+100 µM CAA	+0.003 % MMS	+Mag1p	+100 µM CAA	+0.003 % MMS
Targeted mutation rate (cell^{-1} gen^{-1})	3×10^{-8}	3×10^{-5}	1×10^{-4}	1×10^{-4}	3×10^{-5}	5×10^{-5}	5×10^{-5}
Point mutant fraction[a]	26/48	16/48	40/48	8/12	11/12	35/48	9/12
Background mutation rate (cell^{-1} gen^{-1})	*CAN1*: 1×10^{-8}	*CAN1*: 3×10^{-8} +dox: 2×10^{-6}	+dox: 2×10^{-6}	+dox: 2×10^{-5}	+dox: 3×10^{-6}	+dox: 1×10^{-6}	+dox: 3×10^{-7}
Transitions	22 %	16 %	26 %	–	19 %	18 %	31 %
TA>CG	13 %	6.1 %	0.0 %	–	15 %	4.5 %	1.4 %
CG>TA	8.7 %	10 %	26 %	–	4 %	14 %	30 %
Transversions	48 %	59 %	59 %	–	52 %	82 %	64 %
TA>GC	17 %	0.0 %	3.7 %	–	6.3 %	0.0 %	2.9 %
GC>TA	22 %	27 %	55.6 %	–	25 %	77 %	40 %
TA>AT	4.3 %	18 %	0.0 %	–	16 %	0.0 %	17 %
GC>CG	4.3 %	14 %	0.0 %	–	4.2 %	4.5 %	4.3 %
InDels	30 %	25 %	3.7 %	–	29 %	0.0 %	4.3 %

[a]Measured by ability to PCR *KlURA3*.

of essential genes adjacent to or telomere-proximal to the *HIS3* locus cause the growth defect. We therefore recommend integrating the array at subtelomeric gene-poor regions like ChrI:197000. TaGTEAM has also been successful with the array on a plasmid, with only a moderate growth defect when selecting for the plasmid.

Finally, the choice of mutator impacts both the spectrum of mutations generated and the tendency to favor point mutations as opposed to other mutagenic outcomes (Fig. 7.1). We list results for 240x tetO array at ChrI:197000 in the W303A background in Table 7.1 to serve as a guide (Finney-Manchester and Maheshri 2013). Mutators must be expressed at high levels, either from centromeric plasmids or the genome. Targeted mutagenesis can be toggled by using an inducible promoter or the addition of doxycycline, which prevents binding of the mutator to the tetO array.

In Table 7.2, we provide detailed protocols for deploying TaGTEAM using constructs available on Addgene (http://www.addgene.org), measuring mutation rates, and performing a selection.

7.2.2 Detailed Protocol

7.2.2.1 Materials

- Standard synthetic dropout or complete media with either 2 % glucose (SD) or 2 % galactose (SG) (Guthrie and Fink 2004).
- Canavanine selection plates: 600 mg/L L-canavanine sulfate (Sigma) in SD lacking arginine.
- 5′FOA selection plates: 1 g/L 5′FOA (US biological) in SD complete supplemented with 50 mg/L uracil.
- Methyl-methanosulfonate (MMS) (Sigma).
- Chloroacetaldehyde (CAA) (Sigma).

7.2.2.2 Constructing TaGTEAM-Capable Strains

1. Integrate mutator. (Mutators on centromeric plasmids are transformed after tetO array/target sequence integration.)
   (a) In prototrophic strain, integrate mutator at *URA3* locus.
   - Amplify desired U3KO constructs using primers in Table 7.3.

Table 7.2 Constructs for generating TaGTEAM-capable strains

Plasmid	Addgene #	Integration locus	Marker	Usage notes
PRS415-GAL1pr-MAG1-sctetR	44752	Centromeric	*LEU2*	Galactose-controlled Mag1-sctetR expression.
PRS415-TDH3pr-MAG1-sctetR	53718	"	"	Constitutive Mag1-sctetR expression.
PRS415-GAL1pr-sctetR-FokI	44756	"	"	Galactose-controlled sctetR-FokI expression.
PRS415-TDH3pr-sctetR-FokI	43721	"	"	Constitutive sctetR-FokI expression.
PRS415-GAL1pr-sctetR-FokI-GAL10pr-MAG1	53722	"	"	Galactose-controlled expression of sctetR-FokI and untargeted Mag1p.
PRS303-GAL1pr-MAG1-sctetR	44752	*HIS3*	*HIS3*	Digest with PstI or BsiWI before transformation.
PRS303-GAL1pr-sctetR-FokI	53723	"	"	"
U3KO-EV	53724	*URA3*	–	Inserts a nonfunctional *ura3* allele. Serves as an empty vector control in prototrophic strains.
U3KO-GAL1pr-MAG1-sctetR	53725	"	–	To insert a galactose-controlled Mag1-sctetR at *URA3* resulting in a *ura-* phenotype.
U3KO-TDH3pr-MAG1-sctetR	53726	"	–	To insert constitutive Mag1-sctetR at *URA3*.
U3KO-GAL1pr-sctetR-FokI	53727	"	–	To insert galactose-controlled sctetR-FokI at *URA3*.
U3KO-TDH3pr-sctetR-FokI	53728	"	–	To insert constitutive sctetR-FokI at *URA3*.
PRS303-240xtetO	44754	CHRI:197000	*HIS3*	Digest with AscI for CHRI integration. CEN-XbaI-240x-XhoI-TEL.
PRS316-240xtetO	44755	Centromeric	*URA3*	URA3 marked centromeric plasmid containing 240x array.
PRS316kl-120xtetO-ade2-1		Centromeric	*KlURA3*	*KlURA3* marked centromeric plasmid containing 120x array and gain of function *ade2-1* mutation rate marker.
pCHRI-85xtetO	44757	CHRI:197000	*KlURA3*	Digest with AscI for CHRI integration of 85x array.

Table 7.3 Primers for constructing TaGTEAM-compatible strains

Primer	Sequence	Template
U3KO(+)	GGAGCACAGACTTAGATTGG	U3KO plasmids
U3KO(−)	CTTTGTCGCTCTTCGCAATGTC	"
U3KO-chk(+)	TGCGAGGCATATTTATGGTGAAG	Genomic (g)DNA after ura3KO with mutator
U3KO-gal1chk(−)	CCATCCAAAAAAAAAGTAAGAATTTTTG	gDNA after ura3KO with galactose-controlled mutator
U3KO-tdh3chk(−)	GGCAGTATTGATAATGATAAACTCG	gDNA after ura3KO with constitutive mutator
KlURA3(+)	CATCAAATGGTGGTTATTCGTGG	gDNA of mutants to confirm PMs
KlURA3(−)	CTCTTTTTCGATGATGTAGTTTCTGG	"

- Transform, plate on SD complete, replica plate on 5′FOA next day.[1]
- Confirm integration by PCR using primers in Table 7.3.
(b) For an auxotrophic strain, can directly transform pRS303-based vectors (1–5 μg of PstI-digested vector) with mutators (Table 7.2) at *HIS3*.
2. Clone target genes into desired tetO array plasmid (sequence available from Addgene) and introduce into strain.[2]
  (a) For integration at ChrI:197000, digest 1–5 μg with AscI and transform.
  (b) If the target region is on a plasmid, transform in strain where mutator is integrated.

7.2.2.3 Fluctuation Analysis to Confirm TaGTEAM Functionality

This assay requires *KlURA3* in the target/array region (for integrated targets) or *ade2-1* (for targets on centromeric plasmids) and measures mutation rate as loss (*KlURA3*) or gain (*ade2-1*) of function.

1. Inoculate an overnight culture in appropriate SD dropout media (to maintain any plasmid).
2. Dilute cells to 10,000 cells/mL in appropriate SD dropout or SG dropout for mutator induction.

3. Aliquot 12 replicate 20 μL cultures per strain into a sterile 384 well plate. Seal to prevent evaporation. Incubate at 30 °C for 2–3 days to stationary phase.[3,4,5]
4. Measure final OD_{600} of cultures on a plate reader.
5. Plate a small fraction of each culture on 30 mm YPD plates (to give 10–50 colonies) to estimate the total number of cells per culture.[6]
6. Plate the remaining culture on 30 mm 5′FOA plates (or appropriate media) to estimate the number of mutants per culture.
7. Incubate at 30 °C for 2–3 days. Use colony counts to estimate the mutation rate.[7]
8. Confirm that a sufficient number of mutants represent base pair substitutions.

---

[1] We typically transform 400 μL of PCR product using the LiAc/PEG/ssDNA method (Gietz and Woods 2002).

[2] If integrating target genes, create one version including KlURA3, to measure the mutation rate of a particular strain/mutator combination. If placing on centromeric plasmid, measure mutation rate using ade2-1 reversion.

[3] Shaking has no appreciable effect on mutation rate in these conditions.

[4] Periodically measuring the OD_{600} of a replicate plate can be used to confirm cultures have entered stationary phase.

[5] Background mutation rates can be estimated by loss of function at CAN1. These rates are usually 10^2–10^3-fold below targeted rates (Table 34.2) so 500 μL, rather than 20 μL of culture should be grown in a 96-deep well plate. The same format should be used when measuring reversion rates of ade2-1, because this gain of function mutation requires mutagenesis of an internal stop codon and has similar rates (~10^{-7} revertants per generation). Use canavanine selection or SD ura- ade- plates, respectively.

[6] Establish a calibration between OD_{600} and cell density at the end of growth to eliminate this step if repeated assays with the strain are planned.

[7] We obtain a maximum likelihood estimate of the mutation rate by fitting measurements to the Luria-Delbruck distribution (Foster 2006).

7 In Vivo Targeted Mutagenesis in Yeast Using TaGTEAM

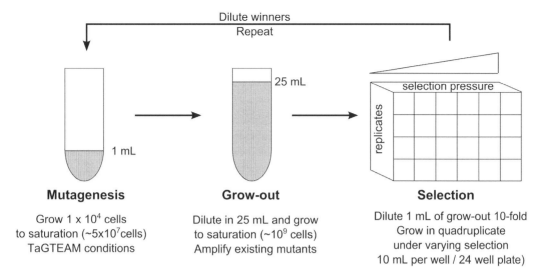

Fig. 7.3 A possible evolutionary protocol using a TaGTEAM-capable strain with a selectable phenotype

(a) Patch one mutant colony per culture onto a fresh 5′FOA plate to confirm mutant phenotype.
(b) Prepare genomic DNA from each mutant.
(c) Verify *KlURA3* marker presence using primers in Table 7.3.

7.2.2.4 Using TaGTEAM to Create Mutant Libraries

While the details will depend on a specific application, we list three guidelines to deploying TaGTEAM. Further discussion is in (Finney-Manchester 2013).

1. *Decouple mutagenesis from selection.* TaGTEAM point mutations are generated by HR which occurs during the S/G2 cell-cycle phase. Most selections involve a metabolic burden that slows growth by increasing the length of G1. Hence target specificity of TaGTEAM decreases because mutational processes in G1 are untargeted. In addition, decoupling mutagenesis and selection allows the use of growth conditions ideal for each process (Fig. 7.3).
2. *Perform mutagenesis under marker selection to minimize rearrangements.* An appreciable fraction of mutational events result in large-scale deletion of the target region. Place a selectable marker (*KlURA3*, *HIS3*, etc.) within the target region, effectively enriching the population for mutants of interest.
3. *Periodically assess TaGTEAM stability.* Repeated DNA damage at the same locus can lead to loss of function. There are number of ways to do so.
   (a) Verify retention of target genes and tetO array via PCR.
   • Isolate and prepare genomic DNA from several clones within the mutating population.
   • PCR with primers to target genes and/or the tetO array. Use genomic DNA from an unmutated clone as a positive control.[8,9]
   (b) Fluorescently tag mutators (to verify their continued expression).[10]
   (c) Fluctuation analysis at target to verify elevated targeted mutation rate.
   (d) Fluctuation analysis at *CAN1* to verify background mutation rate.

---

[8] PCR of arrays longer than 85 copies of tetO can be difficult, but it can detect changes in array size that can occur through HR.

[9] Next-gen sequencing is a viable alternative.

[10] We have C-terminally tagged Mag1-sctetR and N-terminally tagged sctetR-FokI with YFP and see no change in function, provided an N-terminal SV40 NLS remains in the tagged construct (Finney-Manchester and Maheshri 2013).

7.2.2.5 A Suggested Evolutionary Protocol for TaGTEAM Library Generation Coupled to Selection

When selecting for a desired phenotype, implementing TaGTEAM involves periodic switches between TaGTEAM optimal mutagenesis and selection conditions. We describe one way to do so in Fig. 7.3. Assuming a uniform 10^7 bp^{-1} gen^{-1} mutation rate, each single base pair mutation is realized in the target region at 5× coverage, as the final 5×10^7 cells determine the library size. Grow-out after mutagenesis of the 1 mL culture amplifies mutants and ensures cells express mutant proteins prior to selection. When splitting the grow-out into 24×1 mL samples, the 5× coverage ensures at least one of each type of mutant is in each sample. By performing selection at various selection pressures, we can identify the largest leap forward in phenotypic improvement per evolutionary round. We are able to assess how reproducible and likely any phenotypic improvement was due to mutagenesis in the target region by doing the selection in quadruplicate. Volumes are chosen for simple bench-scale manipulation amenable to automation.

Acknowledgements We thank L. Samson, B. Engelward, and K. Prather for useful discussions. Funding sources include a National Science Foundation graduate fellowship to S.F-M and an MIT Reed Research Fund and NIEHS Pilot P30-ES002109 to N.M. Work in N.M.'s laboratory is also funded by NIH GM095733.

References

de Groot MJ, Bundock AP, Hooykaas PJJ, Beijersbergen AGM (1998) Agrobacterium tumefaciens-mediated transformation of filamentous fungi. Nat Biotechnol 16:839–842

Finney-Manchester SP (2013) Harnessing mutagenic homologous recombination for targeted mutagenesis in vivo by TaGTEAM. PhD thesis, Massachusetts Institute of Technology, Cambridge, MA

Finney-Manchester SP, Maheshri N (2013) Harnessing mutagenic homologous recombination for targeted mutagenesis *in vivo* by TaGTEAM. Nucleic Acids Res 41:e99

Foster PL (2006) Methods for determining spontaneous mutation rates. Methods Enzymol 409:195–213

Gietz RD, Woods RA (2002) Transformation of yeast by the Liac/SS carrier DNA/PEG method. Methods Enzymol 350:87–96

Goldsmith M, Tawfik DS (2012) Directed enzyme evolution: beyond the low-hanging fruit. Curr Opin Struct Biol 22:406–412

Guthrie C, Fink GR (2004) Guide to yeast genetics and molecular and cell biology. Part A. Elsevier, Amsterdam, Netherlands

Kim B, Du J, Zhao H (2012) Strain improvement via evolutionary engineering, Ch 4. In: Patnaik R (ed) Engineering complex phenotypes in industrial strains. Wiley, Hoboken

Roberts SA, Sterling J, Thompson C, Harris S, Mav D, Shah R, Klimczak LJ, Kryukov GV, Malc E, Mieczkowski PA, Resnick M, Gordenin DA (2012) Clustered mutations in yeast and human cancers can arise from damaged long single-strand DNA regions. Mol Cell 46:424–35

Yang Y, Gordenin DA, Resnick MA (2010) A single-strand specific lesion drives MMS-induces hypermutability at a double-strand break in yeast. DNA Repair 9:914–921

Part III

Endogenous DNA: Gene Expression Control

RNA Silencing in Filamentous Fungi: From Basics to Applications

8

Nguyen Bao Quoc and Hitoshi Nakayashiki

8.1 Introduction

8.1.1 The Discovery of RNA Silencing in Fungi

Double-stranded RNA (dsRNA)-mediated post-transcriptional gene silencing (PTGS) was independently discovered as RNA interference (RNAi) in *Caenorhabditis elegans*, co-suppression in plants, and quelling in fungi (Napoli et al. 1990; van der Krol et al. 1990; Cogoni et al. 1996; Fire et al. 1998), which here we collectively describe as RNA silencing. In the canonical RNA silencing pathway, exogenous or endogenous dsRNA is the triggering molecule that is subsequently cleaved by an RNase-III-like enzyme, the so-called DICER, into small interfering RNA (siRNA), typically 20–25 bp in length. siRNA is then incorporated into the RNA-induced silencing complex (RISC) to function as a guide molecule to target cognate messenger RNA (mRNA)

for degradation (Bernstein et al. 2001; Hammond et al. 2001; Ketting and Plasterk 2000; Elbashir et al. 2001; Zamore et al. 2000).

The story of RNA silencing in fungi began with the discovery of quelling reported by Romano and Macino in 1992. Quelling is a kind of co-suppression induced by a transgene in the fungus *Neurospora crassa*. Introduction of a coding sequence of the *al-1* gene, which is essential for carotenoid biosynthesis, into *N. crassa* resulted in transformants showing a pale yellow/white color, indicating that the endogenous *al-1* gene was suppressed to varying degrees. Quelling has several interesting features that provide mechanistic insights into RNA silencing: (1) the gene inactivation by quelling is reversible and the reversion is correlated with the release of exogenous DNA; (2) the reduction of the mRNA steady-state level of the duplicated gene is due to a posttranscriptional effect on its accumulation; (3) transgenes containing transcribed regions are able to induce gene silencing, whereas promoter regions are ineffective; (4) quelling is dominant in heterokaryotic strains containing a mixture of transgenic and nontransgenic nuclei, indicating the involvement of a diffusible trans-acting molecule (Cogoni and Macino 1997).

A molecular genetic approach was successfully used to isolate several quelling-defective (qde) genes (Cogoni and Macino 1997, 1999a, b). QDE-1 is an RNA-dependent RNA polymerase (RdRP) possibly involved in the generation of

---

N.B. Quoc, Ph.D. (✉)
Research Institute of Biotechnology and Environment, Nong Lam University, Linh Trung Ward, Thu Duc District, Ho Chi Minh City 70000, Vietnam
e-mail: baoquoc@hcmuaf.edu.vn

H. Nakayashiki, Ph.D.
Graduate School of Agricultural Science, Laboratory of Cell Function and Structure, Kobe University, Kobe, Hyogo, Japan

---

M.A. van den Berg and K. Maruthachalam (eds.), *Genetic Transformation Systems in Fungi, Volume 2*, Fungal Biology, DOI 10.1007/978-3-319-10503-1_8,
© Springer International Publishing Switzerland 2015

dsRNA from repetitive loci through aberrant RNA (aRNA) production and also in subsequent processes for signal amplification (Cogoni and Macino 1999a; Lee et al. 2010a). QDE-2 belongs to the Argonaute (Ago) proteins that typically exhibit the PAZ and PIWI domains and constitute the core components of RISC complexes (Song et al. 2004). PAZ and PIWI domains are proposed to function as a binding site of small RNA and catalytic core of slicer, respectively (Kim et al. 2009). In *N. crassa*, two RNase-III family or so-called DICER enzymes, DCL-1 and DCL-2, were shown to be redundantly involved in the quelling pathway, with a larger contribution by DCL-2 (Catalanotto et al. 2004). These findings indicated that quelling in *N. crassa* belongs to a broad category of RNA-mediated gene silencing mechanisms typical to RNAi.

With recent advances in the fungal genome sequencing, data mining of proteins involved in the RNA silencing machinery has revealed that the mechanisms of RNA silencing are conserved in most fungal species with a few exceptions such as *C. tropicalis*, *C. albicans*, *C. lusitaniae*, *S. cerevisae*, *U. maydis* (Nakayashiki et al. 2006). In fact, RNA silencing is used as a genetic tool in various fungal species now. Interestingly, comparative phylogenetic analysis shows that numbers of Dicer, Argonaute, and RdRP genes vary significantly among fungal species, suggesting that RNA silencing pathways have diversified in the evolution of fungi (Nakayashiki et al. 2006).

8.1.2 Molecular Mechanisms of RNA Silencing in Fungi

It is generally believed that quelling in fungi is equivalent to RNAi in animals or posttranscriptional gene silencing (PTGS) in plants because core RNA silencing components such as Dicer, Argonaute, and RdRP genes are used in all of these pathways (Fagard et al. 2000). However, besides these common components, several additional genes in the quelling pathway have also been identified in *N. crassa*. The third QDE protein, namely QDE-3, is a DNA helicase involved in DNA repair and the production of aRNA that

is subsequently converted to dsRNA by QDE-1 (Cogoni and Macino 1999b; Kato et al. 2005; Lee et al. 2010a). It has been proposed that QDE-3 can recognize aberrant DNA structures in repetitive sequences, and unwind such dsDNA to ssDNA that serves as a template for aRNA production by the DNA-dependent RNA polymerase activity of QDE-1 (Dang et al. 2011). The QDE-2-interacting protein (QIP) is an exonuclease originally isolated through physical interaction with QDE-2 (Maiti et al. 2007). The siRNA duplex loaded into the Argonaute protein must be in a single-stranded form to function as a guide for targeted mRNA degradation. QIP was shown to increase the efficiency of passenger strand removal from the siRNA duplex leading to RISC activation (Maiti et al. 2007).

In addition to quelling, several other RNA silencing-related pathways have been identified in *N. crassa*. Meiotic silencing by unpaired DNA (MSUD) is the second pathway and operates during the sexual phase from an early stage of meiosis after karyogamy to ascospore maturation (Shiu et al. 2001). MSUD was first reported in studies on the meiotic transvection of the ascospore maturation gene (*asm-1*) in *N. crassa* (Aramayo and Metzenberg 1996; Shiu and Metzenberg 2002; Shiu et al. 2001). In the MSUD pathway, it is supposed that some meiotic sensing operates to detect unpaired DNA, i.e., DNA segments that are present in one parental chromosome but not in its pairing partner. Such unpaired DNA triggers gene silencing of all homologous sequences in the genome through the production of unpaired DNA-specific aRNA that is converted into dsRNA and sRNA (Pratt et al. 2004; Shiu et al. 2001; Shiu and Metzenberg 2002). The mechanism of MSUD was elucidated by molecular genetic approaches that identified a set of important genes highly homologous to the components of the quelling pathway (Shiu et al. 2001; Lee et al. 2003). SAD-1 (*s*uppressor of *a*scus *d*ominance 1), a paralog of QDE-1, was shown to localize to the perinuclear region. SAD-1 is supposed to be the RdRP required for production of dsRNA from unpaired DNA-derived aRNA in meiotic silencing (Shiu et al. 2001). SMS-2 (*s*uppressor of *m*eiotic *s*ilencing-2) and SMS-3

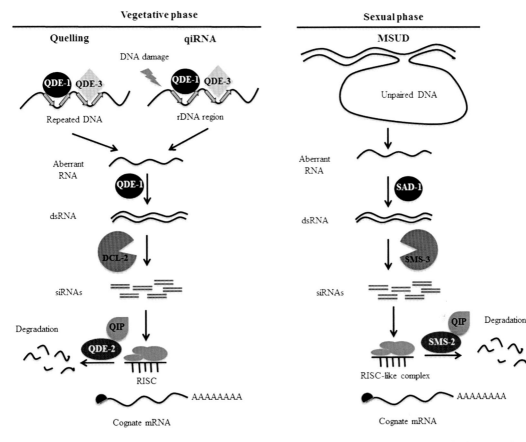

Fig. 8.1 A proposed model of RNA silencing pathways in *Neurospora crassa*. During the vegetative phases of the *N. crassa* life cycle, repeated sequence (quelling) and ribosomal DNA loci (qiRNA) in the genome can induce silencing phenomena. In this pathway, the production of aRNAs is done by QDE-1 and QDE-3, then converted to dsRNA by QDE-1. siRNAs are produced through the procession of dsRNAs by the action of Dcl-2, then loaded onto QDE-2/QIP-based RISC complex to degrade the cognate mRNA. During meiosis, unpaired DNA can trigger the second RNA silencing pathway in *N. crassa*, called MSUD. Mechanisms of silencing in MSUD are supposed to be quite similar to those in quelling except that MSUD uses a different set of silencing protein components (paralogs) from those in the quelling pathway. *qiRNA* QDE-2 interacting sRNA, *QDE* quelling defective, *dsRNA* double-stranded RNA, *SAD-1* suppressor of ascus dominance, *DCL* Dicer-like protein, *QIP* QDE-2 interacting protein, *RISC* RNA-induced silencing complex, *SMS* suppressor of meiotic silencing, *MSUD* meiotic silencing by unpaired DNA

(DCL-1) are Argonaute and Dicer proteins homologous to QDE-2 and DCL-2 in the quelling pathway, respectively (Lee et al. 2003; Alexander et al. 2008). QIP has been shown to play an important role in the MSUD pathway as in the quelling pathway (Hammond et al. 2011; Xiao et al. 2010). Therefore, in *N. crassa*, two RNA silencing pathways operate in different stages of the life cycle, quelling in the vegetative phase and MSUD in the sexual phase, using mostly discrete sets of RNA silencing components (Cogoni 2001; Dang et al. 2011; Nakayashiki 2005; Nakayashiki et al. 2006; Shiu et al. 2001). A simplified model of the quelling and MSUD pathways is shown in Fig. 8.1.

Recently, Dang et al. (2011) identified several novel RNA silencing-related pathways in *N. crassa*. QDE-2-interacting small RNA (qiRNA) is a new type of small interfering RNA induced by DNA damage (Chang et al. 2012). The production of qiRNAs involves major quelling components such as QDE-1, QDE-3, DCL-1, and DCL-2. qiRNA originates mostly from the ribosomal DNA locus and likely plays a role in the DNA-damage response by inhibiting protein translation (Lee et al. 2009). In this pathway, production of aRNA from the rRNA locus requires replication protein A (RPA), which interacts with QDE-1 (Lee et al. 2010a; Nolan et al. 2008). Because RPA forms a single-stranded DNA-binding complex, it is proposed that RPA is the protein that recruits QDE-1 to ssDNA that can be used as a template for aRNA generation (Lee et al. 2010a). In addition, several miRNA-like small RNAs and Dicer-independent siRNAs were identified in *N. crassa* (Lee et al. 2010b). Even though their biological roles largely remain to be elucidated, it has been demonstrated that diverse molecular mechanisms are involved in their biogenesis (Lee et al. 2010b).

8.1.3 The Roles of RNA Silencing in Fungi

8.1.3.1 RNA Silencing in Genome Stability

The possible role of RNA silencing in controlling the integrity of the rDNA locus and in genome stability has been reported in fission yeast and insects (Cam et al. 2005; Peng and Karpen 2007). Similarly in fungi, a significant reduction of the copy number of rDNA repeats was observed in quelling mutants (Cecere and Cogoni 2009), indicating a role for quelling in the maintenance of the integrity and stability of the rDNA locus in *N. crassa*. As described above, the qiRNA-mediated pathway also contributes to the maintenance of genome stability in *N. crassa* because its defective mutants show higher sensitivity to DNA damage (Lee et al. 2009). The role of RNA silencing in heterochromatin formation has also been reported in the fission yeast *S. pombe* through the detection of siRNA of noncoding transcripts derived from the centromere, and the association of Ago1 and RdRP1 with all major heterochromatic loci (Cam et al. 2005; Djupedal et al. 2009). In contrast, it was reported that RNA silencing components were independent of DNA methylation and heterochromatin formation in *N. crassa* (Freitag et al. 2004).

8.1.3.2 RNA Silencing as Viral Defense

RNA silencing has been demonstrated to be a potent defense mechanism against invasive viruses and transposons in plants and animals (Harvey et al. 2011; Wilkins et al. 2005; Zambon et al. 2006; Jeang 2012). In fungi, the crucial role of RNA silencing in antiviral defense was first reported in the chestnut blight fungus, *Cryphonectria parasitica*. The *C. parasitica* mutant of DCL-2, one of the two Dicer proteins in *C. parasitica*, showed higher sensitivity to infection with *Cryphonectria* hypovirus 1 (CHV1) (Seger et al. 2007). Similarly, a drastic increase in viral infections was observed in a knockout mutant of an Argonaute-like gene, *agl-2* (Zhang et al. 2008; Sun et al. 2009). Interestingly, hairpin RNA (hpRNA)-induced GFP silencing was significantly diminished in *C. parasitica* strains infected with CHV1, suggesting that CHV1 carried a suppressor of RNA silencing as previously shown with plant viruses (Seger et al. 2006). The RNA silencing suppressor was the virus-encoded papain-like protease p29, which shares some sequence similarity with HC/Pro, an RNA silencing suppressor encoded by plant viruses belonging to the genus Potyvirus. A crucial role for RNA silencing in antiviral defense has been also demonstrated in *Aspergillus nidulans* and *Magnaporthe oryzae* (Hammond et al. 2008; Himeno et al. 2010).

8.1.3.3 RNA Silencing as Defense Against Transposons

The contribution of RNA silencing to genome defense against transposable elements has also been reported in several fungal species. In *N. crassa*, the LINE1-like retrotransposon Tad was shown to be repressed by a quelling-related pathway (Nolan et al. 2005). QDE-2 and Dicer, but not QDE-1 nor QDE-3, were required for Tad control. The LTR-retrotransposon MAGGY identified in *M. oryzae* was also targeted for RNA silencing. Its mRNA accumulation was drastically increased in knockout mutants of *MoDcl2*, one of the two Dicer genes in the *M. oryzae* genome (Murata et al. 2007). However, the activity of MAGGY transposition decreased as its genomic copy number increased even in the *modcl2* mutant, suggesting that *M. oryzae* possesses some transposon suppression mechanism other than RNA silencing (Murata et al. 2007). In the human fungal pathogen *Cryptococcus neoformans*, transposon activity and genome integrity are controlled by the production of transposon-mediated siRNAs (Wang et al. 2010). This mechanism is highly activated during the sexual phase, and thus has been designated sex-induced silencing (SIS). Interestingly, the efficiency of SIS has been shown to correlate with the copy number of target sequences in the genome (Wang et al. 2010). Thus, it appears to be a common rule that repetitiveness is an important factor to trigger RNA silencing in fungi.

8.2 RNAi Strategies in Fungi

8.2.1 Advantage of RNAi as a Genetic Tool in Fungal Research

Because of recent advances in sequencing technologies, the number of sequenced fungal genomes has been increasing remarkably. In the post-genomics era, RNA silencing (RNAi) can serve as a powerful reverse genetic tool to identify gene function in a wide range of eukaryotes including fungi. Searching for RNAi components in fungal genomes has revealed that RNA silencing pathways are conserved in various fungal species encompassing the major fungal taxa (Nakayashiki et al. 2006). In fact, RNAi-related gene silencing has been demonstrated in Ascomycota (Romano and Macino 1992; Goldoni et al. 2004; Hamada and Spanu 1998; Kadotani et al. 2003; Fitzgerald et al. 2004; Mouyna et al. 2004; Rappleye et al. 2004; Hammond and Keller 2005; McDonald et al. 2005; Spiering et al. 2005; Zheng et al. 1998; Yamada et al. 2007; Ngiam et al. 2000; Oliveira et al. 2008; Janus et al. 2007; Ullan et al. 2008; Cardoza et al. 2006; Seger et al. 2006; Erental et al. 2007; Moriwaki et al. 2007; Engh et al. 2007; Shafran et al. 2008; Patel et al. 2008; Singh et al. 2010; Brotman et al. 2008; Liu et al. 2010; Tinoco et al. 2010; Krajaejun

et al. 2007; Vermout et al. 2007), Basidiomycota (de Jong et al. 2006; Liu et al. 2002; Namekawa et al. 2005; Matiyahu et al. 2008; Salame et al. 2010; Costa et al. 2009; Kemppainen et al. 2009; Caribe dos Santos et al. 2009), Zygomycota (Nicolas et al. 2003; Takeno et al. 2004), fungus-like Oomycota (Latijnhouwers et al. 2004; van West et al. 1999), and Myxomycetes (Martens et al. 2002) (Table 8.1). In some fungal species such as *Saccharomyces cerevisiae*, *Candida albicans*, *and Ustilago maydis*, however, some or all of the RNA silencing components may have been lost during evolution, indicating that RNA silencing and its related pathways are not essential for growth and development in the life cycle of these fungal species (Nakayashiki et al. 2005, 2006; Aravind et al. 2000). Interestingly, a noncanonical Dicer gene was found in the genome of *C. albicans*. This was consistent with the finding that gene silencing of the EFG1 gene was successfully induced through transfection of related 19 nt siRNAs in *C. albicans* (Moazeni et al. 2012). In contrast, however, Staab et al. (2011) reported that hairpin dsRNA did not trigger RNA interference in *C. albicans*. Recently the RNAi pathway was reconstructed in the budding yeast by introducing the missing genes (Argonaute and Dicer) from *S. castellii* or human (Drinnenberg et al. 2009; Suk et al. 2011). This strategy provides a chance for scientists to use RNAi in yeasts including *C. albicans*.

For functional genomics research in fungi, RNAi has become an important alternative to conventional gene knockout strategies. There are three major advantages of RNAi in functional genomics research. First, RNAi can be applied to functional analysis of a multiple gene family because it induces gene suppression in a sequence-specific but not locus-specific manner using a mobile trans-acting signal in the cytoplasm. Functional redundancy is the major obstacle in functional genomics by gene knockout approaches because it can mask the effect of deleterious mutations. Second, RNAi can be used for analyses of lethal genes because it does not completely shut down gene expression. The functions of essential genes remain largely unknown because classical genetic approaches such as mutant screening or targeted gene disruption are not available because of lethality. Imperfect silencing with reduced levels of gene expression could shed light on unexpected roles of essential genes in fundamental biological phenomena. Third, RNAi can be used in fungal species with low gene targeting efficiency. The efficiency of homologous recombination varies considerably among fungal species. As described later, several RNAi strategies are now available, and none of them require targeted integration of an RNAi vector. Thus, RNAi should offer a convenient genetic tool in fungal species where the efficiency of gene targeting is low.

8.2.2   Long Hairpin RNA

RNAi vectors producing hpRNA or intron-containing hairpin RNA (ihpRNA) are common genetic tools in various eukaryotes. In fungi, most RNAi investigations have been done with these types of vectors (Table 8.1). This strategy induces gene silencing at high efficiency even though the silencing level usually differs considerably among the resulting transformants. In *N. crassa*, hpRNA-producing constructs were shown to induce more efficient and stable gene silencing compared to "canonical" quelling induced by promoterless transgenes (Goldoni et al. 2004; Romano and Macino 1992).

To facilitate RNAi experiments in fungi, several vectors have been constructed. pSilent-1 is one such vector available at the Fungal Genetic Stock Center (http://www.fgsc.net/) (Nakayashiki et al. 2005). pSilent-1 carries a hygromycin resistance cassette and a transcriptional unit for hairpin RNA expression with a spacer consisting of a cutinase gene intron from *M. oryzae* (Fig. 8.2a). This vector has been used to induce RNAi in various ascomycete fungi (Table 8.2). Recently, the Gateway system has been introduced to pSilent-1 for high-throughput gene analysis (Shafran et al. 2008). Several RNAi vectors with an inducible promoter have also been constructed *N. crassa*, *A. nidulans*, and *M. oryzae* (Barton and Prade 2008; Goldoni et al. 2004; Vu et al. 2013). The use of an inducible promoter in RNAi experiments

can provide a powerful means to genetically elucidate the function of an essential gene because gene knockout approaches are not applicable to such genes. Inducible RNAi also make it possible to analyze a causal gene for a phenotype, where a particular phenotype can be linked to suppression of a specific gene.

8.2.3   Convergent Transcription

RNAi vectors producing hpRNA and ihpRNA usually require two steps of oriented cloning of a target gene fragment. This process is time-consuming, and thus limits the usefulness of these vectors for small-scale analysis. As mentioned above, the introduction of the Gateway system into the vectors can help with this problem. A less costly alternative is to use a dual promoter system in which sense and antisense transcripts are individually expressed under the control of two opposing RNA polymerase II promoters. Consequently, the sense and antisense transcripts form dsRNA to trigger RNAi in the cell. This type of vector allows single-step cloning for generation of an RNAi construct, and thus is applicable for high-throughput RNAi analyses. However, the main drawback of dual promoter vectors is lower RNAi efficiency compared with hpRNA- or ihpRNA-expressing vectors. In the former vectors, the annealing of two different RNA molecules separately transcribed by two promoters is required to form dsRNA in the cells, while dsRNA can be formed by self-folding of inverted repeats within an RNA molecule in the latter system. This difference in the dsRNA formation process between the two systems may cause different RNAi efficiencies. In fact, in *Histoplasma capsulatum*, an RNAi vector with opposing promoters triggered only moderate silencing of the GFP reporter gene with a 35 % reduction on average (Rappleye et al. 2004). In *M. oryzae*, the RNAi vector pSilent-dual1 (pSD1) with opposing *A. nidulans* trpC and gpd promoters was constructed and shown to induce GFP silencing mostly at moderate levels (Nguyen et al. 2008) (Fig. 8.2b). Approximately 8 % transformants with the pSD1-based GFP silencing

Table 8.1 RNA silencing in fungi and fungus-like organisms

Fungal species	RNAi trigger	RNAi target	Transformation	Reference
Ascomycota				
Neurospora crassa	Homologous transgene	albino-1 (al-1) and albino-3 (al-3)	PEG	Romano and Macino (1992)
	IR[a]			Goldoni et al. (2004)
Cladosporium fulvum	Homologous transgene	Hydrophobin (Hcf-1)	PEG	Hamada and Spanu (1998)
Magnaporthe oryzae	IR	GFP, PKS, MPG1,	PEG	Kadotani et al. (2003)
Venturia inaequalis	IR	GFP, THN	PEG	Fitzgerald et al. (2004)
Aspergillus fumigatus	IR	ALB1/PKSP	PEG	Mouyna et al. (2004)
Aspergillus nidulans	IR	aflR	PEG	Hammond and Keller (2005)
Aspergillus oryzae	Homologous transgene	Cpase O, amyB	PEG	Zheng et al. (1998)
	IR			Yamada et al. (2007)
Aspergillus niger	Homologous transgene	pdiA	Lithium acetate	Ngiam et al. (2000)
	IR	xlnR	PEG	Oliveira et al. (2008)
Histoplasma capsulatum	IR	ADE2	Electroporation	Rappleye et al. (2004)
Fusarium graminearum	IR	tri6	PEG	McDonald et al. (2005)
Fusarium solani	IR	Csn-1	PEG	Liu et al. (2010)
Fusarium verticillioides	IR	Gus	Agrobacterium tumefaciens	Tinoco et al. (2010)
Neotyphodium uncinatum	IR	lolC-2	Electroporation	Spiering et al. (2005)
Acremonium chrysogenum	IR	DsRed	PEG	Janus et al. (2007)
Penicillium chrysogenum	Convergent transcription	pcbC	PEG	Ullan et al. (2008)
Trichoderma harzianum	IR	erg1	PEG	Cardoza et al. (2006)
Trichoderma asperellum	IR	TasSwo	Microprojectile bombardment	Brotman et al. (2008)
Cryphonectria parasitica	IR	GFP	PEG	Seger et al. (2006)
Sclerotinia sclerotiorum	IR	pph-1, rgb-1	PEG	Erental et al. (2007)
Bipolaris oryzae	IR	PKS	PEG	Moriwaki et al. (2007)
Sordaria macrospora	IR	sdh	PEG	Engh et al. (2007)
Collectotrichum gloeosporioides	IR	PAC1	PEG	Shafran et al. (2008)
Botrytis cinerea	IR	bcsod-1	PEG	Patel et al. (2008)

(continued)

Table 8.1 (continued)

Fungal species	RNAi trigger	RNAi target	Transformation	Reference
Verticillium longisporum	IR	*Vlaro-2*	*Agrobacterium tumefaciens*	Singh et al. (2010)
Blastomyces dermatitidis	IR	*CDC11*	*Agrobacterium tumefaciens*	Krajaejun et al. (2007)
Microsporumn canis	IR	*SUB3*	PEG	Vermout et al. (2007)
Basidiomycota				
Schizophyllum commune	IR	*SC15*	PEG	de Jong et al. (2006)
Cryptococcus neoformans	IR	*CAP59*	Electroporation	Liu et al. (2002)
Coprinus cinereus	IR	*LIM15*	Lithium acetate	Namekawa et al. (2005)
Phanerochaete chrysosporium	IR	Mn*SOD1*	Electroporation	Matiyahu et al. (2008)
Pleurotus ostreatus	IR	*mmp-3*	PEG	Salame et al. (2010)
Agaricus bisporus	IR	*URA3/CBX*	*Agrobacterium tumefaciens*	Costa et al. (2009)
Laccaria bicolor	IR	*NR*	*Agrobacterium tumefaciens*	Kemppainen et al. (2009)
Moniliophthora perniciosa	IR	Mp*PRX1*/Mp*HYD3*	PEG	Caribe dos Santos et al. (2009)
Zygomycota				
Mucor circinelloides	Homologous transgene	*carB*	PEG	Nicolas et al. (2003)
Mortierella alpine	IR	delta12-desaturase	Microparticle bombardment	Takeno et al. (2004)
Oomycota[b]				
Phytophthora infestans	Homologous transgene	*INF1*	PEG[c]	van West et al. (1999)
	Homologous transgene	*Pigpa1*	Electroporation	Latijnhouwers et al. (2004)
	dsRNA	*gfp*	Lipofectin	Whisson et al. (2005)
Myxomycete (slime mold)[b]				
Dictyostelium discoideum	IR	*beta-gal* *discoidin* gene family	Electroporation	Martens et al. (2002)

[a]IR, hairpin RNA or inverted repeat RNA expressing plasmid
[b]Fungus-like organisms
[c]Lipofectin was added to increase transformation efficiency

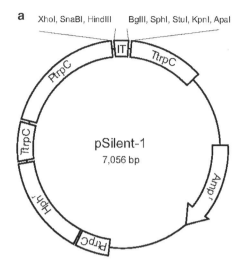
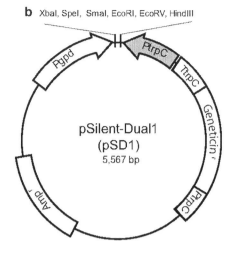

Fig. 8.2 Schematic representation of the silencing vectors pSilent-1, and pSilent-dual1. (**a**) A map of pSilent-1. *Amp*ʳ ampicillin-resistant gene, *Hyg*ʳ hygromycin-resistant gene, *IT* intron 2 of cutinase gene from *Magnaporthe oryzae*, *PtrpC Aspergilus nidulans* trpC promoter, *TtrpC A. nidulans* trpC terminator. (**b**) A map of pSilent-dual1. *Gen*ʳ geneticin-resistant gene, *PgpdA. nidulans* gpdA promoter. Restriction sites with a *asterisk* (*) are unique sites in the vector

Table 8.2 RNAi induction by pSilent-1 and pSilent-dual1 in fungi

Fungal species	RNAi vector	Targets	References
Acremonium chrysogenum	pSilent-1	Cephalosporin C	Janus et al. (2007)
Bipolaris oryzae	pSilent-1	*PKS*, *BLR2*	Moriwaki et al. (2007, 2008)
Magnaporthe oryzae	pSilent-1	*PKS*	Nakayashiki et al. (2005)
	pSilent-dual1	36 Calcium proteins, endoxylanases	Nguyen et al. (2008)
			Nguyen et al. (2011)
		MAGGY transposon, cellulases	Vu et al. (2011, 2012)
Piriformospora indica	pSilent-dual1	*PiPT*	Johri (2010)
Sclerotinia sclerotiorum	pSilent-1	Ss*Nox1*, Sx*Nox2*	Kim et al. (2011)
Sordaria macrospora	pSilent-1	1,8 Dihydroxynaphthalene	Engh et al. (2007)
Trichoderma asperellum T203	pSilent-1	Swollenin, ACC deminase	Brotman et al. (2008)
			Viterbo et al. (2010)
Verticillium longisporum	pSilent-1	*Vlaro2*	Singh et al. (2010)
Fusarium oxysporum	pSilent-1	Hexose transporter (*Hxt*)	Ali et al. (2013)
Alternaria alternata	pSilent-1	*pksJ* and *altR*	Saha et al. (2012)
Neurospora crassa	pSilent-1	*CHR-1*	Flores-Alvarez et al. (2012)
Chaetomium cupreum	pSilent-1	*ERG-2*	Odeph et al. (2011)

vector showed strong GFP silencing (>80 % reduction) in a Dicer-dependent manner (Nguyen et al. 2008). Various *M. oryzae* endogenous genes such as polyketide synthase (PKS), calcium signaling proteins, and cell wall degrading enzymes (CWDEs) have been successfully targeted for RNAi using pSD1 (Nguyen et al. 2008, 2011; Vu et al. 2012). Similarly, in the basidiomycete fungus *C. neoformans*, the RNAi vector pIP48 with two opposing Gal7 promoters was used to efficiently suppress the expression of endogenous genes such as *URA5*, *ADE2*, *LAC1*, and *CAP59* (Liu et al. 2002; Bose and Doering 2011).

8.2.4 Direct dsRNA or siRNA Delivery into Fungal Cells

Active uptake of nucleic acids was first reported in *C. albicans*. The levels of dsDNA and dsRNA accumulation were higher in the intracellular space than in the extracellular space, indicating that *C. albicans* has machinery for active uptake of double-stranded nucleic acids (Disney et al. 2003). To investigate whether this RNA uptake mechanism can be exploited to induce gene silencing in fungi, direct delivery of synthetic dsRNA or siRNA into fungal protoplasts or germinating spores has been examined in many studies. In *A. fumigatus* and *A. nidulans*, introduction of 21 and 23 nt-siRNAs into germinating spores induced sequence-specific gene silencing at least 72 h after treatment (Jochl et al. 2009; Khatri and Rajam 2007; Barnes et al. 2008). In the fungus-like oomycete *Phytophthora infestans*, Lipofectin-mediated transfection of protoplasts with *in vitro* synthesized Cy3-labeled long dsRNA (150–300 bp in size) caused sequence-specific gene silencing of GFP 4 days after transfection (Whisson et al. 2005). The GFP silencing then diminished but remained at a significant level (7–67 % of the controls) until 17 days after transfection. In contrast, a significant reduction (11–44 % of the controls) in mRNA accumulation of the endogenous *inf1* gene was detected only 15 days after transfection in *P. infestans*. Generally, the greatest level of silencing was observed from 12 to 15 days after transfection in *P. infestans* (Whisson et al. 2005). Similarly, in the basidiomycete *Moniliophthora perniciosa*, introduction of synthesized long dsRNA into protoplasts by $CaCl_2$/PEG-mediated transformation or electroporation induced a sequence-specific reduction in gene expression (Caribe dos Santos et al. 2009). The transcript levels of the endogenous *Prx1* and the hydrophobin gene family were reduced by 23–87 % and 90 %, respectively, 28 days after treatment (Caribe dos Santos et al. 2009). The drawbacks of this strategy are the late occurrence and diverse levels of gene silencing. Because the original trigger molecule may not be stable in the longer term nor equally incorporated into every cell, some signal amplification is required to fully activate RNAi induced by exogenous dsRNA or siRNA in fungal cells. This process of signal amplification may result in the late occurrence of RNAi in this strategy. Even though direct dsRNA or siRNA delivery into the cell has so far been reported in limited fungal species, this method can offer a convenient RNAi tool in fungi as is the case in animals.

8.2.5 Simultaneous Silencing of Multiple Genes

Simultaneous silencing of multiple genes can be an efficient method for analyzing the biological functions of a gene family or functionally related genes. Such genes are often functionally redundant, and thus difficult to analyze by gene knockout methods. In this strategy, either a DNA fragment from a conserved region or a chimeric DNA fragment of multiple genes is used as an RNAi trigger. Sometimes, a DNA fragment of a marker gene such as GFP or dsRed is also included in a simultaneous silencing vector as an indicator of gene silencing (Liu et al. 2002; Mouyna et al. 2004; Fitzgerald et al. 2004; Nguyen et al. 2008; Lacroix and Spanu 2009). Simultaneous silencing in fungi was first described in the human pathogenic fungus *C. neoformans* with the two genes *ADE2* and *CAP59*. In the results, more than 80 % of the *ADE2*-silenced transformants also exhibited a *CAP59*-silenced phenotype (Liu et al. 2002). Recently, simultaneous silencing of multiple genes was shown to work efficiently in the ascomycetes *A. fumigatus*, *V. inaequalis*, *C. fulvum*, and *A. oryzae* (Mouyna et al. 2004; Fitzgerald et al. 2004; Lacroix and Spanu 2009; Yamada et al. 2007). In *M. oryzae*, a novel strategy named the "building blocks method" was proposed for simultaneous RNAi in fungi. In this method, an artificial sequence composed of short nucleotide units, each of which has a sequence 40–60 bp long derived from a single target gene, is used as a silencing trigger to achieve the maximum average efficiency of simultaneous gene silencing. As described earlier in more detail, this method was successfully employed to reveal the biological roles of two CWDE families, cellulases and xylanases (Nguyen et al. 2008, 2011; Vu et al. 2012).

8.3 RNA Silencing (RNAi) as a Tool for Functional Genomics in Fungi

8.3.1 RNAi in Fungal Research

RNAi has been successfully used in various fields of fungal research such as biomaterials production, bioconversion, and plant–fungal interactions (reviewed by Salame et al. 2011). The koji mold, *Aspergillus oryzae*, is widely used in fermentative industries for the production of sake, miso, and soy sauce. In *A. oryzae*, RNAi-related techniques have been employed to dissect the biological roles of several genes such as the *xylA* gene encoding a secreted beta-xylosidase responsible for the browning of soy sauce, carboxypeptidase genes encoding proteases, and the *brlA* gene encoding a transcription factor involved in conidium formation (Zheng et al. 1998; Kitamoto et al. 1999; Yamada et al. 2007).

RNAi has been more frequently used in molecular studies on plant pathogenic fungi. The process of fungal infection of plants comprises several key steps such as conidiation, appressoria formation, penetration of the host cuticle, and development of invasive hyphae in plant cells. In the rice blast fungus, several genes involved in the infection process were analyzed by RNAi. The knockdown of *Mpg1* encoding a hydrophobin caused a decrease in the hydrophobicity of mycelia and pathogenicity, suggesting it has a role in spore attachment to plant leaf surfaces that are highly hydrophobic (Nakayashiki et al. 2005). Genes related to Ca^{2+} signaling were systemically examined by RNAi in *M. oryzae* (Nguyen et al. 2008) because the involvement of Ca^{2+} signaling in infection-related morphogenesis was suggested in various phytopathogenic fungi. The results indicated that at least 26, 35, and 15 of the 37 genes examined were involved in hyphal growth, sporulation, and pathogenicity, respectively. These included several novel findings such as

that *Pmc1*-, *Spf1*-, and *Neo1*-like Ca^{2+} pumps, calreticulin, and the calpactin heavy chain were essential for fungal pathogenicity (Nguyen et al. 2008; Nakayashiki and Nguyen 2008). Fungal CWDEs have also been subjected to analysis by RNAi because such enzymes often belong to gene families that are difficult to analyze by gene knockout approaches. In fact, several attempts with targeted gene disruption methods failed to reveal their biological functions, likely because of functional redundancy. An RNAi approach named the "building blocks method" was successfully used to demonstrate that CWDEs such as xylanases and cellulases play significant roles in both vertical penetration and horizontal expansion of *M. oryzae* in infected plants (Nguyen et al. 2011; Vu et al. 2012). Cytological analysis revealed that xylanase RNAi mutants showed a severe defect in cuticle penetration while the ability of cellulase RNAi mutants to penetrate the plant cell wall was likely impaired (Nguyen et al. 2011; Vu et al. 2012).

In the fungus-like oomycete *P. infestans*, which is the causal agent of late blight on potato, knockdown of *Cdc14*, a gene whose ortholog in *S. cerevisiae* is essential and involved in mitosis and the cell cycle, resulted in a defect in sporulation but not hyphal growth (Ah Fong and Judelson 2003). Similarly, the involvement of the G protein α-subunit, *Pigpa1*, and β-subunit, *Pigpb1*, in zoospore formation was supported by RNAi experiments. Zoospore production was reduced by 20–45 % in *Pigpa1*-silenced transformants in comparison with the wild-type strain (Latijnhouwers et al. 2004). The *Pigpa1*-silenced transformants also showed a severe reduction in pathogenicity on the host plants. In addition, the *Pigpb1*-silenced mutants produced many abnormal zoospores, suggesting it has a role in fungal development (Latijnhouwers and Gover 2003).

RNAi has also been used to characterize fungal genes involved in pathogenicity. Genes involved in toxin production are often present in multiple copies in the genome. RNAi was successfully employed to reveal genes involved in the production of host-selective toxins and mycotoxins in *Alternaria alternata* and *Fusarium graminearum* (Ajiro et al. 2010; Miyamoto et al. 2008;

McDonald et al. 2005). In *F. oxysporum* f. sp. *lycopersici*, the role of tomatinase, which detoxifies α-tomatine (a preformed inhibitor of fungal growth in tomato plants), in fungal pathogenicity was demonstrated by an RNAi-related technique (Ito et al. 2002).

In the human pathogenic fungi, *H. capsulatum* and *C. neoformans*, several virulence factors such as *AGS1*, *CAP59*, *ADE2*, and *SEC6* were characterized by RNAi (Rappleye et al. 2004; Liu et al. 2002; Panepinto et al. 2009). These genes encode proteins involved in various aspects of fungal virulence; α (1,3)-glucan (a cell wall component of the fungal pathogen) production, the synthesis of the polysaccharide capsule, phosphoribosyl-aminoimidazole carboxylase, and polarized fusion of exocytic vesicles, respectively.

Although only a few RNAi studies have been reported in homobasidiomycete fungi to date (Schuurs et al. 1997; Namekawa et al. 2005; Walti et al. 2006), RNAi can also serve as a powerful genetic tool in mushrooms. In the model mushroom *Coprinopsis cinerea*, efficient silencing of the GFP gene was observed by introducing a hairpin GFP RNA-expression construct. Interestingly, simultaneous silencing of the CGL gene family was achieved by solely using the CGL2 gene as a trigger (Walti et al. 2006). This strategy was also used to suppress successfully the expression of meiotic recombination related genes, LIM15/DMC1 in C. *cinerea* (Namekawa et al. 2005).

8.3.2 Host-Induced Gene Silencing in Fungi

RNA silencing is a natural defense mechanism against invading viruses in plants (Csorba et al. 2009; Harvey et al. 2011; Hu et al. 2011). Recently, a novel application for RNA silencing named host-induced gene silencing (HIGS) has been developed, where plants are genetically modified to express small RNAs specific to a pathogen gene required for infection structure formation, growth, or pathogenicity on plants. Recent studies have indicated that HIGS can be an efficient tool to protect plants from infection

with root-knot nematodes (Huang et al. 2006; Fairbairn et al. 2007; Dubreuil et al. 2009), bacterial pathogens (Escobar and Dandekar 2003; Escobar et al. 2001, 2002; Viss et al. 2003), and insects (Baum et al. 2007; Mao et al. 2007; Turner et al. 2006). Several attempts at applying HIGS to phytopathogenic fungi are also being made. HIGS was first reported in the pathogenic fungus *F. verticillioides* with the β-glucuronidase (GUS) reporter gene. GUS gene expression was specifically silenced in *F. verticillioides* interacting with transgenic tobacco carrying a GUS hpRNA expression cassette (Tinoco et al. 2010), suggesting that GUS-siRNAs in the transgenic tobacco moved into fungal cells similarly to the cell-to-cell transmission of a specific gene silencing signal between transgenic lettuce and its parasitic plant *Triphysaria versicolor* (Tomilov et al. 2008). Furthermore, Nowara et al. (2010) showed that the expression of dsRNA of *Avra10*, an effector gene in the powdery mildew fungus *Blumeria graminis*, in susceptible barley (*Hordeum vulgaris*) and wheat (*Triticum aestivum*) cultivars reduced the severity of *B. graminis* infection. This strategy was also successfully employed to prevent infection with *Puccinia striiformis* f. sp. *tritici* and *F. oxysporum* f. sp. *lycopersici* (Yin et al. 2010; Singh et al. 2010). These results suggest that small RNA molecules can move from plant cells to fungal cells during infection, and effectively silence target genes in the infecting fungi. Therefore, HIGS will offer a potent strategy to control plant pathogens in the near future.

8.4 Conclusion

RNA silencing (RNAi) has become a fundamental genetic tool in molecular biology. Because the efficiency of gene targeting by homologous recombination is relatively high in most, albeit not all, fungal species, gene knockout methods have been commonly used in fungal genetics research so far. However, as described in this chapter, RNAi can be a useful tool for genetic research even when gene knockout methods cannot be used. In particular, RNAi will facilitate our understanding of the biological roles of essential genes and gene families. In addition, as exemplified by the HIGS and building blocks methods, novel approaches based on the sequence-specific nature of RNAi can also be developed in the future and used for basic and applied research in fungi.

Acknowledgments This work was supported in part by CRP-ICGEB Research Grant (CRP/VIET13-02) from International Centre for Genetic Engineering and Biotechnology, Trieste, Italy.

References

Ah Fong AMV, Judelson HS (2003) Cell cycle regulator Cdc14 is expressed during sporulation but not hyphal growth in the fungus-like oomycete *Phytophthora infestans*. Mol Microbiol 50(2):487–494

Ajiro N, Miyamoto Y, Masunaka A, Tsuge T, Yamamoto M, Ohtani K, Fukumoto T, Gomi K, Peever TL, Izumi Y, Tada Y, Akimitsu K (2010) Role of the host-selective ACT-toxin synthesis gene ACTTS2 encoding an enoyl-reductase in pathogenicity of the tangerine pathotype of *Alternaria alternata*. Phytopathology 100:120–126

Alexander WG, Raju NB, Xiao H, Hammond TM, Perdue TD, Metzenberg RL, Pukkila PJ, Shiu PK (2008) DCL-1 localizes with other components of the MSUD machinery and is required for silencing. Fungal Genet Biol 45:719–727

Ali SS, Nugent B, Mullins E, Doohan FM (2013) Insights from the fungus *Fusarium oxysporum* point to high affinity glucose transporters as targets for enhancing ethanol production from lignocellulose. PLoS One 8(1):e54701

Aramayo R, Metzenberg RL (1996) Meiotic transvection in fungi. Cell 86:103–113

Aravind L, Watanabe H, Lipman DJ, Koonin EV (2000) Lineage-specific loss and divergence of functionally linked genes in eukaryotes. Proc Natl Acad Sci 97: 11319–11324

Barnes SE, Alcocer MJC, Archer DB (2008) siRNA as a molecular tool for used in *Aspergillus niger*. Biotechnol Lett 30:885–890

Barton LM, Prade RA (2008) Inducible RNA interference of brlAβ in *Aspergillus nidulans*. Eukaryot Cell 7:2004–2007

Baum JA, Bogaert T, Clinton W, Heck GR, Feldmann P, Llagan O, Johnson S, Plaetinck G, Munyikwa T, Pleau M, Vaughn T, Robert J (2007) Control of coleopteran insect pests through RNA interference. Nat Biotechnol 25:1322–1326

Bernstein E, Caudy AA, Hammond SM, Hannon GJ (2001) Role for a bidentate ribonuclease in the initiation step of RNA interference. Nature 409:363–366

Bose I, Doering TL (2011) Efficient implementation of RNA interference in the pathogenic yeast *Cryptococcus neoformans*. J Microbiol Methods 86(2):156–159

Brotman Y, Briff E, Viterbo A, Chet I (2008) Role of swollenin, an expansion-like protein from *Trichoderma*, in plant root colonization. Plant Physiol 147:779–6789

Cam HP, Sugiyama T, Chen ES, Chen X, Fitzgerald PC, Grewal SIS (2005) Comprehensive analysis of heterochromatin- and RNAi-mediated epigenetic control of the fission yeast genome. Nat Genet 37(8):809–819

Cardoza RE, Vizcaino JA, Hermosa MR, Sousa S, Gonzalez FJ, Llobell A, Monte E, Gutierrez S (2006) Cloning and characterization of the *erg1* gene of *Trichoderma harzianum*: effect of the *erg1* silencing on ergosterol biosynthesis and resistance to terbinafine. Fungal Genet Biol 43:164–178

Caribe dos Santos AC, Sena JAL, Santos SC, Dias CV, Pirovani CP, Pungartnik C, Valle RR, Cascardo JCM, Vincentz M (2009) dsRNA-induced gene silencing in *Moniliophthora perniciosa*, the causal agent of witches' broom disease of cacao. Fungal Genet Biol 46(11):825–836

Catalanotto C, Pallotta M, ReFalo P, Sachs MS, Vayssie L, Macino G, Cogoni C (2004) Redundancy of the two dicer genes in transgene induced posttranscriptional gene silencing in *Neurospora crassa*. Mol Cell Biol 24:2536–2545

Cecere G, Cogoni C (2009) Quelling targets the rDNA locus and functions in rDNA copy number control. BMC Microbiol 9(1):44–54

Chang SS, Zhang Z, Liu Y (2012) RNA interference pathways in fungi: mechanisms and functions. Annu Rev Microbiol 66:305–323

Cogoni C (2001) Homology-dependent gene silencing mechanism in fungi. Annu Rev Microbiol 55: 381–406

Cogoni C, Macino G (1997) Isolation of quelling-defective (qde) mutants impaired in posttranscriptional transgene-induced gene silencing in *Neurospora crassa*. Proc Natl Acad Sci 94:10233–10238

Cogoni C, Macino G (1999a) Gene silencing in *Neurospora crassa* requires a proteins homologous to RNA-dependent RNA polymerase. Nature 399:166–169

Cogoni C, Macino G (1999b) Posttranscriptional gene silencing in *Neurospora* by a RecQ DNA helicase. Science 286:2342–2344

Cogoni C, Irelan JT, Schumacher M, Schmidhauser TJ, Selker EU, Macino G (1996) Transgene silencing of the al-1 gene in vegetative cells of *Neurospora* is mediated by a cytoplasmic effectors and does not depend on DNA-DNA interactions or DNA methylation. EMBO J 15:3153–3163

Costa ASMB, Thomas DJI, Eastwood D, Cutler SB, Bailey AM, Foster GD, Mills PR, Challen MP (2009) Quantifiable downregulation of endogenous genes in *Agaricus bisporus* mediated by expression of RNA hairpin. J Microbiol Biotechnol 19:271–276

Csorba T, Pantaleo V, Burgyan J (2009) RNA silencing: an antiviral mechanism. In: Loebenstein G, Carr JP (eds) Advances in virus research, vol 75. Academic, New York, pp 35–71

Dang Y, Yang Q, Xue Z, Liu Y (2011) RNA interference in fungi: pathways, functions, and applications. Eukaryot Cell 10:1148–1155 de Jong IF, Deelstra HL, Wosten HA, Lugones LG (2006) RNA mediated gene silencing in monokarryons and dikaryons of *Schizophyllum commune*. Appl Environ Microbiol 72:1267–1269

Disney MD, Haidaris CG, Turner DH (2003) Uptake and antifungal activity of oligonucleotides in *Candida albicans*. Proc Natl Acad Sci 100(4):1530–1534

Djupedal I, Kos-Braun IC, Mosher RA, Soderholm N, Simmer F, Hardcastle TJ, Fender A, Heidrich N, Kagansky A, Bayne E, Wagner EGH, Baulcombe DC, Allshire RC, Ekwall K (2009) Analysis of small RNA in fission yeast: centromeric siRNAs are potentially generated through a structured RNA. EMBO J 28(24): 3832–3844

Drinnenberg IA, Weinberg DE, Xie KT, Mower JP, Wolfe KH, Fink GR, Bartel DP (2009) RNAi in budding yeast. Science 326(5952):544–550

Dubreuil G, Magliano M, Dubrana MP, Lozano J, Lecomte P, Favery B, Abad P, Rosso MN (2009) Tobacco rattle virus mediates gene silencing in a plant parasitic root-knot nematodes. J Exp Bot 60: 4041–4050

Elbashir SM, Lendeckel W, Tuschl T (2001) RNA interference is mediated by 21 and 22 nucleotide RNAs. Genes Dev 15:188–200

Engh I, Nowrousian M, Kuck U (2007) Regulation of melanin biosynthesis via the dihydroxynaphthalene pathway is dependent on sexual development in the ascomycete *Sordaria macrospora*. FEMS Microbiol Lett 275:62–70

Erental A, Harel A, Yarden O (2007) Type 2A phosphoprotein phosphatase is required for asexual development and pathogenesis of *Sclerotinia sclerotiorum*. Mol Plant Microbe Interact 20:944–954

Escobar MA, Dandekar AM (2003) *Agrobacterium tumefaciens* as an agent of disease. Trends Plant Sci 8: 380–386

Escobar MA, Civerolo EL, Summerfelt KR, Dandekar AM (2001) RNAi mediated oncogene silencing confers resistance to crown gall tumorigenesis. Proc Natl Acad Sci 98:13437–13442

Escobar MA, Leslie CA, McGranahan GH, Dandekar AM (2002) Silencing crown gall disease in walnut (*Juglans regina* L.). Plant Sci 163:591–597

Fagard M, Boutet S, Morel JB, Bellini C, Vaucheret H (2000) AGO1, QDE-2 and RDE1 are related proteins required for post-transcriptional gene silencing in plants, quelling in fungi and RNA interference in animals. Proc Natl Acad Sci 97:11650–11654

Fairbairn DL, Cavallaro AS, Bernard M, Mahalinga-Iyer J, Graham MW, Botella JR (2007) Host-delivered RNAi: an effective strategy to silence genes in plant parasitic nematodes. Planta 226:1525–1533

Fire A, Xu S, Montgomery MK, Kostas SA, Driver SE, Mello CC (1998) Potent and specific genetic interference by double-stranded RNA in *Caenorhabditis elegans*. Nature 391:806–811

Fitzgerald A, Kan JA, Plummer KM (2004) Simultaneous silencing of multiple genes in the apple scab fungus, *Venturia inaequalis*, by expression of RNA with chimeric inverted repeats. Fungal Genet Biol 41:963–971

Flores-Alvarez LJ, Corrales-Escobosa AR, Cortes-Penagos C, Martinez-Pacheco M, Wrobel-Zasada K, Wrobel-Kaczmarczyk K, Cervantes C, Gutierrez-Corona F (2012) The *Neurospora crassa chr-1* gene is up-regulated by chromate and its encoded CHR-1 protein causes chromate sensitivity and chromium accumulation. Curr Genet 58(5–6):281–290

Freitag M, Lee DW, Kothe GO, Pratt RJ, Aramayo R, Selker EU (2004) DNA methylation is independent of RNA interference in *Neurospora*. Science 304:1939

Goldoni M, Azzalin G, Macino C, Cogoni G (2004) Efficient gene silencing by expression of double stranded RNA in *Neurospora crassa*. Fungal Genet Biol 41:1016–1024

Hamada W, Spanu PD (1998) Co-suppression of the hydrophobin gene *Hcf-1* is correlated with antisense RNA biosynthesis in *Cladosporium fulvum*. Mol Gen Genet 259:630–638

Hammond TM, Keller NP (2005) RNA silencing in *Aspergillus nidulans* is independent of RNA dependent RNA polymerase. Genetics 169:607–617

Hammond SM, Boettcher S, Caudy AA, Kobayashi R, Hannon GJ (2001) Argaunaute 2, a link between genetic and biochemical analyses of RNA. Science 293:1146–1150

Hammond TM, Andrewski MD, Roossinck MJ, Keller NP (2008) *Aspergillus* mycoviruses are targets and suppressors of RNA silencing. Eukaryot Cell 7: 350–357

Hammond TM, Xiao H, Boone EC, Perdue TD, Pukkila PJ, Shiu PK (2011) SAD-3, a putative helicase required for meiotic silencing by unpaired DNA, interacts with other components of the silencing machinery. G3 (Bethesda) 1:369–376

Harvey JJ, Lewsey MG, Patel K, Westwood J, Heimstadt S, Carr JP, Baulcombe DC (2011) An antiviral defense role of AGO2 in plants. PLoS One 6(1):e14639

Himeno M, Maejima K, Komatsu K, Ozeki J, Hashimoto M, Kagiwada S, Yamaji Y, Namba S (2010) Significantly low level of small RNA accumulation derived from an encapsidated mycovirus with dsRNA genome. Virology 396(1):69–75

Hu Q, Niu Y, Zhang K, Liu Y, Zhou X (2011) Virus-derived transgenes expressing hairpin RNA give immunity to *Tobacco mosaic* virus and *Cucumber mosaic* virus. Virol J 8:1–11

Huang G, Allen R, Davis EL, Baum TJ, Hussey RS (2006) Engineering broad root-knot resistance in transgenic plants by RNA silencing of a conserved and essential root-knot nematode parasitism gene. Proc Natl Acad Sci 103:14302–14306

Ito S, Takahara H, Kawagushi T, Tanaka S, Iwaki-Kameya M (2002) Post-transcriptional silencing of the tomatinase gene in *Fusarium oxysporum* f.sp *lycopersici*. J Phytopathol 150(8–9):474–480

Janus D, Hoff B, Hofmann E, Kuck U (2007) An efficient fungal RNA-silencing system using the DsRed reporter gene. Appl Environ Microbiol 73:962–970

Jeang KT (2012) RNAi in the regulation of mammalian viral infections. BMC Biol 10:58

Jochl C, Loh E, Ploner A, Hass H, Huttenhofer A (2009) Development dependent scavenging of nucleic acids in the filamentous fungus, *Aspergillius fumigatus*. RNA Biol 6(2):178–186

Johri AK (2010) Development of electroporation-mediated transformation system for axenically cultivable root endophyte fungus *Piriformospora indica*. Protocol Exchange. doi:10.1038/nprot.2010.57

Kadotani N, Nakayashiki H, Tosa Y, Mayama S (2003) RNA silencing in the phytopathogenic fungus *Magnaporthe oryzae*. Mol Plant Microbe Interact 16:769–775

Kato H, Goto DB, Martienssen RA, Urano T, Furukawa K, Murakami Y (2005) RNA polymerase II is required for RNAi-dependent heterochromatin assembly. Science 309:467–469

Kemppainen MJ, Duplessis S, Martin F, Pardo AG (2009) RNA silencing in the model mycorrhizal fungus *Laccaria bicolor*: gene knock-down of nitrate reductase results in inhibition of symbiosis with Populus. Environ Microbiol 11:1878–1896

Ketting RF, Plasterk RH (2000) A genetic link between co-suppression and RNA interference in *C. elegans*. Nature 404:296–298

Khatri M, Rajam MV (2007) Targeting polyamines of *Aspergillus nidulans* by siRNA specific to fungal ornithine decarboxylase gene. Med Mycol 45(3):211–220

Kim VN, Han J, Siomi MC (2009) Biogenesis of small RNAs in animals. Nat Rev Mol Cell Biol 10:126–139

Kim H, Chen C, Kabbage M, Dickman MB (2011) Identification and characterization of *Sclerotinia sclerotiorum* NADPH oxidases. Environ Microbiol 77(21):7721–7729

Kitamoto N, Yoshino S, Ohmiya K, Tsukagoshi N (1999) Sequence analysis, overexpression, and antisense inhibition of a beta-xylosidase gene, *xylA*, from *Aspergillus oryzae* KBN616. Appl Environ Microbiol 65:20–24

Krajaejun T, Gauthie GM, Rappleye CA, Sullivan TD, Klei BS (2007) Development and application of a green fluorescent protein sentinel system for identification of RNA interference in *Blastomyces dermatitidis* illuminates the role of septin in morphogenesis and sporulation. Eukaryot Cell 6:1299–1309

Lacroix H, Spanu PD (2009) Silencing of six hydrophobins in *Cladosporium fulvum*: complexities of simultaneously targeting multiple genes. Appl Environ Microbiol 75(2):542–546

Latijnhouwers M, Gover F (2003) A *Phytophthora infestans* G-protein subunit is involved in sporangium formation. Eukaryot Cell 2(5):971–977

Latijnhouwers M, Ligterink W, Vleeshouwers VGAA, van West P, Gover F (2004) A G-alpha subunit controls zoospore motility and virulence in the potato late blight pathogen *Phytopthora infestans*. Mol Microbiol 51:925–936

Lee DW, Pratt RJ, McLaughlin M, Aramayo R (2003) An argonaute-like protein is required for meiotic silencing. Genetics 164:821–828

Lee HC, Chang SS, Choudhary S, Aalto AP, Maiti M, Bamford DH, Liu Y (2009) qiRNA is a new type of small interfering RNA induced by DNA damage. Nature 459:274–277

Lee HC, Aalto AP, Yang Q, Chang S, Huang G, Fisher D, Cha J, Poranen MM, Bamford DH, Liu Y (2010a) The DNA/RNA-dependent RNA polymerase QDE-1 generates aberrant RNA and dsRNA for RNAi in a process requiring replication protein A and a DNA helicase. PLoS Biol 8(10):e1000496. doi:10.1371/journal.pbio.1000496

Lee HC, Li L, Gu W, Xue Z, Crosthwaite SK, Pertsemlidis A, Lewis ZA, Freitag M, Selker EU, Mello CC, Liu Y (2010b) Diverse pathways generate microRNA-like RNAs and Dicer-independent small interfering RNAs in fungi. Mol Cell 38:803–814

Liu H, Cottrell TR, Pierini LM, Goldman WE, Doering TL (2002) RNA interference in the pathogenic fungus *Cryptococcus neoformans*. Genetics 160:463–470

Liu H, Zhang B, Li C, Bao X (2010) Knock down of chitosanase expression in phytopathogenic fungus *Fusarium solani* and its effect on pathogenicity. Curr Genet 56:275–281

Maiti M, Lee HC, Liu Y (2007) QIP, a putative exonuclease, interacts with the *Neurospora* Argonaute protein and facilitates conversion of duplex siRNA into single strands. Genes Dev 21:590–600

Mao YB, Cai WJ, Wang JW, Hong GJ, Tao XY, Wang LJ, Huang YP, Chen XY (2007) Silencing a cotton bollworm P450 monooxygenase gene by plant-mediated RNAi impairs larval tolerance of gossypol. Nat Biotechnol 25:1307–1313

Martens H, Novotny J, Oberstrass J, Steck TL, Postlethwait P, Nellen W (2002) RNAi in *Dictyostelium*: the role of RNA-directed RNA polymerases and double-stranded RNase. Mol Biol Cell 13:445–453

Matiyahu A, Hadar Y, Dosoretz CG, Belinky PA (2008) Gene silencing by RNA interference in the white rot fungus *Phanerochaete chrysosporium*. Appl Environ Microbiol 74:5359–5365

McDonald T, Brown D, Keller NP, Hammond TM (2005) RNA silencing of mycotoxin production in *Aspergillus* and *Fusarium* species. Mol Plant Microbe Interact 18(6):539–545

Miyamoto Y, Masunaka A, Tsuge T, Yamamoto M, Ohtani K, Fukumoto T, Gomi K, Peever TL, Akimitsu K (2008) Functional analysis of a multicopy host-selective ACT-toxinbiosynthesis gene in the tangerine pathotype of *Anternaria alternata* using RNA silencing. Mol Plant Microbe Interact 21:1591–1599

Moazeni M, Khoramizadeh MR, Kordbacheh P, Sepehrizadeh Z, Zeraati H, Noorbakhsh F, Teimoori-Toolabi L, Rezaie S (2012) RNA-mediated gene silencing in Candida albicans: inhibition of hyphae formation by use of RNAi technology. Mycopathologia 174(3):177–185

Moriwaki A, Ueno M, Arase S, Kihara J (2007) RNA mediated gene silencing in the phytopathogenic fungus *Bipolaris oryzae*. FEMS Microbiol Lett 269: 85–89

Moriwaki A, Katsube H, Ueno M, Arase S, Kihara J (2008) Cloning and characterization of the *BLR2*, the homologue of the blue-light regulator of *Neurospora crassa* WC-2, in the phytopathogenic fungus *Bipolaris oryzae*. Curr Microbiol 56:115–121

Mouyna I, Henry C, Doering TL, Latge JP (2004) Gene silencing with RNA interference in the human pathogenic fungus *Aspergillus fumigatus*. FEMS Microbiol Lett 237:317–324

Murata T, Kadotani N, Yamaguchi M, Tosa Y, Mayama S, Nakayashiki H (2007) siRNA-dependent and independent post-transcriptional cosuppression of the LTR-retrotransposon MAGGY in the phytopathogenic fungus *Magnaporthe oryzae*. Nucleic Acids Res 35(18):5987–5994

Nakayashiki H (2005) RNA silencing in fungi: mechanisms and applications. FEBS Lett 579:5950–5957

Nakayashiki H, Hanada S, Quoc NB, Kadotano N, Tosa Y, Mayama S (2005) RNA silencing as a tool for exploring gene function in Ascomycete fungi. Fungal Genet Biol 42:257–283

Nakayashiki H, Kadotani N, Mayama S (2006) Evolution and diversification of RNA silencing proteins in fungi. J Mol Evol 63(1):127–135

Nakayashiki H, Nguyen QB (2008) RNA interference: roles in fungal biology. Curr Opin Microbiol 11:494–502

Namekawa SH, Iwabata K, Sugawara H, Hamada FN, Koshiyama A, Chiku H, Kamada T, Sakaguchi K (2005) Knock-down of LIM15/DMC1 in the mushroom Coprinus cineteus by doublestranded RNA-mediated gene silencing. Microbiology 151:3669–3678

Napoli C, Lemieux C, Jorgensen R (1990) Introduction of a chimeric chalcone synthase gene into petunia results in reversible co-suppression of homologous genes in trans. Plant Cell 2:279–289

Ngiam C, Jeenes DJ, Punt PJ, Van Den Hondel CA, Archer DB (2000) Characterization of a foldase, protein disulfide isomerase A, in the protein secretory pathway of *Aspergillus niger*. Appl Environ Microbiol 66:775–782

Nguyen QB, Kadotani N, Kasahara S, Tosa Y, Mayama S, Nakayashiki H (2008) Systemic functional analysis of calcium signaling proteins in the genome of the rice blast fungus, *Magnaporthe oryzae*, using a high-throughput RNA silencing system. Mol Microbiol 68:1348–1365

Nguyen QB, Itoh K, Vu BV, Tosa Y, Nakayashiki H (2011) Simultaneous silencing of endo-β-1,4 xylanase genes reveals their roles in the virulence of *Magnaporthe oryzae*. Mol Microbiol 81(4):1008–1019

Nicolas FF, Torres-Martinez S, Ruiz-Vazquez RM (2003) Two classes of small antisense RNAs in fungal RNA silencing triggered by non-integrative transgenes. EMBO J 22:3983–3991

Nolan T, Braccini L, Azzalin G, De Toni A, Macino G, Cogoni C (2005) The post-transcriptional gene silencing machinery functions independently of DNA methylation to repress a LINE1-like retrotransposon in *Neurospora crassa*. Nucleic Acids Res 33:1564–1573

Nolan T, Cecere G, Mancone C, Alzoni T, Tripodi M, Catalanotto C, Cogoni C (2008) The RNA dependent RNA polymerase essential for post transcriptional gene silencing in *Neurospora crassa* interacts with replication protein A. Nucleic Acids Res 36:532–538

Nowara D, Gay A, Lacomme C, Shaw J, Ridout C, Douchkov D, Hensel G, Kumlehn J, Schweizer P (2010) HIGS: Host-induced gene silencing in the obligate biotrophic fungal pathogen *Blumeria graminis*. Plant Cell 22:3130–3141

Odeph M, Yang Q, Phiri J (2011) Molecular cloning of Erg2 gene to effect knockdown in Chaetomium cupreum. In: Proceedings of the 6th International Conference, Broadband and Biomedical Communications (IB2Com), Melbourne, VIC, Australia. IEEEXplore

Oliveira JM, van der Veen D, de Graaff LH, Qui L (2008) Efficient cloning system for construction of gene silencing vectors in *Aspergillus niger*. Appl Microbiol Biotechnol 80:917–924

Panepinto J, Komperda K, Frases S, Park YD, Djordjevic JT, Casadevall A, Williamson PR (2009) Sec6-dependent sorting of fungal extracellular exosomes and laccase of *Cryptococcus neoformans*. Mol Microbiol 71:1165–1176

Patel RM, van Kan JA, Bailey AM, Foster GD (2008) RNA mediated gene silencing of superoxide dimutase (*bcsod1*) in *Botrytis cinerea*. Phytopathology 98:1334–1339

Peng JC, Karpen GH (2007) H3K9 methylation and RNA interference regulate nucleolar organization and repeated DNA stability. Nat Cell Biol 9:25–35

Pratt RJ, Lee DW, Aramayo R (2004) DNA methylation affects meiotic trans-sensing, not meiotic silencing in *Neurospora*. Genetics 168:1925–1935

Rappleye CA, Engle JT, Goldman WE (2004) RNA interference in *Histoplasma capsulatum* demonstrates a role for α-(1,3)-glucan in virulence. Mol Microbiol 53:153–165

Romano N, Macino G (1992) Quelling: transient inactivation of gene expression in *Neurospora crassa* by transformation with homologous sequences. Mol Microbiol 6:3343–3353

Saha D, Fetzner R, Burkhardt B, Podlech J, Metzler M, Dang H, Lawrence C, Fischer R (2012) Identification of a polyketide synthase required for alternariol (AOH) and alternariol-9-methyl ether (AME) formation in *Alternaria alternata*. PLoS One 7(7):e40564

Salame TM, Yarden O, Hadar Y (2010) *Pleurotus ostreatus* manganese-dependent peroxidase silencing impairs decolourization of Orange II. J Microbial Biotechnol 3:93–106

Salame TM, Ziv C, Hadar Y, Yarden O (2011) RNAi as a potential tool for biotechnological applications in fungi. Appl Microbiol Biotechnol 89:501–512

Schuurs TA, Schaeffer EAM, Wessels JGH (1997) Homology-dependent silencing of the SC3 gene in *Schizophyllum commune*. Genetics 147:589–596

Seger GC, van Wezel R, Zhang X, Hong Y, Nuss DL (2006) Hypovirus papain-like protease p29 suppresses RNA silencing in the natural fungal host and in a heterologous plant system. Eukaryot Cell 6:896–904

Seger GC, Zhang X, Deng F, Sun Q, Nuss DL (2007) Evidence that RNA silencing functions as an antiviral defense mechanism in fungi. Proc Natl Acad Sci 104:12902–12906

Shafran H, Miyara I, Eshed R, Prusky D, Sherman A (2008) Development of new tools for studying gene function in fungi based on the Gateway system. Fungal Genet Biol 45:1147–1154

Shiu PK, Metzenberg RL (2002) Meiotic silencing by unpaired DNA: properties, regulation and suppression. Genetics 161:1483–1495

Shiu PK, Raju NB, Zichler D, Matzenberg RL (2001) Meiotic silencing by unpaired DNA. Cell 107: 905–916

Singh S, Braus-Stromeyer SA, Timpner C, Tran VT, Lohaus G, Reusche M, Knufer J, Teichmann T, von Tiedemann A, Braus GH (2010) Silencing of Vlaro2 for chorismate synthase revealed that the phytopathogen Verticillium longisporum induces the cross-pathway control in the xylem. Appl Microbiol Biotechnol 85:1961–1976

Song JJ, Smith SK, Hannon GJ, Joshua-Tor L (2004) Crystal structure of Argonaute and its implications for RISC slicer activity. Science 305(5689):1434–1437

Spiering MJ, Moon CD, Wilkinson HH, Schardl CL (2005) Gene clusters for insecticidal loline alkaloids in the grass-endophytic fungus Neotyphodlum uncinatum. Genetics 169:1403–1414

Staab JF, White TC, Marr KA (2011) Hairpin dsRNA does not trigger RNA interference in Candida albicans cells. Yeast 28:1–8

Suk K, Choi J, Suzuki Y, Ozturk SB, Mellor JC, Wong KH, Mackay JL, Gregory RI, Roth FP (2011) Reconstitution of human RNA interference in budding yeast. Nucleic Acids Res 39(7):e43

Sun Q, Choi GH, Nuss DL (2009) A single Argonaute gene is required for induction of RNA silencing antiviral defense and promotes viral RNA recombination. Proc Natl Acad Sci 42:17927–17932

Takeno S, Sakuradani F, Murata S, Inohara-Ochiai M, Kawashima H, Ashikari T, Shimizu S (2004) Establishment of an overall transformation system for an oil-producing filamentous fungus, Mortierrlla alpina 1S-4. Appl Microbiol Biotechnol 65:419–425

Tinoco MLP, Dias BBA, Dall'Astta RC, Pamphile JA, Aragao FJL (2010) In vivo tran-specific gene silencing in fungal cells by in planta expression of a double-stranded RNA. BMC Biol 8:27. doi:10.1186/1741-7007-8-27

Tomilov AA, Tomilova NB, Wroblewski T, Michelmore R, Yoder JL (2008) Trans-specific gene silencing between host and parasitic plants. Plant J 56:389–397

Turner CT, Davy MW, MacDiarmid RM, Plummer KM, Birch NP, Newcomb RD (2006) RNA interference in the light brown apple moth, Epiphyas postvittana (Walker) induced by double-stranded RNA feeding. Insect Mol Biol 15:383–391

Ullan RV, Godio RP, Teijeira F, Vaca I, Garcia-Estrada C, Feltrer R, Kosalkova K, Martin JF (2008) RNA silencing in Penicillium chrysogenum and Acremonium chrysogenum: validation studies using beta-lactam genes expression. J Microbiol Methods 75:209–218

van der Krol AR, Mur LA, Beld M, Mol JN, Stuitje AR (1990) Flavonoid genes in Petunia: addition of a limited number of gene copies may lead to a suppression of gene expression. Plant Cell 2:291–299

van West P, Kamoun S, van't Klooster JW, Govers F (1999) Internuclear gene silencing in Phytophthora infestans. Mol Cell 3:339–348

Vermout S, Tarbart J, Baldo A, Monod M, Losson B, Mignon B (2007) RNA silencing in the dermatophyte Microsporum canis. FEMS Microbiol Lett 275:38–45

Viss WJ, Pitrak J, Humann J, Cook M, Driver J, Ream W (2003) Crown-gall-resistant transgenic apple trees that silence Agrobacterium tumefaciens oncogenes. Mol Breed 12:283–295

Viterbo A, Landau U, Kim S, Chernin L, Chet I (2010) Characterization of ACC deaminase from the biocontrol and plant growth-promoting agent Trichoderma asperrellum T203. FEMS Microbiol Lett 305:42–48

Vu BV, Takino M, Murata T, Nakayashiki H (2011) Novel vectors for retrotransposon-induced gene silencing in Magnaporthe oryzae. J Gen Plant Pathol 77:147–151

Vu BV, Itoh K, Nguyen QB, Tosa Y, Nakayashiki H (2012) Cellulases belonging to glycoside hydrolase families 6 and 7 contribute to the virulence of Magnaporthe oryzae. Mol Plant Microbe Interact 25(9):1135–1141

Vu BV, Pham KT, Nakayashiki H (2013) Substrate-induced transcriptional activation of the MoCel7C cellulase gene is associated with methylation of histone H3 at lysine 4 in the rice blast fungus Magnaporthe oryzae. Appl Environ Microbiol 79:6823–6832

Walti MA, Villalba C, Buser RM, Grunler A, Aebi M, Kunzler M (2006) Targeted gene silencing in the model mushroom Coprinopsis cinerea (Coprinus cinereus) by expression of homologous hairpin RNAs. Eukaryot Cell 5(4):732

Wang X, Hsueh YP, Li W, Floyd A, Skalsky R, Heitman J (2010) Sex-induced silencing defends the genome of Cryptococcus neoformans via RNAi. Genes Dev 24:2566–2582

Whisson SC, Avrova AO, Van West P, Jones JT (2005) A method for double-stranded RNA mediated transient gene silencing in Phytophthora infestans. Mol Plant Pathol 6:153–163

Wilkins C, Dishongh R, Moore SC, Whitt MA, Chow M, Machaca K (2005) RNA interference is an antiviral defense mechanism in Caenorhabditis elegans. Nature 436(7053):1044–1047

Xiao H, Alexander WG, Hammond TM, Boone EC, Perdue TD, Pukkila PJ, Shiu PK (2010) QIP, a protein that converts duplex siRNA into single strands, is required for meiotic silencing by unpaired DNA. Genetics 186:119–126

Yamada O, Ikeda R, Ohkita Y, Hayashi R, Sakamoto K, Akita O (2007) Gene silencing by RNA interference in the koji mold Aspergillus oryzae. Biosci Biotechnol Biochem 71:138–144

Yin C, Jurgenson J, Hulbert S (2010) Development of a host-induced RNAi system in the wheat stripe rust fungus

Puccinia striiformis f. sp. *tritici*. Mol Plant Microbe Interact 24:554–561

Zambon RA, Vakharia VN, Wu LP (2006) RNAi is an antiviral immune response against a dsRNA virus in Drosophila melanogaster. Cell Microbiol 8:880–889

Zamore PD, Tuschl T, Sharp PA, Bartel DP (2000) RNAi: double-stranded RNA directs the ATP-dependent cleavage of mRNA at 21 to 23 nucleotide intervals. Cell 101:25–33

Zhang X, Seger GC, Sun Q, Deng F, Nuss DL (2008) Characterization of hypovirus-derived small RNAs generated in the chestnut blight fungus by an inducible DCL-2-dependent pathway. J Virol 6:2613–2619

Zheng XF, Kobayashi Y, Takeuchi M (1998) Construction of low-serine-type-carboxypeptidase-producing mutant of *Aspergillus oryzae* by the expression of antisense RNA and its use as a host for heterologous protein secretion. Appl Microbiol Biotechnol 49:39–44

RNAi-Mediated Gene Silencing in the Beta-Lactam Producer Fungi *Penicillium chrysogenum* and *Acremonium chrysogenum*

9

Carlos García-Estrada and Ricardo V. Ullán

9.1 Introduction

Targeted gene disruption in ascomycete fungi has been the technique of choice for the generation of null mutants (knock-out), a process that is useful for the characterization of gene function. However, in most ascomycete fungi, the ectopic integration of the transforming DNA makes this process difficult and time consuming due to the low frequency of homologous recombination. Clear examples of this experimental problem are the filamentous fungi *Penicillium chrysogenum* and *Acremonium chrysogenum* (Casqueiro et al. 1999; Hoskins et al. 1990; Chaveroche et al. 2000; Ullán et al. 2002a, b; Colot et al. 2006). Even with species like *Aspergillus nidulans* and *Neurospora crassa*, which show a relatively high targeting efficiency (Asch and Kinsey 1990; Miller et al. 1985), the screening of the desired homologous recombination is tedious. An alternative method that overcomes some of the problems associated with the knock-out technique is the silencing of gene expression at post-transcriptional level (gene knock-down) by

targeting the mRNA with small interfering RNA (siRNA) molecules in a mechanism known as RNA interference (RNAi).

9.2 Methods

The silencing mechanism involves the specific degradation of target mRNA molecules that have sequence homology with the inducer RNA. This process requires the formation of double-stranded RNA (dsRNA) molecules that have homologous sequences to the target RNA. The activation of the silencing process takes place when the dsRNA molecules are produced and recognized by the Dicer RNase (member of the RNase III family of nucleases), which degrade them in short molecules of dsRNA of 21–26 nucleotides, referred to as siRNA. The siRNA molecules are then integrated in the RISC (RNA Induced Silencing Complex) multicomponent system, which is able to degrade target mRNA strands that are complementary to the RISC-associated siRNA (reviewed in Dang et al. 2011 and Chang et al. 2012).

Several RNAi systems have been developed for functional genomic analysis in a broad range of ascomycete fungi such as *N. crassa* (Goldoni et al. 2004), *A. nidulans* (Hammond and Keller 2005), *Aspergillus fumigatus* (Mouyna et al. 2004), *Magnaporthe oryzae* (Kadotani et al. 2003), *A. chrysogenum* (Janus et al. 2007; Ullán et al. 2008),

---

C. García-Estrada, D.V.M., Ph.D. (✉)
R.V. Ullán, Ph.D.
INBIOTEC (Instituto de Biotecnología de León),
León 24006, Spain
e-mail: carlos.garcia@inbiotec.com; c.gestrada@unileon.es

M.A. van den Berg and K. Maruthachalam (eds.), *Genetic Transformation Systems in Fungi, Volume 2*, Fungal Biology, DOI 10.1007/978-3-319-10503-1_9,
© Springer International Publishing Switzerland 2015

and *P. chrysogenum* (Ullán et al. 2008). The majority of the systems for gene silencing are based on a promoter that controls the expression of sense and antisense sequences separated by a spacer that generates a hairpin dsRNA (Nakayashiki et al. 2005; Shafran et al. 2008). These sequences should be derived from exons and not from introns or promoter regions. Although these systems are stable and highly efficient, the cloning of hairpin constructs is often time consuming and difficult. An alternative method to unleash the RNAi response consists of the formation of dsRNA molecules through the co-expression of sense and antisense RNA strands from two divergent strong promoters in the pJL43-RNAi vector (Ullán et al. 2008). Advantages of this expression system include the rather easy construction of the vector and the high number of genes that can be tested.

9.3 Detailed Procedure

Plasmid pJL43-RNAi (Fig. 9.1) is an alternative tool for gene silencing. It includes two divergently orientated strong promoters; the *P. chrysogenum* *pcbC* gene promoter (P*pcbC*) (Gutiérrez et al. 1991) and the *A. nidulans* glyceraldehyde 3-phosphate dehydrogenase gene promoter (P*gpd*) (Punt et al. 1992). Between both promoters there is a single *Nco*I restriction site for the cloning of an exon DNA fragment of the target gene. The exon fragment, whose size ranges from 250 to 600 bp, is obtained by PCR using primers with *Nco*I sites at the 5′ end of the sequence. After digestion of the PCR product with *Nco*I, it is subcloned into the *Nco*I site of the pJL43-RNAi vector. The two convergent promoters and the subcloned DNA fragment constitute the silencing cassette (Fig. 9.1), which generates the sense and antisense mRNA strands that form dsRNA molecules selectively inducing gene knock-down.

Plasmid pJL43-RNAi also contains the phleomycin resistance cassette (Fig. 9.1) for selection of *A. chrysogenum* and *P. chrysogenum* (or other ascomycetes that exhibit sensitivity to this antibiotic) transformants. This cassette consists of the *ble* gene of *Streptoalloteichus hindustanus* under the control of the *Aspergillus awamori* glutamate dehydrogenase gene promoter (P*gdh*) and the *Saccharomyces cerevisiae* cytochrome c gene transcriptional terminator (T*cyc1*).

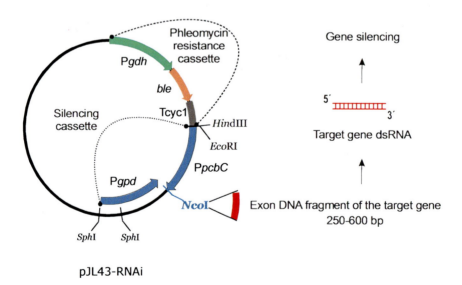

Fig. 9.1 Map of the pJL43-RNAi silencing vector, which contains the phleomycin resistance cassette and the silencing cassette with a single *Nco*I cloning site. The formation of dsRNA molecules from a target gene is shown. Abbreviations: *ble*, phleomycin resistance gene of *S. hindustanus*; P*gdh*, promoter of the *A. awamori* glutamate dehydrogenase gene; P*gpd*, promoter of the *A. nidulans* glyceraldehyde-3-phosphate dehydrogenase gene; P*pcbC*, promoter of the *P. chrysogenum pcbC* gene; T*cyc1*, transcriptional terminator of the cytochrome c gene of *S. cerevisiae*

Once the silencing vector has been built, it is transformed into the filamentous fungus of interest. The transformation of *P. chrysogenum* and *A. chrysogenum* can be performed according to the procedures described previously by Fierro et al. (1993), Cantoral et al. (1987), Díez et al. (1987) and Gutiérrez et al. (1991). The pJL43-RNAi vector is an integrative vector that allows the selection of positive transformants (using phleomycin at concentrations of 10 μg mL^{-1} for *A. chrysogenum* and 30 μg mL^{-1} for *P. chrysogenum*) that exhibit a phleomycin-resistant phenotype.

It is necessary to perform a PCR analysis as an initial screening to determine the complete integration of the silencing cassette in the genome of the phleomycin-resistant transformants. Primers should be designed to amplify the full silencing cassette. The untransformed parental strain is used as a negative control in the PCR amplifications.

Afterwards, total DNA from those transformants with a positive result in the PCR experiment is extracted and digested with specific restriction enzymes to release the silencing cassette. Then, Southern hybridization analysis must be performed to confirm the presence of the silencing cassette in the transformants and its absence in the control untransformed strain. The exon DNA fragment of the silencing vector can be used as a probe.

Finally, different techniques can be employed to test the target gene expression rate after knocking-down. Semiquantitative PCR, Northern blot hybridization, or quantitative PCR are the techniques of choice. The *actA* gene (*P. chrysogenum*) or the γ-actin (*A. chrysogenum*) can be used as internal controls for normalization.

9.4 Results

Plasmid pJL43-RNAi has been successfully utilized for the generation of knock-down mutants in different filamentous fungi (Table 9.1). This plasmid was tested firstly for the silencing of the *P. chrysogenum pcbC* and *A. chrysogenum cefEF* genes. It was found that 15–20 % of the selected transformants were knock-down mutants with reduced penicillin or cephalosporin C production, indicating that this process is effective (Ullán et al. 2008).

Another gene that was silenced through this approach was the *P. chrysogenum laeA* gene (Kosalková et al. 2009), which encodes a global regulator of secondary metabolism. Only one transformant was tested and showed a dramatic decrease in the steady-state levels of the *PclaeA* transcript. Consequently, this knock-down mutant exhibited drastically reduced levels of penicillin and showed pigmentation and sporulation defects.

Table 9.1 Examples of gene silencing in filamentous fungi using the pJL43-RNAi plasmid

Filamentous fungus	Target gene	Reference
Acremonium chrysogenum	*cefEF*	Ullán et al. (2008)
Penicillium chrysogenum	*pcbC*	Ullán et al. (2008)
Penicillium chrysogenum	*laeA*	Kosalková et al. (2009)
Trichoderma longibrachiatum	*cmt1*	Feltrer et al. (2010)
Penicillium chrysogenum	*rds*	García-Estrada et al. (2011)
	rpt	
	gmt	
	Pc21g15460	
	Pc21g15470	
Penicillium chrysogenum	*Pcrfx1*	Domínguez-Santos et al. (2012)
Penicillium chrysogenum	*penT*	Yang et al. (2012)
Penicillium chrysogenum	*paaT*	Fernández-Aguado et al. (2013a)
Penicillium chrysogenum	*penV*	Fernández-Aguado et al. (2013b)
Penicillium chrysogenum	*chs4*	Liu et al. (2013a, b)

Later, another transcriptional regulator, PcRFX1, was characterized generating loss-of-function mutants after the transformation of *P. chrysogenum* with the pJL43-RNAi plasmid (Domínguez-Santos et al. 2012). Up to seven knock-down transformants were analyzed and four out of those seven showed a significant reduction in *Pcrfx1* gene expression (between 3.39-fold and 2.67-fold). The rest of transformants did not vary in the expression of the target gene. *Pcrfx1* gene silencing led to a reduction in the expression of the penicillin biosynthetic genes *pcbAB*, *pcbC*, and *penDE* and to a decrease in the penicillin titers, thus confirming the involvement of this transcription factor in the regulation of the penicillin biosynthetic process.

RNAi-mediated gene silencing was used to partially characterize the roquefortine C/meleagrin biosynthetic pathway in *P. chrysogenum* (García-Estrada et al. 2011). Five different genes out of seven from the biosynthetic cluster were knocked-down using plasmid pJL43-RNAi. Two transformants were tested for each gene and the reduction of gene expression ranged from 20 to 72 %. Silencing of the *rds* gene (encoding a dipeptide synthetase) led to a significant decrease in the production of roquefortine C and meleagrin as compared to the parental strain. These results suggested that this dipeptide synthetase was involved in the early steps of the roquefortine C biosynthesis and that the same protein was required for the biosynthesis of meleagrin. When the *rpt* gene (encoding a prenyltransferase) was knocked-down, similar results were obtained, thus indicating that this protein was also part of the roquefortine C/meleagrin biosynthetic pathway. Silencing of the oxidoreductases Pc21g15460 and Pc21g15470 reduced the production of both mycotoxins, confirming that these proteins were involved in the biosynthesis of roquefortine C and meleagrin. Finally, when the methyltransferase gene (*gmt*) was silenced, the production of roquefortine C in the silenced transformants did not vary regarding control values, unlike meleagrin, whose levels decreased in comparison with the parental strain. These results, together with the accumulation of the intermediate glandicoline B in the knock-down

mutant, indicated that the *gmt* gene encoded a protein involved in the last step of the biosynthetic pathway, namely the methylation of glandicoline B and the formation of meleagrin (García-Estrada et al. 2011). Recent studies using knock-out mutants of the genes of the roquefortine C/meleagrin cluster confirmed the above-mentioned results and allowed the elucidation of the entire biosynthetic pathway (Ali et al. 2013). This confirms that RNAi-mediated gene silencing is an accurate tool for the characterization of gene function.

The characterization of some transporters has been achieved using plasmid pJL43-RNAi. The expression of Pc21g01300, which was renamed *paaT* (Fernández-Aguado et al. 2013a) was silenced using siRNA molecules. Six transformants were analyzed, which showed reductions in the *paaT* expression level from 20 to 90 %. The silenced mutants showed a clear reduction in the benzylpenicillin specific production (from 25 to 40 %), whereas isopenicillin N (IPN) production remained similar to that observed in the control strain. Interestingly, the toxicity of phenylacetic acid (the benzylpenicillin side chain precursor) increased in the knock-down mutants, which led to the conclusion that the transport of phenylacetic acid across the peroxisomal membrane is mediated by the PaaT protein in *P. chrysogenum*. Another research group also characterized the Pc21g01300 gene (named in this case *penT*) in *P. chrysogenum* using a similar silencing system (Yang et al. 2012). Nine transformants were analyzed, all of them showing a reduction in the levels of penicillin. Authors concluded that PenT stimulates penicillin production probably through enhancing the translocation of penicillin precursors (e.g., phenylacetic acid) across the fungal cellular membrane.

The function of another transporter from *P. chrysogenum* was analyzed using the RNAi approach. Pc22g22150, which was renamed *penV* (Fernández-Aguado et al. 2013b) was silenced with plasmid pJL43-RNAi. Three transformants were randomly selected and one of them showed a reduction of around 70 % in the expression of *penV*. Interestingly, this led to a reduction in the expression of the second and third genes of the penicillin biosynthetic pathway (*pcbC* and *penDE*). However, the expression of the *pcbAB* gene encoding the δ-(L-α-aminoadipyl)-L-cysteinyl-D-valine (ACV) synthetase (the first enzyme in the pathway) did not vary. Knockdown also gave rise to a reduction in the production of benzylpenicillin, IPN, and ACV. In light of these results, authors concluded that PenV is a vacuolar membrane protein related to the transport of the precursors L-cysteine and L-valine from the vacuolar lumen to the cytosol supplying these amino acids to the ACV synthetase.

The silencing of the class III chitin synthase gene (*chs4*) was also reported. Eight transformants were tested and up to 91 % decreases in the mRNA levels were achieved with this approach. Mutants exhibited a slow growth rate and shorter but highly branched hyphae (Liu et al. 2013a). In a similar work, the same group also tested the relationship between *chs4* and penicillin production (Liu et al. 2013b). In this work, four transformants were analyzed. Gene expression strongly decreased in two of them and these transformants showed an increase in penicillin titers.

The plasmid pJL43-RNAi has not been tested only in *P. chrysogenum* and *A. chrysogenum*. In *Trichoderma longibrachiatum* silencing of the *cmt1* gene (encoding an enzyme with chlorophenol O-methyltransferase activity) was analyzed in six different transformants (Feltrer et al. 2010). Five out of the six transformants showed reduced levels of CMT1 activity in comparison to the wild-type strain. This reduction was especially relevant (48.9 ± 5.2 %) in one of those transformants, which exhibited a correlated decrease in the *cmt1* mRNA transcript levels (more than 30 %).

9.5   Concluding Remarks

The RNAi-mediated silencing process triggered by the pJL43-RNAi vector represents a relatively fast and efficient way to analyze gene function in those filamentous fungi that exhibit sensitivity to phleomycin. Random integration of the silencing cassette into the target fungal genome and variability in the number of integrated copies give rise to a specific gene silencing pattern for each transformant. Results provided by different experiments have confirmed this situation; i.e. some knock-down transformants with high silencing rates, others with low rates, and some others with no silencing at all. Therefore, several transformants should be tested for each specific gene knock-down experiment in order to obtain complete information about gene function. Even when the analysis of several transformants is necessary, the RNAi-mediated silencing approach described above is still advantageous in comparison with the generation of knock-out mutants in filamentous fungi, which due to the low frequency of homologous recombination requires a tedious and time-consuming screening process.

Acknowledgements The development of the pJL43-RNAi plasmid was part of the research activity supervised by Professor Juan-Francisco Martín (University of León, Spain) and was supported by grants of the European Union (Eurofung QLRT-1999-00729I) and DSM (Delft, The Netherlands).

References

Ali H, Ries MI, Nijland JG, Lankhorst PP, Hankemeier T, Bovenberg RA, Vreeken RJ, Driessen AJ (2013) A branched biosynthetic pathway is involved in production of roquefortine and related compounds in *Penicillium chrysogenum*. PLoS One 8:e65328

Asch DK, Kinsey JA (1990) Relationship of vector insert size to homologous integration during transformation of *Neurospora crassa* with the cloned am (GDH) gene. Mol Gen Genet 221:37–43

Cantoral JM, Díez B, Barredo JL, Álvarez E, Martín JF (1987) High-frequency transformation of *Penicillium chrysogenum*. Nat Biotechnol 5:494–497

Casqueiro J, Gutiérrez S, Bañuelos O, Hijarrubia MJ, Martín JF (1999) Gene targeting in *Penicillium chrysogenum*: disruption of the *lys2* gene leads to penicillin overproduction. J Bacteriol 181:1181–1188

Chang SS, Zhang Z, Liu Y (2012) RNA interference pathways in fungi: mechanisms and functions. Annu Rev Microbiol 66:305–323

Chaveroche MK, Ghigo JM, d'Enfert C (2000) A rapid method for efficient gene replacement in the filamentous fungus *Aspergillus nidulans*. Nucleic Acids Res 28:E97

Colot HV, Park G, Turner GE, Ringelberg C, Crew CM, Litvinkova L, Weiss RL, Borkovich KA, Dunlap JC (2006) A high-throughput gene knockout procedure for *Neurospora* reveals functions for multiple transcription factors. Proc Natl Acad Sci U S A 103: 10352–10357

Dang Y, Yang Q, Xue Z, Liu Y (2011) RNA interference in fungi: pathways, functions, and applications. Eukaryot Cell 10:1148–1155

Díez B, Álvarez E, Cantoral JM, Barredo JL, Martín JF (1987) Isolation and characterization of pyrG mutants of *Penicillium chrysogenum* by resistance to 5′-fluoroorotic acid. Curr Genet 12:277–282

Domínguez-Santos R, Martín JF, Kosalková K, Prieto C, Ullán RV, García-Estrada C (2012) The regulatory factor PcRFX1 controls the expression of the three genes of β-lactam biosynthesis in *Penicillium chrysogenum*. Fungal Genet Biol 49:866–881

Feltrer R, Alvarez-Rodríguez ML, Barreiro C, Godio RP, Coque JJ (2010) Characterization of a novel 2,4,6-trichlorophenol-inducible gene encoding chlorophenol O-methyltransferase from *Trichoderma longibrachiatum* responsible for the formation of chloroanisoles and detoxification of chlorophenols. Fungal Genet Biol 47:458–467

Fernández-Aguado M, Ullán RV, Teijeira F, Rodríguez-Castro R, Martín JF (2013a) The transport of phenylacetic acid across the peroxisomal membrane is mediated by the PaaT protein in *Penicillium chrysogenum*. Appl Microbiol Biotechnol 97:3073–3084

Fernández-Aguado M, Teijeira F, Martín JF, Ullán RV (2013b) A vacuolar membrane protein affects drastically the biosynthesis of the ACV tripeptide and the beta-lactam pathway of *Penicillium chrysogenum*. Appl Microbiol Biotechnol 97:795–808

Fierro F, Gutiérrez S, Díez B, Martín JF (1993) Resolution of four chromosomes in penicillin-producing filamentous fungi: the penicillin gene cluster is located on chromosome II (9.6 Mb) in *Penicillium notatum* and chromosome I (10.4 Mb) in *Penicillium chrysogenum*. Mol Gen Genet 241:573–578

García-Estrada C, Ullán RV, Albillos SM, Fernández-Bodega MA, Durek P, von Döhren H, Martín JF (2011) A single cluster of coregulated genes encodes the biosynthesis of the mycotoxins roquefortine C and meleagrin in *Penicillium chrysogenum*. Chem Biol 18:1499–1512

Goldoni M, Azzalin G, Macino G, Cogoni C (2004) Efficient gene silencing by expression of double stranded RNA in *Neurospora crassa*. Fungal Genet Biol 41:1016–1024

Gutiérrez S, Díez B, Álvarez E, Barredo JL, Martín JF (1991) Expression of the *penDE* gene of *Penicillium chrysogenum* encoding isopenicillin N acyltransferase in *Cephalosporium acremonium*: production of benzylpenicillin by the transformants. Mol Gen Genet 225:56–64

Hammond TM, Keller NP (2005) RNA silencing in *Aspergillus nidulans* is independent of RNA-dependent RNA polymerases. Genetics 169:607–617

Hoskins JA, O'Callaghan N, Queener SW, Cantwell CA, Wood JS, Chen VJ, Skatrud PL (1990) Gene disruption of the *pcbAB* gene encoding ACV synthetase in *Cephalosporium acremonium*. Curr Genet 18:523–530

Janus D, Hoff B, Hofmann E, Kuck U (2007) An efficient fungal RNA-silencing system using the *DsRed* reporter gene. Appl Environ Microbiol 73:962–970

Kadotani N, Nakayashiki H, Tosa Y, Mayama S (2003) RNA silencing in the phytopathogenic fungus *Magnaporthe oryzae*. Mol Plant Microbe Interact 16:769–776

Kosalková K, García-Estrada C, Ullán RV, Godio RP, Feltrer R, Teijeira F, Mauriz E, Martín JF (2009) The global regulator LaeA controls penicillin biosynthesis, pigmentation and sporulation, but not roquefortine C synthesis in *Penicillium chrysogenum*. Biochimie 91:214–225

Liu H, Wang P, Gong G, Wang L, Zhao G, Zheng Z (2013a) Morphology engineering of *Penicillium chrysogenum* by RNA silencing of chitin synthase gene. Biotechnol Lett 35:423–429

Liu H, Zheng Z, Wang P, Gong G, Wang L, Zhao G (2013b) Morphological changes induced by class III chitin synthase gene silencing could enhance penicillin production of *Penicillium chrysogenum*. Appl Microbiol Biotechnol 97:3363–3372

Miller BL, Miller KY, Timberlake WE (1985) Direct and indirect gene replacements in *Aspergillus nidulans*. Mol Cell Biol 5:1714–1721

Mouyna I, Henry C, Doering TL, Latge JP (2004) Gene silencing with RNA interference in the human pathogenic fungus *Aspergillus fumigatus*. FEMS Microbiol Lett 237:317–324

Nakayashiki H, Hanada S, Nguyen BQ, Kadotani N, Tosa Y, Mayama S (2005) RNA silencing as a tool for exploring gene function in ascomycete fungi. Fungal Genet Biol 42:275–283

Punt PJ, Kramer C, Kuyvenhoven A, Pouwels PH, van den Hondel CA (1992) An upstream activating sequence from the *Aspergillus nidulans gpdA* gene. Gene 120:67–73

Shafran H, Miyara I, Eshed R, Prusky D, Sherman A (2008) Development of new tools for studying gene function in fungi based on the Gateway system. Fungal Genet Biol 45:1147–1154

Ullán RV, Liu G, Casqueiro J, Gutiérrez S, Bañuelos O, Martín JF (2002a) The *cefT* gene of *Acremonium chrysogenum* C10 encodes a putative multidrug efflux pump protein that significantly increases cephalosporin C production. Mol Genet Genomics 267:673–683

Ullán RV, Casqueiro J, Banuelos O, Fernandez FJ, Gutierrez S, Martin JF (2002b) A novel epimerization system in fungal secondary metabolism involved in the conversion of isopenicillin N into penicillin N in *Acremonium chrysogenum*. J Biol Chem 277:46216–46225

Ullán RV, Godio RP, Teijeira F, Vaca I, García-Estrada C, Feltrer R, Kosalková K, Martín JF (2008) RNA-silencing in *Penicillium chrysogenum* and *Acremonium chrysogenum*: validation studies using β-lactam genes expression. J Microbiol Methods 75:209–218

Yang J, Xu X, Liu G (2012) Amplification of an MFS transporter encoding gene *penT* significantly stimulates penicillin production and enhances the sensitivity of *Penicillium chrysogenum* to phenylacetic acid. J Genet Genomics 39:593–602

Controlling Fungal Gene Expression Using the Doxycycline-Dependent Tet-ON System in *Aspergillus fumigatus*

10

Michaela Dümig and Sven Krappmann

10.1 Introduction

Pathogenic fungi have gained steadily increasing attention as they pose a serious threat to the immunocompromised individual. Among the estimated 1.5 million fungal species (Hawksworth 2001), only few have evolved to infect susceptible human beings, with about ten species causing infections predominantly. Among these, the saprobe and ubiquitous mould *Aspergillus fumigatus* has emerged in recent decades to cause a severe mycosis, termed aspergillosis, in distinct clinical settings, with more than 200,000 life-threatening infections per year worldwide that are linked to mortality rates of 30–95 %.

The general resistance of humans against fungal infections reflects the fact that the human immune system is well equipped to fight fungal invaders and that the immune status of a host determines the outcome of the intimate interplay with a fungal pathogen. Yet, *A. fumigatus* has evolved to infect and severely harm susceptible individuals that may suffer from one of several risk factors, such as leukaemia, prolonged

M. Dümig • S. Krappmann, Prof. Dr. rer. nat. (✉)
Mikrobiologisches Institut-Klinische
Mikrobiologie, Immunologie und Hygiene
Universitätsklinikum Erlangen, Friedrich-Alexander-
Universität Erlangen-Nürnberg Wasserturmstr. 3/5,
91054 Erlangen, Germany
e-mail: sven.krappmann@uk-erlangen.de

neutropenia, or immune suppression after stem cell transplantation. Virulence of *A. fumigatus* is a multifactorial trait, and its determinants are encoded in the fungal genome in a polygenic fashion (Tekaia and Latgé 2005; Krappmann 2008). To address putative factors that contribute to fungal pathogenicity it is advisable to follow the rationale behind the molecular interpretation of Koch's postulates, that is that (1) the gene of interest confers a specific phenotype like contributing to virulence of the pathogen, (2) inactivation of the gene eliminates this phenotype and therefore abolishes virulence, and (3) reintroduction of the gene in the mutant reinstates the wild-type phenotype to restore virulence capacities (Falkow 1988, 2004). According to this straight-forward approach, the generation of a mutant strain is crucial for the identification and validation of a microbial virulence determinant.

Defined mutants are generated by means of gene targeting to eliminate or modify any gene of interest. Genes for which no *null* mutation can be isolated are considered essential and represent prime candidates for antifungal therapy, as they are required for growth of the pathogen in general. Essentiality of a given gene can easily be tested in the filamentous coenocyte *A. fumigatus* by heterokaryon rescue (Osmani et al. 2006), but the absence of growth of the corresponding *null* mutant interferes with any further characterization of the presumably essential factor. Infection experiments, for instance, cannot be executed as

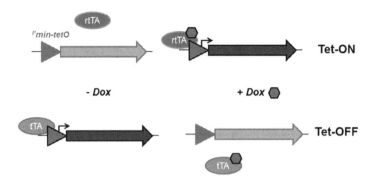

Fig. 10.1 Schematic outline of Tet-ON and Tet-OFF systems for conditional gene expression. The gene of interest is placed behind a minimal promoter *Pmin* containing *tetO* binding sites for the synthetic tetracycline-binding transcriptional activator TA. The TA nature determines whether the *Pmin-tetO* sites are targeted upon binding of the tetracycline derivative doxycycline (Dox) to activate transcription: rtTA attaches to the promoter in its Dox-bound form (Tet-ON), whereas the reverse version tTA is excluded from the promoter to result in silenced transcription (Tet-OFF)

no viable spores can be collected, and therefore, any impact of the factor in question on virulence cannot be addressed in a definitive manner. This is especially true for mutants that are impaired in biosynthetic pathways: the absence of growth even at full supplementation under in vitro conditions might not translate into any essential nature of the gene for growth inside a susceptible host, as the conditions differ significantly. To circumvent this drawback, conditional gene expression is the method of choice by replacing the promoter region of the gene in question for a controllable one (conditional promoter replacement, CPR) (Roemer et al. 2003; Hu et al. 2007). Hence, expression of the gene is tuned by defined environmental conditions to achieve overexpression or, on the other side, complete silencing. Such adjustable promoters that are functional in *Aspergillus* species have been characterized only to a limited extent, with the majority derived from metabolic genes (Zadra et al. 2000; Romero et al. 2003; Grosse and Krappmann 2008; Monteiro and De Lucas 2010). In a more sophisticated manner, artificial and heterologous systems based on the human estrogen receptor or the bacterial Tet repressor have been successfully employed for tunable gene expression in *Aspergillus* (Pachlinger et al. 2005; Vogt et al. 2005). Especially the latter was extensively exploited to become a highly versatile tool in molecular biology for a number of organisms. It is derived from the tetracycline resistance-conferring operon of *Escherichia coli*, which comprises a repressor protein TetR that detects the presence of tetracycline and binds to conserved operator sequences *tetO* to prevent transcription of the operon (Hillen and Berens 1994). Binding of the antibiotic prevents TetR from binding to *tetO* elements in the native setting, whereas it is the prerequisite for TetR–*tetO* association in the reverse derivative of the system (Fig. 10.1). By fusing the tetracycline repressor to functional transcriptional activation domains, the system was successfully adapted for a variety of eukaryotes (Gossen and Bujard 1992; Gossen et al. 1993, 1995; Weinmann et al. 1994; Stebbins and Yin 2001; Knott et al. 2002), among them fungal hosts (Nakayama et al. 1998, 2000; Stoyan et al. 2001; Park and Morschhäuser 2005; Vogt et al. 2005; Weyler and Morschhäuser 2012). The gene regulation modules are made up by two functional elements, the tetracycline-responsive DNA-binding protein that activates transcriptional initiation and a promoter that includes *tetO* sequences. When the presence of tetracycline prevents the synthetic tetracycline transcriptional activator tTA from *tetO* binding, the so-called Tet-OFF system is implemented; in case tetracycline association enables the reverse transactivator rtTA to bind to *tetO* and trigger transcription, the system is denominated as Tet-ON (Gossen et al. 1995). In practice, additional functional elements are necessary to support any Tet system in a eukaryotic host, such as promoters and transcriptional termination sequences for constitutive tTA or rtTA expression, accompanied by a minimal promoter containing the *tetO* elements that is shut off when the transcriptional activator is not bound to them. Nowadays the tetracycline analogue doxycycline is commonly used due to its superior stability, yet, its interference with iron homeostasis needs to be taken into account (Fiori and Van Dijck 2012).

Doxycycline-responsive systems were successfully implemented for controlled gene expression in several fungi, including *Aspergillus* species such as the human pathogen *A. fumigatus*. Doxycycline per se did not influence growth of *A. niger* at concentrations of up to 125 µg/mL (Meyer et al. 2011), and any further effects of the antibiotic itself on the physiology of aspergilli have not been described so far. In the initial study of Vogt et al. (2005), both functional elements, a minimal promoter containing seven *tetO* copies and expression cassettes for tTA and the rtTA2^S-M2 transcriptional regulators, were located on distinct plasmids and therefore require co-transformation. By placing a hygromycin resistance-conferring gene under positive (Tet-ON) or negative (Tet-OFF) doxycycline control, functionality of each system was successfully demonstrated. The Tet-ON system could further be streamlined by integrating all functional modules in one cassette, for which functionality was shown by expressing several genes in a doxycycline-dependent manner in the biotechnological workhorse *A. niger* (Meyer et al. 2011). Conditional expression of a presumably essential gene could also be achieved with this Tet-ON construct in *A. fumigatus* (Dichtl et al. 2012). However, based on the fact that both the constitutive promoter driving rtTA transcription and the minimal promoter sequence preceding the gene of interest stem from the *A. nidulans* *gpdA* promoter, recombination events were detected that interfered with the doxycycline-dependent regulation of gene expression. To circumvent this drawback, two derivative constructs had been generated in which either promoter sequence was replaced by alternative elements (Helmschrott et al. 2013).

The prime application of doxycycline-dependent expression systems in *A. fumigatus* molecular biology is the CPR strategy to identify genes that are required for virulence and therefore represent promising targets for antifungal therapy. The CPR approach allows functional characterization of a gene of interest under in vitro conditions, and by virtue of the Tet-systems also in living systems such as an infected animal. Its application is most appropriate for essential genes that are refractory to deletion by direct gene replacement, and this chapter shall demonstrate the validity of the Tet-ON system for essential gene verification in *A. fumigatus* and its feasibility for infection models of aspergillosis.

10.2 Methods

The methods described below aim at generating an *A. fumigatus* strain that expresses a given gene in a conditional manner, based on the doxycycline-inducible Tet-ON system. For the respective conditional promoter replacement strategy, assembly of a suitable expression cassette and isolation of the replacement module is necessary, and the transformation procedure according to Punt and van den Hondel (1992) to exchange the endogenous copy of the gene in the *A. fumigatus* genome by the recombinant one will be described in detail. Primary transformants need to be selected and purified for clonality to become screened and eventually validated for the correct genotype by diagnostic PCR (Saiki et al. 1985) and/or Southern hybridization analyses (Southern 2006). Phenotypic characterization of the resulting *A. fumigatus* isolate is then carried out with respect to doxycycline-dependent growth in vitro and virulence in animal models of invasive aspergillosis (Smith et al. 1994) with the aim to assess any essentiality of the candidate gene for pathogenicity of this fungal pathogen.

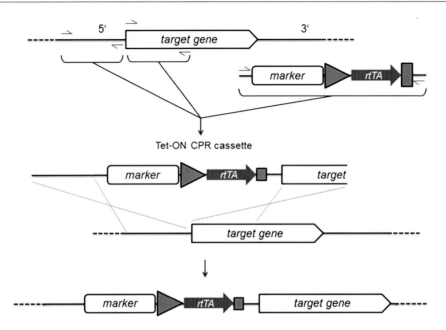

Fig. 10.2 Workflow for generating a doxycycline-dependent CPR cassette based on the Tet-ON system. The amplified components (5′ and 3′ homology arms together with the Tet-ON module linked to genetic marker gene for selection) can be assembled by sequence and ligation-independent cloning (SLIC) approaches. The resulting cassette is transformed in the *A. fumigatus* recipient for gene replacement by homologous recombination

10.2.1 Detailed Procedure

10.2.1.1 Construction and Isolation of Conditional Promoter Replacement Cassette

Generation of a suitable replacement cassette that contains the Tet-ON module fused to the gene of interest and linked to a selectable marker gene can be achieved by standard cloning methods using ligation of respective fragments in a common plasmid backbone. Yet, modern approaches omit the activity of restriction endonucleases and ligation reactions and allow simultaneous and seamless assembly of several DNA fragments. The desired construct is produced in a one-step procedure starting from PCR amplicons that comprise three functional elements: the 5′ upstream region of the target gene, the Tet-ON module comprising the selection marker, and the gene's coding sequence (Fig. 10.2). In one preferred system, the GeneArt® Seamless Cloning and Assembly Kit, recombination of the three fragments is mediated by homology arms of 15 bp that are incorporated in the individual amplicons via the priming oligonucleotides. Furthermore, applicable restriction sites might be included in the distal primers for convenient release of the CPR cassette from the respective plasmid.

Purification of the replacement module prior to transformation is achieved by standard procedures such as separation by agarose gel electrophoresis and subsequent extraction via spin columns as implemented in a variety of commercial kits.

10.2.1.2 Transformation of *Aspergillus fumigatus*

Integration of the CPR cassette can be achieved for *A. fumigatus* by protoplast-mediated transformation followed by selection on appropriate culture media that are supplemented with the respective antibiotic and doxycycline as well to ensure expression of the gene in question. The transformation procedure itself is executed as follows:

1. *Buffers and Solutions*
   (a) Citrate buffer: 150 mM KCl, 580 mM NaCl, 50 mM sodium citrate (dissolve in water and titrate pH to 5.5 with citric acid, store at room temperature after autoclaving)

10 Controlling Fungal Gene Expression Using the Doxycycline-Dependent Tet-ON System...

(b) STC1700 buffer: 1.2 M sorbitol, 10 mM Tris–HCl (pH 5.5), 50 mM CaCl₂, 35 mM NaCl (dissolve in H_2O, autoclave, and store at 4 °C)

(c) PEG4000 buffer 60 % (w/v) polyethylene glycol 4000, 50 mM CaCl₂, 10 mM Tris–HCl (pH 7.5)

2. *Generation and harvesting of A. fumigatus protoplasts*

   (a) Inoculate a 50 mL overnight (o/n) culture in complete media or a 200 mL culture of minimal medium in a baffled Erlenmeyer flask

   (b) Let grow while shaking at about 130–150 rpm at 30 °C or 37 °C

   (c) Filter the o/n culture through a sterile Miracloth filter and wash three times with citrate buffer

   (d) Transfer the mycelial biomass to an Erlenmeyer flask without baffles using a sterile rounded spatula

   (e) Add 200 mg Glucanex (Novozymes) dissolved in 10 mL citrate buffer and suspend the mycelium

   (f) Shake carefully with about 70–80 rpm at 30 °C

   (g) Check after 45 min whether protoplasting has started by transferring ca. 10 μL of the mycelial suspension onto a glass slide, place cover slip on top and inspect with a microscope if protoplasts emerge on hyphal tips; check every 15–20 min for sufficient amounts of protoplasts (about 1×10^8/g wet weight)

   (h) Filter the mixture through a fresh Miracloth filter into a 50 mL reaction tube in ice

   (i) Add a maximum of 30 mL ice-cold STC1700 buffer to a maximum of 20 mL protoplast solution and leave on ice for 10 min

   (j) Spin 12 min with 1,200 g at 4 °C, discard supernatant

   (k) Add 30–50 mL of fresh ice-cold STC1700 buffer and re-suspend the pellet by flicking carefully

   (l) Spin 12 min with 1,200 g at 4 °C, discard supernatant

   (m) Dissolve the pellet in a maximum of 500 μL STC1700 buffer by flicking carefully

   (n) Quantify the amount of protoplasts from a 1:10 dilution in a haemocytometer and use 100–200 μL per transformation

   Note: Handle protoplasts carefully and keep them over or in ice, use cut-off tips to avoid shearing!

3. *Transformation*

   (a) Add 5–10 μg of the DNA fragment in a maximum of 30 μL 5 mM Tris buffer (pH 8.0) to the protoplast solution in a 15 mL reaction tube

   (b) Incubate 20–30 min on ice

   (c) Add 250 μL PEG4000 and mix gently, add another 250 μL PEG4000 and mix gently, finally add 850 μL PEG4000 and mix gently until the solution appears homogenous without smearing, incubate 15–20 min over(!) ice to avoid PEG precipitation

   (d) Fill up with STC1700 buffer to 15 mL and mix gently

   (e) Centrifuge for 15 min at 1,200 g at 4 °C

   (f) Re-suspend the protoplast pellet in max. 500 μL STC1700 buffer

   (g) Make dilutions or aliquots of the mixture according to preferences

   (h) Add aliquots to 5 mL top agar (containing 0.7 % agar in supplemented sorbitol medium, heated and stored in 5–6 mL aliquots in a 50 °C water bath until usage), mix gently by inverting and pour it on a 1.2 M sorbitol-containing culture media dish

   (i) Place in incubator o/n at 30 °C, on the next day, plates can be incubated at 37 °C

   (j) After 2–5 days single colonies can arise, which need to be transferred repeatedly to selective media to avoid heterokaryon formation

10.2.1.3 Validation and Phenotyping of Recombinant Tet-ON Isolates

After generating clonal descendants from the primary transformants, validity of the recombinant strain can be analyzed by conventional means of

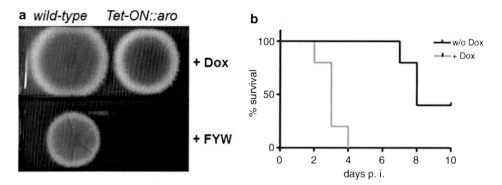

Fig. 10.3 Illustrative data for conditional gene expression in *A. fumigatus* by the Tet-ON system. (**a**) A CPR allele of an essential *aro* gene that encodes a biosynthetic activity of the shikimate pathway allows conditional growth only in the presence of doxycycline but not on minimal medium even in the presence of all three aromatic amino acids phenylalanine, tyrosine, and tryptophan. (**b**) Virulence of the *Tet-ON::aro* strain in a murine model of aspergillosis strongly depends on doxycycline supplementation, demonstrating essentiality of the gene for in vivo growth and, therefore, virulence

genotyping, such as diagnostic PCR or Southern analysis, after isolation of sample genomic DNA. Confirmed isolates are then further inspected in the absence of doxycycline with respect to any phenotypes emerging from reduced expression of the Tet-ON promoter allele. Special attention is set on virulence assessment in immunosuppressed mice using various routes of infection such as inhalation, intranasal instillation, or intravenous injection of conidia to model pulmonary or systemic aspergillosis, respectively. To achieve conditional expression of the *A. fumigatus* gene of interest in the infected host animals, doxycycline can be omitted from or supplied to the drinking water at concentrations of 0.2 %. This allows testing any essentiality for a given gene under in vivo conditions of infection.

10.3 Results

The *A. fumigatus* aromatic amino acid biosynthesis pathway may serve as an illustrative example for the described approach. This anabolic route was targeted with the aim to generate mutants auxotrophic for tyrosine, phenylalanine, and tryptophan (our unpublished results). While single deletion mutants requiring tryptophan or tyrosine/phenylalanine could be isolated successfully, attempts to generate a double deletant requiring all the aromatic amino acids were fruitless. In order to probe the presumably essential nature of the biosynthetic route, an integral *aro* gene of the shikimate pathway was replaced by a conditional promoter allele carrying the Tet-ON module. The resulting strain displays a conditional auxotrophy that indeed cannot be rescued by supplementation with the aromatic amino acids (Fig. 10.3a). Furthermore, the *Tet-ON::aro* strain was tested for virulence in a murine model of invasive aspergillosis (Fig. 10.3b). After systemic infection of immunosuppressed animals disease progression was monitored for two cohorts, one served with doxycycline-containing drinking water while for the other this supplement was omitted. In agreement with the in vitro data, the absence of doxycycline does correlate with increased survival rates, indicating that expression of the corresponding *aro* gene is required for growth and virulence of *A. fumigatus* in the respective disease model. Control animals that received doxycycline in their drinking water but had not been infected did not show any signs of disease or other matters, indicating the absence of relevant side effects from the antibiotic. Altogether, these data demonstrate the overall feasibility of conditional gene expression by the doxycycline-dependent Tet-ON system with the aim to define novel target pathways within the *A. fumigatus* virulome.

10.4 Conclusion

Conditional gene expression after promoter replacement by doxycycline-dependent modules represents the method of choice when studying essential genes in fungi. This is of special relevance for pathogenic ones in the context of infection, as exemplified for the major mould pathogen of human *A. fumigatus*.

References

Dichtl K, Helmschrott C, Dirr F, Wagener J (2012) Deciphering cell wall integrity signalling in *Aspergillus fumigatus*: identification and functional characterization of cell wall stress sensors and relevant Rho GTPases. Mol Microbiol 83:506–519

Falkow S (1988) Molecular Koch's postulates applied to microbial pathogenicity. Rev Infect Dis 10(suppl 2): S274–S276

Falkow S (2004) Molecular Koch's postulates applied to bacterial pathogenicity—a personal recollection 15 years later. Nat Rev Microbiol 2:67–72

Fiori A, Van Dijck P (2012) Potent synergistic effect of doxycycline with fluconazole against *Candida albicans* is mediated by interference with iron homeostasis. Antimicrob Agents Chemother 56:3785–3796

Gossen M, Bujard H (1992) Tight control of gene expression in mammalian cells by tetracycline-responsive promoters. Proc Natl Acad Sci U S A 89:5547–5551

Gossen M, Bonin AL, Bujard H (1993) Control of gene activity in higher eukaryotic cells by prokaryotic regulatory elements. Trends Biochem Sci 18:471–475

Gossen M, Freundlieb S, Bender G, Muller G, Hillen W, Bujard H (1995) Transcriptional activation by tetracyclines in mammalian cells. Science 268:1766–1769

Grosse V, Krappmann S (2008) The asexual pathogen *Aspergillus fumigatus* expresses functional determinants of *Aspergillus nidulans* sexual development. Eukaryot Cell 7:1724–1732

Hawksworth DL (2001) The magnitude of fungal diversity: the 1.5 million species estimate revisited. Mycol Res 105:1422–1432

Helmschrott C, Sasse A, Samantaray S, Krappmann S, Wagener J (2013) Upgrading fungal gene expression on demand: improved systems for doxycycline-dependent silencing in *Aspergillus fumigatus*. Appl Environ Microbiol 79:1751–1754

Hillen W, Berens C (1994) Mechanisms underlying expression of Tn*10* encoded tetracycline resistance. Annu Rev Microbiol 48:345–369

Hu W, Sillaots S, Lemieux S, Davison J, Kauffman S, Breton A, Linteau A, Xin C, Bowman J, Becker J, Jiang B, Roemer T (2007) Essential gene identification and drug target prioritization in *Aspergillus fumigatus*. PLoS Pathog 3:e24

Knott A, Garke K, Urlinger S, Guthmann J, Muller Y, Thellmann M, W. Hillen (2002) Tetracycline-dependent gene regulation: combinations of transregulators yield a variety of expression windows. Biotechniques 32:796–806

Krappmann S (2008) Pathogenicity determinants and allergens. In: Osmani SA, Goldman GH (eds) The Aspergilli: genomics, medical aspects, biotechnology, and research methods. CRC Press, Boca Raton, pp 377–400

Meyer V, Wanka F, van Gent J, Arentshorst M, van den Hondel CA, Ram AF (2011) Fungal gene expression on demand: an inducible, tunable, and metabolism-independent expression system for *Aspergillus niger*. Appl Environ Microbiol 77:2975–2983

Monteiro MC, De Lucas JR (2010) Study of the essentiality of the *Aspergillus fumigatus triA* gene, encoding RNA triphosphatase, using the heterokaryon rescue technique and the conditional gene expression driven by the *alcA* and *niiA* promoters. Fungal Genet Biol 47:66–79

Nakayama H, Izuta M, Nagahashi S, Sihta EY, Sato Y, Yamazaki T, Arisawa M, Kitada K (1998) A controllable gene-expression system for the pathogenic fungus *Candida glabrata*. Microbiology 144(Pt 9):2407–2415

Nakayama H, Mio T, Nagahashi S, Kokado M, Arisawa M, Aoki Y (2000) Tetracycline-regulatable system to tightly control gene expression in the pathogenic fungus *Candida albicans*. Infect Immun 68:6712–6719

Osmani AH, Oakley BR, Osmani SA (2006) Identification and analysis of essential *Aspergillus nidulans* genes using the heterokaryon rescue technique. Nat Protoc 1:2517–2526

Pachlinger R, Mitterbauer R, Adam G, Strauss J (2005) Metabolically independent and accurately adjustable *Aspergillus* sp. expression system. Appl Environ Microbiol 71:672–678

Park YN, Morschhäuser J (2005) Tetracycline-inducible gene expression and gene deletion in *Candida albicans*. Eukaryot Cell 4:1328–1342

Punt PJ, van den Hondel CA (1992) Transformation of filamentous fungi based on hygromycin B and phleomycin resistance markers. Methods Enzymol 216: 447–457

Roemer T, Jiang B, Davison J, Ketela T, Veillette K, Breton A, Tandia F, Linteau A, Sillaots S, Marta C, Martel N, Veronneau S, Lemieux S, Kauffman S, Becker J, Storms R, Boone C, Bussey H (2003) Large-scale essential gene identification in *Candida albicans* and applications to antifungal drug discovery. Mol Microbiol 50:167–181

Romero B, Turner G, Olivas I, Laborda F, De Lucas JR (2003) The *Aspergillus nidulans alcA* promoter drives tightly regulated conditional gene expression in *Aspergillus fumigatus* permitting validation of essential genes in this human pathogen. Fungal Genet Biol 40:103–114

Saiki RK, Scharf S, Faloona F, Mullis KB, Horn GT, Erlich HA, Arnheim N (1985) Enzymatic amplification of β-globin genomic sequences and restriction site analysis for diagnosis of sickle cell anemia. Science 230:1350–1354

Smith JM, Tang CM, Van Noorden S, Holden DW (1994) Virulence of *Aspergillus fumigatus* double mutants lacking restriction and an alkaline protease in a low-dose model of invasive pulmonary aspergillosis. Infect Immun 62:5247–5254

Southern E (2006) Southern blotting. Nat Protoc 1:518–525

Stebbins MJ, Yin JC (2001) Adaptable doxycycline-regulated gene expression systems for *Drosophila*. Gene 270:103–111

Stoyan T, Gloeckner G, Diekmann S, Carbon J (2001) Multifunctional centromere binding factor 1 is essential for chromosome segregation in the human pathogenic yeast *Candida glabrata*. Mol Cell Biol 21:4875–4888

Tekaia F, Latgé J-P (2005) *Aspergillus fumigatus*: saprophyte or pathogen? Curr Opin Microbiol 8:385–392

Vogt K, Bhabhra R, Rhodes JC, Askew DS (2005) Doxycycline-regulated gene expression in the opportunistic fungal pathogen *Aspergillus fumigatus*. BMC Microbiol 5:1

Weinmann P, Gossen M, Hillen W, Bujard H, Gatz C (1994) A chimeric transactivator allows tetracycline-responsive gene expression in whole plants. Plant J 5: 559–569

Weyler M, Morschhäuser J (2012) Tetracycline-inducible gene expression in *Candida albicans*. Methods Mol Biol 845:201–210

Zadra I, Abt B, Parson W, Haas H (2000) *xylP* promoter-based expression system and its use for antisense downregulation of the *Penicillium chrysogenum* nitrogen regulator NRE. Appl Environ Microbiol 66: 4810–4816

Part IV

Tools and Applications: Selection Markers and Vectors

Expanding the Repertoire of Selectable Markers for *Aspergillus* Transformation

11

Khyati Dave, V. Lakshmi Prabha, Manmeet Ahuja, Kashyap Dave, S. Tejaswini, and Narayan S. Punekar

11.1 Introduction

Aspergilli are a unique group of saprophytic filamentous fungi that dwell in wide spectrum of habitat such as soil, decaying organic matter, and moist indoor environments. Enormous nutritional flexibility underlies their successful lifestyle. The genus *Aspergillus* is gifted with rich ability to secrete degradative enzymes, organic acids and secondary metabolites. These fungi are capable of growth in a wide range of pH (between pH 2.0 and pH 11.0), temperature (10–50 °C) and high osmolarity (Kis-Papo et al. 2003). These endowments have fascinated both academic and industrial researchers for over a century. While *A. nidulans* is an extensively used genetic model organism and an academic favorite, many other Aspergilli are industrial workhorses (Meyer 2008). For instance, *A. niger* is an avid citric acid producer (Karaffa et al. 2001); *A. oryzae* is used to brew sake and make soy sauce (Barbesgaard et al. 1992); *A. terreus* is employed for the production of itaconic acid (Okabe et al. 2009) and lovastatin (Bizukojc and Ledakowicz 2009). Some of them

K. Dave, M.Sc. • V.L. Prabha, Ph.D.
M. Ahuja, Ph.D. • K. Dave, Ph.D.
S. Tejaswini, M.Sc. • N.S. Punekar, Ph.D. (✉)
Department of Bioscience and Bioengineering,
Indian Institute of Technology Bombay,
Powai, Mumbai, Maharashtra 400 076, India
e-mail: khyatimehta@iitb.ac.in; nsp@iitb.ac.in

like *A. niger* and *A. oryzae* are choice platforms for heterologous protein production due to their exceptional secretion capacity (Su et al. 2012). In contrast, the genus also includes pathogens like *A. fumigatus*, *A. parasiticus*, and *A. flavus*.

Many whole genome sequences now available for Aspergilli are indicative of their acknowledged value. Genome sequences of several industrially and medically important *Aspergillus* species are freely accessible (Arnaud et al. 2012; Gibbons and Rokas 2013). This has enabled a better appreciation of gene expression, metabolism, and its regulation. Understanding the organization, regulation, and manipulation of fungal genes requires the development of various genetic tools. Due to the industrial importance of these fungi many of such tools may have remained trade secrets. Low transformation efficiency, limited choice of selection markers, and poor frequency of targeted gene insertions are largely responsible for the delay in development of genetic engineering toolkit for these organisms.

A. nidulans was the first *Aspergillus* to be successfully transformed and this involved the use of protoplasts (Tilburn et al. 1983). Subsequently other DNA delivery methods were described along with modifications to suit individual fungal strains (for details see other chapters of this book). Each of these methods has its own advantages and disadvantages (Prabha and Punekar 2004). Regardless of these variations, the transforming DNA either integrates into the host

M.A. van den Berg and K. Maruthachalam (eds.), *Genetic Transformation Systems in Fungi, Volume 2*, Fungal Biology, DOI 10.1007/978-3-319-10503-1_11,
© Springer International Publishing Switzerland 2015

genome or replicates autonomously. With rare exceptions, integrative transformation is predominant in filamentous fungi including Aspergilli. Selectable markers are necessary and are used to score transformants (and DNA integration events) in a background of the recipient host. The repertoire of convenient selection markers available for *Aspergillus* transformation is limited. While some markers work well across Aspergilli others are species specific. Several earlier reviews have stressed on various transformation techniques and development of strains with different genetic backgrounds, to study filamentous fungal biology (Fleissner and Dersch 2010; Jiang et al. 2013; Kuck and Hoff 2010; Lubertozzi and Keasling 2009; Meyer 2008; Meyer et al. 2011; Ruiz-Diez 2002; Su et al. 2012; Prabha and Punekar 2004; Weld et al. 2006). However, the description of selectable markers for *Aspergillus* transformation has attracted limited attention. A comprehensive and up-to-date account on various selection markers and different strategies exploiting these markers in manipulating Aspergilli is given here.

11.2 Selectable Markers for *Aspergillus* Transformation

Selectable markers are one of the important molecular tools that distinguish transformed cells from the untransformed host. They generally fall into three categories: resistance markers, nutritional markers, and bidirectional markers. Besides, a few reporter genes also serve as convenient visual markers to select transformants. Selectable markers functionally tested in Aspergilli are described below. Three specific markers, namely *bar*, *agaA*, and *sC* as examples relating to *A. niger* transformation, are particularly emphasized.

11.2.1 Resistance Markers

Resistance markers allow the growth of an organism in the presence of inhibitors (antibiotic or antimetabolite). These are routinely used to transform wild type/natural isolates of fungal strains; they circumvent the need to create an auxotrophic host background. The only requirement is that the recipient organism is sensitive to the selection pressure applied. These markers are particularly useful for organisms with little genetic information available. Resistance markers by and large are dominant and provide a tool for positive selection—the gene imparting a survival phenotype to the recipient on the respective selection medium. Table 11.1 provides a list of such selectable markers reported till date for Aspergilli; of these, *hph*, *bar*, and *benA* are more commonly used. Interestingly, *hsv-1 tk* (coding for herpes simplex virus type 1 thymidine kinase) provides for a negative selection—its expression is lethal to the host organism when grown in the presence of ganciclovir. Lethality is due to the conversion of ganciclovir into a cytotoxic nucleoside analogue. In combination with *ble*, *hsv-1 tk* was used to provide both positive and negative selection in *A. fumigatus* (Krappmann et al. 2005).

Resistance to L-phosphinothricin (PPT) (through the *bar/pat* gene) was first used as selection marker in *Neurospora crassa* (Avalos et al. 1989). Inhibition of glutamine synthetase and subsequent accumulation of ammonia are thought to be the major mode of PPT action. Other cellular targets of PPT comprise membrane depolarization (Ullrich et al. 1990) and membrane transport process (Trogisch et al. 1989). The *bar* gene imparts resistance by acetylating PPT; it serves as a dominant selection marker as many fungi are sensitive to PPT (Ahuja and Punekar 2008). Selection of *A. niger* transformants using *bar* gene constructs was achieved with both a homologous promoter (*A. niger PcitA*, Dave and Punekar 2011) and a heterologous promoter (*A. nidulans PtrpC*, Ahuja and Punekar 2008). Besides *bar* transformants, spontaneously resistant *A. niger* colonies were obtained. Analysis of this spontaneous resistance, caused by mutations in the *A. niger* glutamate uptake system, was helpful in fine-tuning *bar* selection conditions. While host sensitivity to PPT may be an issue, the virtues of this marker are many. PPT is an active ingredient of several commercial herbicides (such as Basta, Bayer Crop-Science, India) and is readily accessible. Due to lower toxicity to

11 Expanding the Repertoire of Selectable Markers for *Aspergillus* Transformation

Table 11.1 Resistance (dominant) markers in Aspergilli

Selectable marker	Marker gene function (CDS length)	Metabolic target	Species transformed
bar	Phosphinothricin acetyltransferase (552 bp)	Glutamine synthetase	*A. niger*[a], *A. fumigatus*[b], *A. nidulans*[b]
benA	Benomyl resistant β-tubulin (1344 bp)	Mitosis	*A. flavus*, *A. parasiticus*, *A. nidulans*
ble	Phleomycin-binding protein (381 bp)	DNA scission	*A. nidulans*, *A. oryzae*
blmB	Bleomycin *N*-acetyltransferase (900 bp)	DNA scission	*A. oryzae*[c]
cbx	Carboxin-resistant succinic dehydrogenase mutant (567 bp)	TCA cycle	*A. oryzae*[d], *A. parasiticus*[d]
gliT	Gliotoxin sulfhydryl oxidase (1005 bp)	Not known	*A. fumigatus*[e]
hph	Hygromycin B phosphotransferase (1020 bp)	Translation	*A. niger*, *A. nidulans*, *A. sydowii*, *A. giganteus*, *A. terreus*, *A. ficcum*
hsv-1 tk	Thymidine kinase (1131 bp)	Nucleotide metabolism	*A. fumigatus*[f]
oliC	Oligomycin resistant mitochondrial ATP synthase subunit (432 bp)	ATP synthase	*A. nidulans*, *A. niger*
ptrA	Thiamine biosynthetic enzyme (984 bp)	Thiamine antagonist	*A. oryzae*, *A. nidulans*, *A. kawachii*, *A. terreus*, *A. fumigatus*

Table modified and updated from Prabha and Punekar (2004)
[a]Ahuja and Punekar (2008)
[b]Nayak et al. (2006)
[c]Suzuki et al. (2009)
[d]Shima et al. (2009)
[e]Carberry et al. (2012)
[f]Krappmann et al. (2005)

humans, affordable cost and PPT availability, *bar* selection is attractive. The compact size of *bar* (CDS around 0.5 kb) allows construction of expression vectors with larger inserts for genome engineering. The *bar* marker was successfully employed in *A. niger*: to select for EGFP expression (Dave and Punekar 2011), for deletion analysis of *A. niger citA* promoter (Dave and Punekar 2011), and to disrupt *agaA* from *A. niger* (see next section for details). It is also applied in introducing *ldh* (lactate dehydrogenase), *gbuA* (4-guanidinobutyrase), and *adc* (arginine decarboxylase) expression constructs in this fungus (unpublished work).

Several other resistance markers (not listed in Table 11.1) are available for filamentous fungi, but so far they have not been tested in Aspergilli. These include *bsd* (blasticidin S resistance in *Rhizopus niveus*; Yanai et al. 1991), *sur* (sulfonylurea resistance in *Penicillium chrysogenum*

and *Magnaporthe grisea*, Sweigard et al. 1997) markers and more recently *nat* (nourseothricin resistance in *N. crassa*, *Cryphonectria parasitica* and *P. chrysogenum*; Smith and Smith 2007; Hoff et al. 2010; de Boer et al. 2013) and *ergA* (terbinafine resistance in *P. chrysogenum*; Sigl et al. 2010).

Some of the disadvantages of resistant markers which may limit their utility include poor selection with few markers due to background growth, random integration of the marker gene into the host genome, some of the markers like *oliC* are species specific, some agents are mutagenic to the fungus (i.e., phleomycin), dominant selection can be too harsh requiring a nonselective phase before cultivating under selective pressure, and restricted availability and toxicity of the antibiotic/antimetabolite required for selection. Most importantly, dominant marker genes are susceptible to horizontal transfer in the environment.

It is therefore highly desirable to eliminate the resistance gene from the genetically modified fungus before its large scale use.

11.2.2 Nutritional Markers

Nutritional markers are usually developed from metabolic pathways. The selection involves the complementation of an auxotrophic strain with the corresponding wild type allele. Development of auxotrophic recipient *Aspergillus* strains is often tricky due to the nutritional versatility of these organisms, poor understanding of their metabolism, and restricted range of available genetic tools. The nutritional markers described so far in Aspergilli are listed in Table 11.2. Markers like *argB* and *riboB* are commonly deployed in *Aspergillus* transformation. Disruption of *argB* (coding for ornithine carbamyltransferase of arginine biosynthesis pathway) leads to arginine auxotrophy and this is exploited as a transformation marker by complementing a matching *argB⁻* strain (Table 11.2). Either an *argB* mutant or a deliberately *argB* disrupted recipient is used (Lenouvel et al. 2002). Insertion of multiple copies of the bacterial *aspA* gene relieved the nutritional deficiency in NAD-glutamate dehydrogenase negative *A. nidulans* (Hunter et al. 1992). Thus, the *aspA* marker is unique in providing an efficient means of selecting multi-copy transformants.

Arginase (encoded by *agaA*) catabolizes L-arginine to L-ornithine and urea; this defines the only route for arginine utilization in most fungi. The *A. niger agaA* gene when disrupted using a *PtrpC-bar* cassette (by homologous recombination) results in an arginase negative phenotype. The D-42 strain (*agaA::bar*) is therefore unable to grow on arginine as sole nitrogen source (Dave et al. 2012). While *agaA* mutants are reported from *N. crassa* (Morgan 1970) and *A. nidulans* (Bartnik et al. 1977), the *A. niger* D-42 strain is the first example of targeted *agaA* disruption in filamentous fungi. Combination of D-42 strain and arginase expression construct (*PcitA-agaA*) defines *agaA* as a novel nutritional marker for *A. niger* transformation. Details of applying this

selection marker may be found in an accompanying protocol (see Chap. 39), while the strategy to identify *agaA* transformants is shown in Fig. 11.1. Depending on the integration event, the transformants obtained on arginine plates may either be *agaA⁺bar⁺* (ectopic) or *agaA⁺bar⁻* (homologous). Such an *agaA/bar* combination describes a two-way selection with a potential for marker reuse.

The inability of *A. niger* D-42 strain (*agaA::bar*) to grow on arginine is consistent with a single pathway for arginine utilization. Both biochemical and bioinformatics approaches support the absence of a functional arginine decarboxylase in *A. niger*. But a pathway to catabolize agmatine was shown to exist in this fungus (Kumar 2013). Introducing a functional arginine decarboxylase is expected to confer the ability to utilize L-arginine to the D-42 mutant. Therefore, arginine decarboxylase could become another novel marker based on fungal arginine catabolism. Initial attempts to express bacterial arginine decarboxylase in *A. niger* D-42 strain for this purpose, have been unsuccessful (unpublished). However, as in *S. cerevisiae*, an alternative route for polyamine biosynthesis was reconstituted by introducing *E. coli* arginine decarboxylase (Klein et al. 1999); we still hope to develop this gene as a successful marker for *A. niger*.

11.2.3 Bidirectional Markers

At times it is advantageous if the same marker can be used for both gain and loss of function. All such two-way selection markers reported to date happen to be nutritional markers (Table 11.3). Such markers are called bidirectional markers as they provide a two-way selection. *pyrG* and *amdS* are frequently used bidirectional markers. As is typical for bidirectional markers, *pyrG* allows both positive and negative selection—*pyrG* mutants are auxotrophic for uridine/uracil but are resistant to 5-fluoroorotic acid (5-FOA). The cloned *amdS* gene was first used as homologous transformation marker in *A. nidulans* (Tilburn et al. 1983). The selection is based on utilization of acetamide by *amdS* expressing transformants. For many Aspergilli that cannot utilize acetamide,

11 Expanding the Repertoire of Selectable Markers for *Aspergillus* Transformation

Table 11.2 Nutritional markers in Aspergilli

Selection marker	Marker gene function (CDS length)	Selection	Species transformed
acuD	Isocitrate lyase (1617 bp)	Acetate utilization	*A. nidulans*
adeA	Phosphoribosylaminoimidazolesuccinocarboxamide synthase (2184 bp)	Adenine prototrophy	*A. oryzae*
adeB	Phosphoribosylaminoimidazolecarboxylase (3000 bp)	Adenine prototrophy	*A. oryzae*
agaA	Arginase (975 bp)	Arginine utilization	*A. niger*[a]
argB	Ornithine carbamyltransferase (1194 bp)	Arginine prototrophy	*A. nidulans*, *A. niger*, *A. oryzae*, *A. terreus*
aspA	Aspartase (1437 bp)	Aspartate utilization	*A. nidulans*[b]
bioDA	DAPA[f] synthase and dethiobiotin synthetase (2364 bp)	Biotin prototrophy	*A. nidulans*[c]
hemA	5-Aminolevulinate synthase (1911 bp)	5-Aminolevulinate prototrophy	*A. oryzae*
hoa	Homoserine *O*-acetyltransferase (1572 bp)	Methionine prototrophy	*A. oryzae*[d]
pkiA	Pyruvate kinase (1581 bp)	Fermentable carbon utilization	*A. nidulans*
prn	Proline catabolism (2457 bp)	Proline utilization	*A. nidulans*
pyroA	Not known (915 bp)	Pyridoxine prototrophy	*A. nidulans*[e], *A. fumigatus*[e]
qutE	Catabolic quinate dehydrogenase (462 bp)	Quinate utilization	*A. nidulans*
riboB	Putative GTP cyclohydrolase (1230 bp)	Riboflavin prototrophy	*A. nidulans*, *A. fumigatus*[e]
trpC	Trifunctional enzyme of tryptophan biosynthesis (1995 bp)	Tryptophan prototrophy	*A. nidulans*, *A. niger*

Table modified and updated from Prabha and Punekar (2004)

[a]Dave et al. (2012)
[b]Hunter et al. (1992)
[c]Magliano et al. (2011)
[d]Iimura et al. (1987)
[e]Nayak et al. (2006)
[f]DAPA 7,8-Diaminopelargonic acid

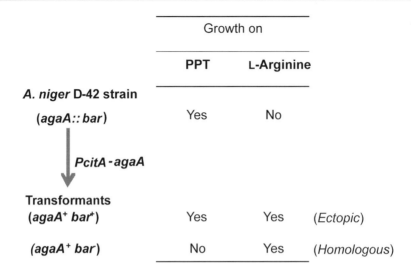

Fig. 11.1 Strategy for *agaA* based selection of transformants in *A. niger*. The *PcitA-agaA* gene construct complements and confers growth phenotype on arginine to D-42 strain (*agaA::bar*). Two types of transformants are expected depending upon the nature of integration event

amdS based selection works directly in the wild type background. In this sense, *amdS* represents a nutritional marker which is dominant. Since *amdS* transformants become sensitive to fluoroacetamide, *amdS* was also exploited as a bidirectional marker (Michielse et al. 2005). Bidirectional markers are convenient tools as they can be repeatedly exploited in the same host. This feature is particularly useful when the range of available markers is restricted. The list of bidirectional markers available for fungi, and for Aspergilli in particular, however is limited (Table 11.3).

As a bidirectional marker *pyrG* is convenient and is frequently used. The availability and cost of 5-FOA is sometimes a concern. *sC* is superior as a marker in this regard. ATP sulfurylase (encoded by *sC*) activates inorganic sulfate to form adenosine 5-phosphosulfate (APS) (Fig. 11.2a). The *sC* gene sequence is highly conserved in Aspergilli (Varadarajalu and Punekar 2005) making it easy to use across species. Since ATP sulfurylase is the first committed step in sulfate assimilation, *sC* mutants are incapable of using sulfate as source of sulfur. At the same time these mutants display selenate (a toxic analogue of sulfate) resistance. Both sC^- (selenate resistant) and sC^+ (sulfate utilization) phenotypes can thus be selected for. Homologous transformation with the *sC* marker was first reported in *A. nidulans* (Buxton et al. 1989) and later in *A. fumigatus* (De Lucas et al. 2001). A heterologous *sC* gene was used to transform *A. niger* (Buxton et al. 1989) and *A. oryzae* (Yamada et al. 1997). A mutant (sC^-) background is a prerequisite to use *sC* marker. Spontaneous *sC* mutants are easily isolated on selenate media as they are resistant to selenate. Mutations in the *sB* gene (encoding sulfate permease) could also result in selenate resistance. However, only *sC* mutants are chromate sensitive and can be distinguished from *sB* mutants (Fig. 11.2b, c). Although *sC* has a long CDS, the high degree of sequence conservation (across Aspergilli) at this genomic locus (Varadarajalu and Punekar 2005) could in principle be used for marker assisted homologous integration events.

11.2.4 Visual Marker Systems

Some reporter genes may also serve as selection markers since they allow the transformants to be visually distinguished. Examples of well-known

Table 11.3 Bidirectional markers from Aspergilli

Selection marker	Marker gene function (CDS length)	Selection	Species transformed
acuA	Acetyl CoA synthase (2013 bp)	– Acetate utilization – Fluoroacetate resistance	A. nidulans
amdS	Acetamidase (1647 bp)	– Acetamide utilization – Fluoroacetamide resistance	A. nidulans, A. niger, A. oryzae, A. ficcum, A. terreus
niaD	Nitrate reductase (2622 bp)	– Nitrate utilization – Chlorate resistance	A. nidulans, A. parasiticus, A. oryzae, A. niger, A. alliaceus, A. flavus
pyr4/pyrG	OMP decarboxylase (843 bp)	– Uridine/uracil prototrophy – 5-FOA resistance	A. niger, A. nidulans, A. oryzae, A. parasiticus, A. aculeatus, A. sojae, A. fumigatus, A. awamori
sC	ATP sulfurylase (1725 bp)	– Sulfate utilization – Selenate resistance	A. nidulans, A. oryzae[a], A. fumigatus[b], A. niger[c]

Table modified and updated from Prabha and Punekar (2004)
[a]Yamada et al. (1997)
[b]De Lucas et al. (2001)
[c]Varadarajalu and Punekar (2005)

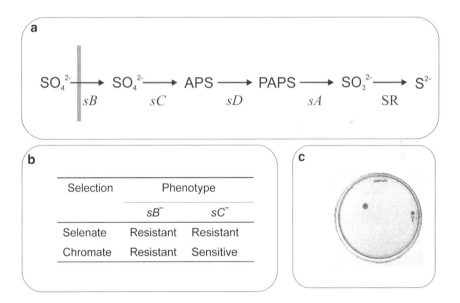

Fig. 11.2 Selection of *sC* mutants. (**a**) Sulfate assimilation pathway in *Aspergillus*, (**b**) strategy to distinguish *sC*⁻ and *sB*⁻ mutants, and (**c**) spontaneous *A. niger sC*⁻ mutant (one out of 10⁶ spores spread per plate)

reporters include *lacZ* (β-D-galactosidase acting on X-Gal), *gusA* (β-glucuronidase acting on X-Gluc), and laccase genes. They were used as reporters in *A. nidulans* (Kanemori et al. 1999) and *A. parasiticus* (Miller et al. 2005). Heterologous laccases were used as reporters in *A. nidulans* and *A. niger* (Mander et al. 2006). Since then various laccase genes from *A. niger* genome have been cloned and functionally annotated (Ramos et al. 2011). Laccases oxidize artificial substrates like ABTS (2,2-azino-di-(3-ethylbenzthiazoline sulfonate)), ADBP/DMA (4-amino-2,6-dibromophenol/3,5-dimethylaniline), and DMPPDA (*N,N*-dimethyl-*p*-phenylenediamine sulfate) thereby help locate the transformants on agar plates containing these substrates. The green fluorescence protein (GFP) and its variants have been extensively used as reporters in Aspergilli (Jiang et al. 2013; Nitsche et al. 2013). However, utility of GFP as a visual marker is constrained because scoring transformants requires their exposure to UV light. In a rare example, the *bar-egfp* protein fusion gene was created to combine selectable and visible markers for *Beauveria bassiana* (Jin et al. 2008).

11.3 Selection Markers in Genome Manipulation Strategies

Transformation with the help of different selectable markers enables genetic manipulation of fungi to study various cellular metabolic processes. Gene knock-in (gene insertion, replacement, or over-expression) and knock-out (gene deletion) approaches have led to the elucidation of many metabolic/gene functions. Industrial strain development may require introduction of one or more steps (genes) of a metabolic pathway. Multiple selectable markers or multiple use of the same marker often becomes a necessity. For instance, a quadruple auxotrophic host ($niaD^-sC^-$ $\Delta argBadeA^-/adeB^-$) was developed to construct more auxotrophic strains of *A. oryzae* (as also other deuteromycetes wherein sexual crossing is impossible) (Jin et al. 2004). Transformation in *Aspergillus* is almost always an integrative event. Therefore one is quickly restricted by the availability of fresh markers. This limitation may be overcome by devising suitable strategies and judicious use of available markers. Aspects of selectable marker exploitation are highlighted below and include locus specific (homologous) DNA integration events for gene disruption, deletion, or insertion as well as marker rescue.

Homologous recombination is necessary for targeted integration of the DNA along with the selectable marker. Homologous recombination occurs by a single crossover event (type I integration) resulting in the insertion of foreign DNA (and the marker) into the target locus or by a double crossover event (type III integration) leading to replacement of target gene by selection marker.

Both these strategies are extensively employed with minor modifications. However, homologous recombination events are rare while ectopic integrations (illegitimate recombination, type II integration) are more common in Aspergilli. Frequency of homologous recombination increases when longer homologous flanking sequences, corresponding to the target locus, are used. While 30–50 bp flanking sequences serve well in *S. cerevisiae* (Hua et al. 1997), much longer flanking sequences (0.5–2.0 kb) are required in Aspergilli (Meyer 2008). The frequency of homologous integration was significantly improved when the components of the nonhomologous end-joining (NHEJ) pathway (of DNA recombination) were deleted in *N. crassa* (Ninomiya et al. 2004). This was extended to Aspergilli by marker-assisted knock out in *A. nidulans* (*nkuA::argB*; Nayak et al. 2006), *A. niger* (*kusA::amdS* and *kueA::pyrG*; Honda et al. 2011; Meyer et al. 2007), *A. oryzae* (*ku70/80::ptrA*; Takahashi et al. 2006), *A. fumigatus* (*ku80::pyrG*; Krappmann et al. 2006), and *A. sojae* (*ku70/80::ptrA*;Takahashi et al. 2006). This approach has also enabled the genome-wide deletion project in *A. nidulans* (Meyer et al. 2011).

11.3.1 Split Marker

The split marker technique, initially developed in *S. cerevisiae* (Fairhead et al. 1996), has found application in many filamentous fungi (Kuck and Hoff 2010). In this strategy, instead of a complete marker sequence, two partially overlapping fragments of the marker are used. The fragments are so designed that the functional marker is generated only after a recombination event (Fig. 11.3a). This technique requires two PCR products (generated either by two step PCR or by ligation fusion PCR), each comprising a fusion of flanking sequence of the target gene and an appropriate partial–inactive fragment of the marker DNA. The two PCR products have internal sequence overlaps such that a complete marker CDS is generated after recombination. On transformation, three crossover events are required for homologous integration of the split marker replacing the

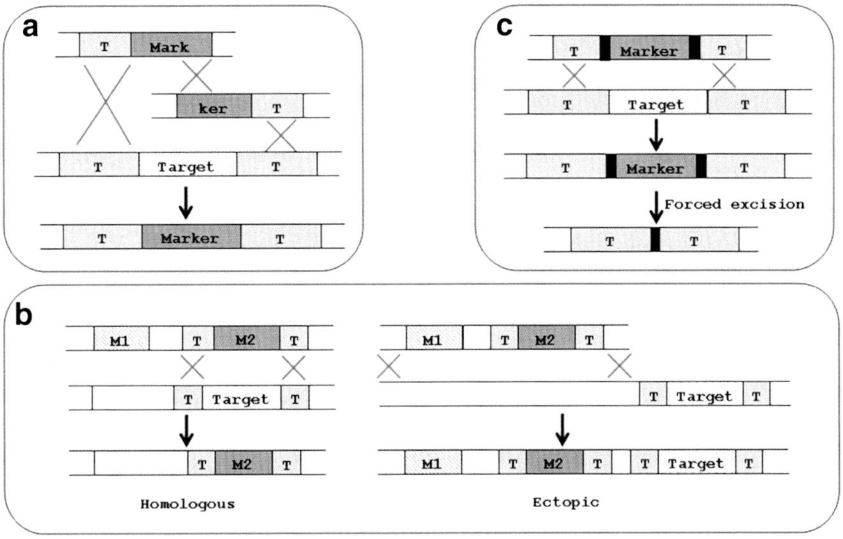

11 Expanding the Repertoire of Selectable Markers for *Aspergillus* Transformation

Fig. 11.3 Use of selection markers in fungal genome manipulation. Strategies for split marker (**a**), dual selection (**b**), and marker rescue (**c**). [*T* flanking homologous regions of the target gene, *M1* positive selectable marker, *M2* negative selectable marker, *black bars* direct DNA repeats]

target gene (Kuck and Hoff 2010). The probability of this event increases when the fragments are in close proximity and when they are at the desired locus. Ectopic integration of either of the fragments of such a split marker will be nonfunctional and hence those transformants are not selected in the screen. The split marker approach significantly improves the chances of homologous over ectopic integrations; but the transformation frequency decreases notably as three recombination events are required. An *hph* split marker was used to disrupt *pyrG* and *acuB* genes of *A. niger* (Nielsen et al. 2007); interestingly, *pyrG* was later rescued (also see Sect. 3.3 below) by eliminating the *hph* marker. Disruption of *A. fumigatus* genes involved in trehalose biosynthetic pathway (*tpsA* and *tpsB*, Al-Bader et al. 2010) and also of photoreceptors (*lreA* and *fphA*, Fuller et al. 2013) was achieved through the split marker strategy.

11.3.2 Dual Selection

Dual selection strategy makes use of two markers— one for positive and another for negative selection. This allows one to distinguish between homologous (positive selection) and ectopic (negative selection) integrations of a gene-cassette. The positive selection marker (often resistance or prototrophy) is designed to be locus specific while the other marker (often lethality or auxotrophic) provides for negative selection. The ectopic transformants express both the markers whereas a targeted gene replacement results in the loss of negative marker and is amenable to positive selection (Fig. 11.3b). Although tested in many fungi (such as Gardiner and Howlett 2004; Khang et al. 2005), dual selection is reported in only two species of Aspergilli. The *niaD*, *areA*, and the tannase genes of *A. sojae* were individually disrupted by dual selection using *pyrG* and *oliC* markers (Takahashi et al. 2004). Homologous integration at *pyrG* locus of *A. awamori* was achieved through the *hph* and *amdS* twin marker system (Michielse et al. 2005).

11.3.3 Marker Rescue

The list of markers at our disposal is not extensive and therefore recycling/reusing a marker is of essence. The issue is particularly acute when the same recipient has to be transformed repeatedly.

The problem was cleverly addressed by the advent of a "blaster cassette" tool for sequential deletion of multiple genes in yeast. This method allows the rescue and reuse of a single selectable marker, mostly a bidirectional marker. The yeast *URA*-blaster cassettes are tripartite sequences consisting of *URA3* gene flanked on both sides by direct DNA repeats (Alani et al. 1987; Fonzi and Irwin 1993). Transformants become prototrophic for uracil upon *URA3* integration. Forced excision of *URA3*, through recombination between the two direct repeats, is achieved in the presence of 5-FOA. The desired locus continues to remain disrupted as one copy of the direct repeat is left behind (Fig. 11.3c). This rescues the *URA* marker and the organism is rendered auxotrophic for uracil. The *URA*-blaster cassette is again available for the next round of gene integration/disruption.

A *URA3* homolog was used to develop the corresponding blaster cassette (*pyrG*-blaster) for Aspergilli. The *pyrG*-blaster was successfully used to disrupt *rodA* gene of *A. fumigatus* (d'Enfert 1996) and *lacA* and *glaA* genes of *A. niger* (Storms et al. 2005). Deletion of *A. nidulans aroC* was achieved with a *pyrG*-blaster module that combines ET cloning (RecE and RecT mediated recombinogenic engineering approach) to rescue the marker (Krappmann and Braus 2003). Marker rescue based on the *cre/loxP* recombination system is also available for Aspergilli. Here, a *loxP* direct DNA repeat is placed on both sides of *pyrG* marker. The Cre recombinase recognizes and catalyzes reciprocal recombination between the pair of *lox* repeats, thereby rescuing the marker. Marker rescue using *cre/loxP* system was used to disrupt *pabaA* and *veA* loci of *A. fumigatus* (Krappmann et al. 2005) and *ligD* gene of *A. oryzae* (Mizutani et al. 2012). Successive disruption of *yA* and *wA* genes was achieved in *A. nidulans* using *cre/loxP* blaster cassettes (Forment et al. 2006). As an improvement, a self-excising marker cassette that employs the prokaryotic β-rec/*six* site-specific recombination system was adopted to recover the *ptrA* marker after disrupting *abr2* and *pksP* genes of *A. fumigatus* (Hartmann et al. 2010).

Marker rescue strategy may also be used to transiently disrupt a gene. As noted above, NHEJ disruption favors homologous recombination. However, such NHEJ disrupted strains display increased sensitivity towards DNA damaging conditions such as γ irradiation (Meyer et al. 2007). Restoring the NHEJ function after the required genetic manipulation is therefore desirable. Transient disruption and subsequent recovery of NHEJ function was demonstrated in *A. nidulans*. The *pyrG*-blaster module served to disrupt *nkuA*; required genetic manipulation was done and finally *nkuA* function was regained by recombining the flanking direct repeats of *nkuA* itself (Nielsen et al. 2008).

11.4 Conclusions

The importance of Aspergilli as an industrially and medically important group of fungi is well established. A growing interest in both basic and applied research on Aspergilli acknowledges this fact. Many whole genome sequences are now available and await exploitation. Both genome annotation and strain development require genetic manipulations. Selectable markers are central to these objectives. A range of selection markers are now available to engineer *Aspergillus* genomes. More are being developed based on the knowledge of the fungal genomes. Selectable markers find direct applications in strain improvement programs for industrially important Aspergilli. Strategies for safe and judicious use of available markers through locus specific integration, marker rescue, and self-excision subsequent to site-specific recombination continue to emerge.

Acknowledgements We acknowledge Department of Biotechnology, Government of India (DBT); Board of Research in Nuclear Science—Department of Atomic Energy (BRNS-DAE), Government of India and New Millennium Indian Technology Leadership Initiative of Council of Scientific and Industrial Research (NMITLI-CSIR), India for funding this research. We also thank Bayer Crop-Science for providing technical grade PPT (glufosinate ammonium).

References

Ahuja M, Punekar NS (2008) Phosphinothricin resistance in *Aspergillus niger* and its utility as a selectable transformation marker. Fungal Genet Biol 45:1103–1110

Alani E, Cao L, Kleckner N (1987) A method for gene disruption that allows repeated use of *URA3* selection in the construction of multiply disrupted yeast strains. Genetics 116:541–545

Avalos J, Geever RF, Case ME (1989) Bialaphos resistance as a dominant selectable marker in *Neurospora crassa*. Curr Genet 16:369–372

Al-Bader N, Vanier G, Liu L, Gravelat FN, Urb M, Hoareau CM, Campoli P, Chabot J, Filler SG, Sheppard DC (2010) Role of trehalose biosynthesis in *Aspergillus fumigatus* development, stress response, and virulence. Infect Immun 78:3007–3018

Arnaud MB, Cerqueira GC, Inglis DO, Skrzypek MS, Binkley J, Chibucos MC, Crabtree J, Howarth C, Orvis J, Shah P, Wymore F, Binkley G, Miyasato SR, Simison M, Sherlock G, Wortman JR (2012) The *Aspergillus* Genome Database (AspGD): recent developments in comprehensive multispecies curation, comparative genomics and community resources. Nucleic Acids Res 40:653–659

Barbesgaard P, Heldt-Hansen HP, Diderichsen B (1992) On the safety of *Aspergillus oryzae*: a review. Appl Microbiol Biotechnol 36:569–572

Bartnik E, Guzewska J, Klimczuk J, Piotrowska M, Weglenski P (1977) Regulation of arginine catabolism in *Aspergillus nidulans*. In: Smith JE, Pateman JA (eds) Genetics and physiology of *Aspergillus*. Academic, London, pp 243–254

Bizukojc M, Ledakowicz S (2009) Physiological, morphological and kinetic aspects of lovastatin biosynthesis by *Aspergillus terreus*. Biotechnol J 4:647–664

Buxton FP, Gwynne DI, Davies RW (1989) Cloning of a new bidirectionally selectable marker for *Aspergillus* strains. Gene 84:329–334

Carberry S, Molloy E, Hammel S, O'Keeffe G, Jones GW, Kavanagh K, Doyle S (2012) Gliotoxin effects on fungal growth: mechanisms and exploitation. Fungal Genet Biol 49:302–312

d'Enfert C (1996) Selection of multiple disruption events in Aspergillus fumigatus using the orotidine-5′-decarboxylase gene, pyrG, as a unique transformation marker. Curr Genet 30:76–82

Dave K, Ahuja M, Jayashri TN, Sirola RB, Punekar NS (2012) A novel selectable marker based on *Aspergillus niger* arginase expression. Enzyme Microb Technol 51:53–58

Dave K, Punekar NS (2011) Utility of *Aspergillus niger* citrate synthase promoter for heterologous expression. J Biotechnol 155:173–177

de Boer P, Bronkhof J, Dukić K, Kerkman R, Touw H, van den Berg M, Offringa R (2013) Efficient gene targeting in *Penicillium chrysogenum* using novel Agrobacterium-mediated transformation approaches. Fungal Genet Biol 61:9–14

De Lucas JR, Dominguez AI, Higuero Y, Martinez O, Romero B, Mendoza A, Garcia-Bustos JF, Laborda F (2001) Development of a homologous transformation system for the opportunistic human pathogen *Aspergillus fumigatus* based on the *sC* gene encoding ATP sulfurylase. Arch Microbiol 176:106–113

Gibbons JG, Rokas A (2013) The function and evolution of the *Aspergillus* genome. Trends Microbiol 21:14–22

Fairhead C, Llorente B, Denis F, Soler M, Dujon B (1996) New vectors for combinatorial deletions in yeast chromosomes and for gap-repair cloning using 'split-marker' recombination. Yeast 12:1439–1457

Fleissner A, Dersch P (2010) Expression and export: recombinant protein production systems for *Aspergillus*. Appl Microbiol Biotechnol 87:1255–1270

Fonzi WA, Irwin MY (1993) Isogenic strain construction and gene mapping in *Candida albicans*. Genetics 134:717–728

Forment JV, Ramon D, MacCabe AP (2006) Consecutive gene deletions in *Aspergillus nidulans*: application of the *Cre/loxP* system. Curr Genet 50:217–224

Fuller KK, Ringelberg CS, Loros JJ, Dunlap JC (2013) The fungal pathogen *Aspergillus fumigatus* regulates growth, metabolism, and stress resistance in response to light. mBio 4:e00142-13

Gardiner DM, Howlett BJ (2004) Negative selection using thymidine kinase increases the efficiency of recovery of transformants with targeted genes in the filamentous fungus *Leptosphaeria maculans*. Curr Genet 45:249–255

Hartmann T, Dumig M, Jaber BM, Szewczyk E, Olbermann P, Morschhauser J, Krappmann S (2010) Validation of a self-excising marker in the human pathogen *Aspergillus fumigatus* by employing the beta-rec/*six* site-specific recombination system. Appl Environ Microbiol 76:6313–6317

Hoff B, Kamerewerd J, Sigl C, Zadra I, Kück U (2010) Homologous recombination in the antibiotic producer *Penicillium chrysogenum*: strain *ΔPcku70* shows up-regulation of genes from the HOG pathway. Appl Microbiol Biotechnol 85(4):1081–1094

Honda Y, Kobayashi K, Kirimura K (2011) Increases in gene-targeting frequencies due to disruption of *kueA* as a *ku80* homolog in citric acid-producing *Aspergillus niger*. Biosci Biotechnol Biochem 75:1594–1596

Hua SB, Qiu M, Chan E, Zhu L, Luo Y (1997) Minimum length of sequence homology required for in vivo cloning by homologous recombination in yeast. Plasmid 38:91–96

Hunter GD, Bailey CR, Arst HN (1992) Expression of a bacterial aspartase gene in *Aspergillus nidulans*: an efficient system for selecting multicopy transformants. Curr Genet 22:377–383

Iimura Y, Gomi K, Uzu H, Hara S (1987) Transformation of *Aspergillus oryzae* through plasmid-mediated complementation of the methionine-auxotrophic mutation. Agric Biol Chem 51:323–328

Jiang D, Zhu W, Wang Y, Sun C, Zhang KQ, Yang J (2013) Molecular tools for functional genomics in filamentous fungi: recent advances and new strategies. Biotechnol Adv 31:1562–1574

Jin FJ, Maruyama J, Juvvadi PR, Arioka M, Kitamoto K (2004) Development of a novel quadruple auxotrophic host transformation system by *argB* gene disruption using *adeA* gene and exploiting adenine auxotrophy in *Aspergillus oryzae*. FEMS Microbiol Lett 239:79–85

Jin K, Zhang Y, Luo Z, Xiao Y, Fan Y, Wu D, Pei Y (2008) An improved method for *Beauveria bassiana* transformation using phosphinothricin acetlytransferase and green fluorescent protein fusion gene as a selectable and visible marker. Biotechnol Lett 30:1379–1383

Kanemori Y, Gomi K, Kitamoto K, Kumagai C, Tamura G (1999) Insertion analysis of putative functional elements in the promoter region of the *Aspergillus oryzae* Taka-amylase A gene (*amyB*) using a heterologous *Aspergillus nidulans amdS-lacZ* fusion gene system. Biosci Biotechnol Biochem 63:180–183

Karaffa L, Sandor E, Fekete E, Szentirmai A (2001) The biochemistry of citric acid accumulation by *Aspergillus niger*. Acta Microbiol Immunol Hung 48:429–440

Khang CH, Park SY, Lee YH, Kang S (2005) A dual selection based, targeted gene replacement tool for *Magnaporthe grisea* and *Fusarium oxysporum*. Fungal Genet Biol 42:483–492

Kis-Papo T, Oren A, Wasser SP, Nevo E (2003) Survival of filamentous fungi in hypersaline Dead Sea water. Microb Ecol 45:183–190

Klein RD, Geary TG, Gibson AS, Favreau MA, Winterrowd CA, Upton SJ, Keithly JS, Zhu G, Malmberg RL, Martinez MP, Yarlett N (1999) Reconstitution of a bacterial/plant polyamine biosynthesis pathway in *Saccharomyces cerevisiae*. Microbiology 145:301–307

Krappmann S, Bayram O, Braus GH (2005) Deletion and allelic exchange of the *Aspergillus fumigatus veA* locus via a novel recyclable marker module. Eukaryot Cell 4:1298–1307

Krappmann S, Braus GH (2003) Deletion of *Aspergillus nidulans aroC* using a novel blaster module that combines ET cloning and marker rescue. Mol Genet Genomics 268:675–683

Krappmann S, Sasse C, Braus GH (2006) Gene targeting in *Aspergillus fumigatus* by homologous recombination is facilitated in a nonhomologous end- joining-deficient genetic background. Eukaryot Cell 5:212–215

Kuck U, Hoff B (2010) New tools for the genetic manipulation of filamentous fungi. Appl Microbiol Biotechnol 86:51–62

Lenouvel F, van de Vondervoort P, Visser J (2002) Disruption of the *Aspergillus niger argB* gene: a tool for transformation. Curr Genet 41:425–432

Lubertozzi D, Keasling JD (2009) Developing *Aspergillus* as a host for heterologous expression. Biotechnol Adv 27:53–75

Magliano P, Flipphi M, Sanglard D, Poirier Y (2011) Characterization of the *Aspergillus nidulans* biotin biosynthetic gene cluster and use of the *bioDA* gene as a new transformation marker. Fungal Genet Biol 48:208–215

Mander GJ, Wang H, Bodie E, Wagner J, Vienken K, Vinuesa C, Foster C, Leeder AC, Allen G, Hamill V, Janssen GG, Dunn-Coleman N, Karos M, Lemaire HG, Subkowski T, Bollschweiler C, Turner G, Nusslein B, Fischer R (2006) Use of laccase as a novel, versatile reporter system in filamentous fungi. Appl Environ Microbiol 72:5020–5026

Meyer V, Arentshorst M, El-Ghezal A, Drews AC, Kooistra R, van den Hondel CA, Ram AF (2007) Highly efficient gene targeting in the *Aspergillus niger kusA* mutant. J Biotechnol 128:770–775

Meyer V (2008) Genetic engineering of filamentous fungi—progress, obstacles and future trends. Biotechnol Adv 26:177–185

Meyer V, Wu B, Ram AF (2011) *Aspergillus* as a multipurpose cell factory: current status and perspectives. Biotechnol Lett 33:469–476

Michielse CB, Arentshorst M, Ram AF, van den Hondel CA (2005) Agrobacterium-mediated transformation leads to improved gene replacement efficiency in *Aspergillus awamori*. Fungal Genet Biol 42:9–19

Miller MJ, Roze LV, Trail F, Linz JE (2005) Role of cis-acting sites *NorL*, a TATA box, and *AflR1* in *nor*-1 transcriptional activation in *Aspergillus parasiticus*. Appl Environ Microbiol 71:1539–1545

Mizutani O, Masaki K, Gomi K, Iefuji H (2012) Modified *Cre-loxP* recombination in *Aspergillus oryzae* by direct introduction of *Cre* recombinase for marker gene rescue. Appl Environ Microbiol 78:4126–4133

Morgan DH (1970) Selection and characterisation of mutants lacking arginase in *Neurospora crassa*. Mol Gen Genet 108:291–302

Nayak T, Szewczyk E, Oakley CE, Osmani A, Ukil L, Murray SL, Hynes MJ, Osmani SA, Oakley BR (2006) A versatile and efficient gene-targeting system for *Aspergillus nidulans*. Genetics 172:1557–1566

Nielsen ML, de Jongh WA, Meijer SL, Nielsen J, Mortensen UH (2007) Transient marker system for iterative gene targeting of a prototrophic fungus. Appl Environ Microbiol 73:7240–7245

Nielsen JB, Nielsen ML, Mortensen UH (2008) Transient disruption of non-homologous end-joining facilitates targeted genome manipulations in the filamentous fungus *Aspergillus nidulans*. Fungal Genet Biol 45:165–170

Ninomiya Y, Suzuki K, Ishii C, Inoue H (2004) Highly efficient gene replacements in *Neurospora* strains deficient for nonhomologous end-joining. Proc Natl Acad Sci U S A 101:12248–12253

Nitsche BM, Burggraaf-van Welzen A-M, Lamers G, Meyer V, Ram AF (2013) Autophagy promotes survival in aging submerged cultures of the filamentous fungus *Aspergillus niger*. Appl Microbiol Biotechnol 97:8205–8218

Okabe M, Lies D, Kanamasa S, Park EY (2009) Biotechnological production of itaconic acid and its biosynthesis in *Aspergillus terreus*. Appl Microbiol Biotechnol 84:597–606

Prabha VL, Punekar NS (2004) Genetic transformation in Aspergilli: tools of trade. Indian J Biochem Biophys 41:205–215

Ramos JA, Barends S, Verhaert RM, de Graaff LH (2011) The *Aspergillus niger* multicopper oxidase family: analysis and overexpression of laccase-like encoding genes. Microb Cell Fact 10:78–89

Ruiz-Diez B (2002) Strategies for the transformation of filamentous fungi. J Appl Microbiol 92:189–195

Shima Y, Ito Y, Kaneko S, Hatabayashi H, Watanabe Y, Adachi Y, Yabe K (2009) Identification of three mutant loci conferring carboxin-resistance and development of a novel transformation system in *Aspergillus oryzae*. Fungal Genet Biol 46:67–76

Sigl C, Handler M, Sprenger G, Kurnsteiner H, Zadra I (2010) A novel homologous dominant selection marker for genetic transformation of *Penicillium chrysogenum*: overexpression of squalene epoxidase-encoding *ergA*. J Biotechnol 151:307–311

Smith RP, Smith ML (2007) Two yeast plasmids that confer nourseothricin-dihydrogen sulfate and hygromycin B resistance in *Neurospora crassa* and *Cryphonectria parasitica*. Fungal Genet Newsl 54:12–13

Storms R, Zheng Y, Li H, Sillaots S, Martinez-Perez A, Tsang A (2005) Plasmid vectors for protein production, gene expression and molecular manipulations in *Aspergillus niger*. Plasmid 53:191–204

Su X, Schmitz G, Zhang M, Mackie RI, Cann IK (2012) Heterologous gene expression in filamentous fungi. Adv Appl Microbiol 81:1–61

Kumar S (2013) Metabolism of guanidinium compounds in *Aspergillus niger*: role of ureohydrolases. Ph.D. thesis, Indian Institute of Technology Bombay, India

Suzuki S, Tada S, Fukuoka M, Taketani H, Tsukakoshi Y, Matsushita M, Oda K, Kusumoto K, Kashiwagi Y, Sugiyama M (2009) A novel transformation system using a bleomycin resistance marker with chemosensitizers for *Aspergillus oryzae*. Biochem Biophys Res Commun 383:42–47

Sweigard JA, Carroll AM, Farrall L, Valent B (1997) A series of vectors for fungal transformation. Fungal Genet Newsl 44:52–53

Takahashi T, Hatamoto O, Koyama Y, Abe K (2004) Efficient gene disruption in the koji-mold *Aspergillus sojae* using a novel variation of the positive-negative method. Mol Genet Genomics 272:344–352

Takahashi T, Masuda T, Koyama Y (2006) Enhanced gene targeting frequency in *ku70* and *ku80* disruption mutants of *Aspergillus sojae* and *Aspergillus oryzae*. Mol Genet Genomics 275:460–470

Tilburn J, Scazzocchio C, Taylor GG, Zabicky-Zissman JH, Lockington RA, Davies RW (1983) Transformation by integration in *Aspergillus nidulans*. Gene 26:205–221

Trogisch GD, Kocher H, Ullrich WR (1989) Effects of glufosinate on anion uptakein *Lemna gibba* G1. Z Naturforsch 44:33–38

Ullrich WR, Ullricheberius CI, Kocher H (1990) Uptake of glufosinate andconcomitant membrane-potential changes in *Lemna-Gibba* G1. Pestic Biochem Physiol 37:1–11

Varadarajalu LP, Punekar NS (2005) Cloning and use of *sC* as homologous marker for *Aspergillus niger* transformation. J Microbiol Methods 61:219–224

Weld RJ, Plummer KM, Carpenter MA, Ridgway HJ (2006) Approaches to functional genomics in filamentous fungi. Cell Res 16:31–44

Yamada O, Lee BR, Gomi K (1997) Transformation system for *Aspergillus oryzae* with double auxotrophic mutations, *niaD* and *sC*. Biosci Biotechnol Biochem 61:1367–1369

Yanai K, Horiuchi H, Takagi M, Yano K (1991) Transformation of *Rhizopus niveus* using bacterial blasticidin S resistance gene as a dominant selectable marker. Curr Genet 19:221–226

Arginase (*agaA*) as a Fungal Transformation Marker

12

Kashyap Dave, Manmeet Ahuja, T.N. Jayashri,
Rekha Bisht Sirola, Khyati Dave,
and Narayan S. Punekar

12.1 Introduction

Filamentous fungi, especially Aspergilli, are exploited for the expression of homologous and heterologous proteins (Lubertozzi and Keasling 2006; Jiang et al. 2013). Different transformation methods for introducing the gene of interest have been described and modified for individual fungi. Various genetic tools are in place for manipulating the fungal genome wherein markers and reporters are the key components (Varadarajalu and Punekar 2004); transformation selection markers enable us to pick up transformants in the parent background. The selectable markers come in two varieties: the nutritional markers and resistance markers that rescue cells from inhibitors (such as antibiotics and antimetabolites). Nutritional markers require an appropriate auxotrophic background in the recipient. The transformants are selected by complementing the nutritional deficiency when the corresponding wild type allele (could be homologous or heterologous) is introduced. Despite the need for a suitable auxotrophic recipient, nutritional markers score over resistance markers as the likely horizontal transfer of resistance (usually dominant) is avoided. The nutritional markers may involve an essential anabolic (such as *argB* and *pyrG*) or a suitable catabolic (such as *prn* and *pkiA*) gene (Durrens et al. 1986; Goosen et al. 1987; de Graaff et al. 1988; Lenouvel et al. 2002). Some nutritional markers can be conveniently operated in both the directions (examples of bidirectional markers include *pyrG*, *sC*, and *niaD*).

Arginase (encoded by *agaA*) hydrolyzes arginine to ornithine and urea and defines the only route for L-arginine utilization in fungi (Dave et al. 2012). Disruption of *agaA* therefore creates a conditional auxotroph—normal growth on a medium containing any nitrogen source except arginine. Such *agaA⁻* strains regain their ability to grow on L-arginine medium upon complementation by an arginase expression construct. An *agaA⁻* host strain along with *agaA* expression construct thus defines a novel nutritional selection for fungal transformation. The protocol to demonstrate *agaA* as a fungal transformation marker is described with *Aspergillus niger* as the model.

12.2 Materials

12.2.1 Chemicals and Supplies

1. Double distilled water (DDW)
2. L-Arginine, Tween 20, and bovine serum albumin (Sigma Aldrich, Gillingham, UK)

K. Dave, Ph.D. • M. Ahuja, Ph.D. • T.N. Jayashri, Ph.D.
R.B. Sirola, Ph.D. • K. Dave, M.Sc.
N.S. Punekar, Ph.D. (✉)
Department of Bioscience and Bioengineering,
Indian Institute of Technology Bombay,
Powai, Mumbai, Maharashtra 400 076, India
e-mail: kashyapdavv@gmail.com; khyatimehta@iitb.ac.in;
nsp@iitb.ac.in

M.A. van den Berg and K. Maruthachalam (eds.), *Genetic Transformation Systems in Fungi, Volume 2*, Fungal Biology, DOI 10.1007/978-3-319-10503-1_12,
© Springer International Publishing Switzerland 2015

3. Sucrose, NaCl, $CaCl_2$, sorbitol, Tris, PEG 8000, Triton X-100, Agar (USB chemicals, Affymetrix, Inc. Ohio, USA)
4. Dextrose, KH_2PO_4, Na_2HPO_4, $MgSO_4 \cdot 7H_2O$, NH_4NO_3, $ZnSO_4 \cdot 7H_2O$, $MnSO_4 \cdot 7H_2O$, $Na_2MoO_4 \cdot H_2O$, $FeCl_3 \cdot 6H_2O$, and $CuSO_4 \cdot H_2O$ (Analytical grade from Merck, India)
5. Agarose (SeaKem LE, Lonza Cologne GmbH, USA)
6. Enzymes: Lysing enzyme (Sigma Aldrich, L-1412), restriction enzymes, DNA ligase, *Taq* polymerase (New England Biolabs, Ipswich, USA or MBI Fermentas, St. Leon-Rot, Germany)
7. Molecular biology kits: GenJET plasmid miniprep kit (MBI Fermentas, St. Leon-Rot, Germany), Nucleobond AX plasmid midiprep kit (for extraction of high purity DNA for transformation; from Macherey-Nagal, Duren, Germany), and GenJET gel extraction kit (from QIAGEN Gmbh, Hilden, Germany)
8. Mira cloth (CalBiochem, Merck KGaA, Germany)
9. Miscellaneous: Muslin cloth, Hettich/Corex glass tubes, micro-centrifuge tubes (1.5 mL capacity), micropipettes, disposable tips, glass Petri-plates (washed with DDW)
10. Water bath shaker
11. Light microscope
12. Neubauer counting chamber
13. Refrigerated centrifuge (Hettich, Germany) with swing out rotor

12.2.2 Organism

The *agaA* disrupted strain (strain D-42; *agaA::bar*) was derived from *A. niger* NCIM 565 (National Collection of Industrial Microorganisms, National Chemical Laboratories, Pune, India) (Dave et al. 2012). The *agaA* (GenBank Acc. No. AF242315.2) flanking sequences were used for targeted disruption by *bar* marker.[1]

---

[1] The *agaA* selection requires a recipient strain that has arginase-less phenotype. This could be achieved by disruption or mutation in the host *agaA* gene. The *agaA*

12.2.3 Media Components and Reagents[2]

1. Potato dextrose agar (PDA)—Weigh peeled and ground potato 200 g, dextrose 25 g, and agar 20 g and make the total volume to one liter with tap water (do not adjust pH). Maintain the fungal culture on PDA.
2. Minimal medium (MM)—Weigh dextrose 10.0 g, KH_2PO_4 3.0 g, Na_2HPO_4 6.0 g, $MgSO_4 \cdot 7H_2O$ 0.5 g, NH_4NO_3 2.25 g, $ZnSO_4 \cdot 7H_2O$ 10.0 mg, $MnSO_4 \cdot 7H_2O$ 3.0 mg, $Na_2MoO_4 \cdot H_2O$ 1.5 mg, $FeCl_3 \cdot 6H_2O$ 20.0 mg, and $CuSO_4 \cdot H_2O$ 1.0 mg. Dissolve these in minimal volume of DDW. Adjust the pH between 5.5 and 6.0 with 0.1 N HCl and make the volume up to 1.0 L. Whenever required, add agar at 2.0 % (w/v) for solid medium.
3. Medium to select *agaA*[+] transformants (MM + Arg + S)—Include 2 mM of L-arginine as the nitrogen source instead of NH_4NO_3 in MM.[3] Also add sucrose to this MM at a final concentration of 1.0 M, for osmotic stability of the protoplasts. Whenever required, add agar at 2.0 % (w/v) for solid medium.
4. Medium to passage *agaA*[+] transformants (MM + Arg)—Include 2 mM of L-arginine as the nitrogen source in place of NH_4NO_3 in MM. Whenever required, add agar at 2.0 % (w/v) for solid medium.
5. Tween solution (DDW + T)—Solution containing 0.005 % (v/v) Tween 20 in DDW.
6. Reagents for transformation:
   (a) Osmotic medium (OM): Solution containing 0.27 M $CaCl_2$ and 0.6 M NaCl prepared in DDW.

---

mutants of fungal recipient strains can be isolated and are reported for *Neurospora crassa* (Morgan 1970) as well as *A. nidulans* (Bartnik et al. 1977). However, a recipient strain with *agaA* deletion or disruption is desirable; this circumvents complications due to spontaneous reversion possible with point mutations.

[2] Sterilize all growth media and transformation reagents by autoclaving. Store the transformation reagents at 4 °C. Each Petri plate should be dispensed with 20 mL of uniformly spread agar medium.

[3] Sterilize L-arginine solution separately by filtration and then add it to the autoclaved MM. Ensure that the final pH is between 5.5 and 6.0.

12 Arginase (*agaA*) as a Fungal Transformation Marker

(b) Double strength Sorbitol/Tris/Calcium chloride (2×STC): Solution containing 2.4 M sorbitol, 20 mM Tris, 100 mM CaCl$_2$, and 70 mM NaCl prepared in DDW. Adjust the pH of this solution to 7.5 with HCl. When required, dilute 2×STC with DDW to prepare 1×STC.

(c) PEG(t): Solution containing 25 % PEG 8000 (w/v) in 1×STC.[4]

(d) Triton X-100: 10 % Triton X-100 solution (v/v) in DDW.

12.3 Methods

The procedures given below describe general protocols for *agaA* vector construction, protoplast preparation and transformation and selection of *agaA* transformants.

12.3.1 agaA Vector Construction

All cloning and propagation of plasmids are done in *Escherichia coli* XL1 Blue (Stratagene, CA, USA) according to standard protocols (Sambrook and Russell 2001). The functional *A. niger citA* promoter (Dave and Punekar 2011) is cloned in frame with *A. niger agaA* cDNA to construct the *PcitA-agaA* expression cassette (plasmid pΔXCA; Dave et al. 2012).[5]

12.3.2 Preparation of Protoplasts[6]

1. Inoculate MM with 10^6–10^7 spores of *A. niger* D-42.[7] Grow for 20 h at 30 °C in a shaker and harvest just when mycelial branching has started.

2. Harvest and wash mycelia on wet muslin cloth with DDW and let the water drain off. Wash these mycelia (on the cloth itself) with about 60–80 mL of OM.[8]

3. Mix lysing enzyme (20 mg/mL) gently into OM.[9]

4. Suspend harvested mycelia in OM containing lysing enzyme. For every spatula full of mycelia add 5 mL of OM containing lysing enzyme. Do not exceed 40 mL suspension in a 100 mL flask.

5. Keep the mycelial suspension on ice until a fresh stock solution of bovine serum albumin (BSA) is prepared and added to a final concentration of 2 mg/mL.[10]

6. Place the above mycelial suspension in a water bath at 37 °C for 4 h, while gentle shaking. Monitor the formation of protoplasts every hour, under a light microscope.

---

[4]Prepare PEG(t) solution freshly before transformation.

[5]Using *agaA* as a nutritional selection marker in *A. niger* was demonstrated with two different promoters namely, the citrate synthase promoter (*PcitA*) of *A. niger* and a truncated tryptophan synthase promoter (*PtrpC*) of *A. nidulans* (Dave et al. 2012). The promoter of *agaA* itself is under nitrogen metabolite regulation and has regulatory elements that respond to L-arginine induction. Therefore, it is desirable to construct a marker that expresses *agaA* cDNA from a constitutive promoter. Heterologous expression of *A. niger agaA* cDNA could complement a *car1* (yeast *agaA* homolog) deletion in *S. cerevisiae* (Jayashri 2006). A constitutive yeast promoter (*Pgpd*) was used for this purpose. Any functional combination of a promoter and arginase cDNA would work well as arginase marker for selection.

---

[6]Carry out all the steps for the preparation and transformation of protoplasts under aseptic conditions.

[7]Maintain the fungal stock cultures as PDA slants. Inoculate the spores from the stock cultures on PDA plates to generate seed cultures. Allow fungal growth on these plates at 37 °C for initiation of conidiation (3–4 days) and then transfer them to room temperature for the spores to mature (10–12 days). Harvest spores from fresh seed plates (12–15 days old) to use as inoculums. Prepare spore suspensions for inoculation in DDW with 0.005 % Tween 20. Count the spores using Neubauer counting chamber and inoculate with approximately 10^8 spores per 100 mL of medium. For preparing protoplasts, inoculate *A. niger* spores into four 1 L flasks (each containing 400 mL of MM).

[8]Pre-cool DDW and OM at 4 °C for washing mycelia. Wash the harvested mycelia with OM till it looks shiny/slimy.

[9]The amount of lysing enzyme used for making protoplasts needs standardization depending upon its strength/efficiency. (You may need only 10 mg/mL if fresh enzymes are used).

[10]Pre-dissolve BSA powder in a small volume of OM. Ensure that the solution does not froth while dissolving BSA.

Table 12.1 Transformation reaction

Protoplast suspension	Addition[a]	2×STC	1×STC	Total reaction volume	Sample
140 μL	X μL DNA	X μL	(60-2X) μL	200 μL	Plus DNA
140 μL	X μL water	X μL	(60-2X) μL	200 μL	Minus DNA

[a]10 μg of DNA in X μL (plus DNA) or X μL of water (minus DNA; as a control)

7. When many protoplasts are visible, stop the incubation by placing the mixture on ice for 10 min.
8. Swirl the suspension vigorously to free protoplasts from mycelial debris. Filter out protoplasts by passing the suspension through a double layer of Mira cloth. Divide and dispense filtrate into two Hettich/Corex glass tubes (about 15 mL in each) and dilute with equal volume of 1×STC. Seal the tubes with Parafilm™, mix by gentle inversion and place the tubes on ice for 10 min.
9. Weight-balance the tubes and centrifuge at 4,000 g for 10 min (use a swing out rotor).
10. Decant the supernatant carefully to avoid the loss of pelleted protoplasts. Wash the pellet gently by layering with 5 mL of 1×STC, followed by centrifugation for 10 min at 4,000 g.
11. Decant the supernatant carefully. Re-suspend the pellet in 1×STC (using the Neubauer counting chamber) to about 10^7 protoplasts per mL. Each individual transformation requires 120–140 μL protoplast suspension.

12.3.3 Transformation and Selection of Transformants

Transform protoplasts with approximately 10 μg of *Sca*I linearized pΔXCA DNA. Set up following transformation reactions.

1. Transformation reaction shown in Table 12.1.
2. To each reaction add 50 μL of PEG(t) and mix gently by pipetting several times. Incubate this reaction (250 μL total) on ice for 30 min.
3. Add 1.0 mL of PEG(t), mix gently, and incubate for 30 min at room temperature.
4. Plate the above transformation reactions (1,250 μL total) on different plates as shown below:

Plus DNA sample:
  (a) Spread 250 μL each on five MM + Arg + S plates (Selection medium).

Minus DNA sample:
  (b) Spread 250 μL on MM + Arg + S plates (Selection medium), in duplicates.
  (c) 100 μL directly spread on MM + S plates.
  (d) 100 μL + 900 μL of 1×STC; spread 100 μL of this 1:10 diluted sample (S1) on MM + S plates, in duplicates.
  (e) 100 μL S1 + 900 μL of 1×STC; spread 100 μL of this 1:100 diluted sample (S2) plate on MM + S plates, in duplicates.
  (f) 100 μL directly spread on MM plates.
  (g) 100 μL + 900 μL of DDW + T; spread 100 μL of this 1:10 diluted sample (W1) on MM, in duplicates.
  (h) 100 μL of W1 + 900 μL of DDW + T; spread 100 μL of this 1:100 diluted sample (W2) on MM, in duplicates.

Viable spore count:
  (i) 100 μL of 10^{-3} and 10^{-4} diluted sample of spore suspension spread on MM (with 0.05 % Triton X-100), in duplicates.

5. Incubate plates (a) and (b) at 37 °C. Count the number of prominent colonies between 48 and 72 h. Pick these putative transformants growing on L-arginine medium and passage them on MM + Arg plates for five generations.[11]

---

[11] Visible growth can be observed after 48 h of incubation on arginine selection plates. True transformants grow better/faster than others upon prolonged incubation (colony size increases). The false positives, if any, do not grow further and die out. Passage the putative transformants on arginine selection plates for five generations. Culture these transformants on MM (i.e. without selection pressure) for the sixth generation and subsequently transfer them back on MM + Arg plates to ensure their genetic stability. Single spore these transformants on MM + Arg plates to obtain a genetically homogenous culture.

6. Incubate plates from (c) through (h), at 37 °C. Count the number of prominent colonies between 36 and 48 h.

12.4 Interpretation of Results

A. niger D-42 strain (*agaA*⁻) is unable to grow on media with L-arginine as sole nitrogen source (MM + Arg + S). This strain regains the ability to utilize L-arginine when transformed with the *PcitA-agaA* expression cassette (linearized pΔXCA). The putative *agaA*⁺ transformants are scored after 48–72 h as prominent colonies directly on MM + Arg + S plates (plus DNA). No visible growth or colonies are observed when protoplasts were incubated with water alone (minus DNA).

Various controls are included to assess the protoplast quality and to quantify the transformation frequency. The number of colonies on MM + S plates indicates the number of total viable cells including viable protoplasts. The S1 and S2 colonies data is used to calculate the Regeneration Count (this is total viable cells including viable protoplasts). The number of colonies on MM plates is used to calculate the Water Count (this accounts for viable cells other than protoplasts; protoplasts burst in water). From the number of protoplasts used per transformation reaction (i.e., Total Protoplast Count in 140 µL; see table above) the Regeneration Frequency is calculated.

$$\mathrm{Regeneration\,Frequency} = \frac{\mathrm{Regeneration\,Count - Water\,Count}}{\mathrm{Total\,Protoplast\,Count}}$$

The regeneration frequency indicates the number of viable protoplasts in the preparation available for transformation. Regeneration frequency varies, even for the same strain, in the range of 5–50 %. The number of transformants obtained on selection plates when compared with protoplast Regeneration Frequency provides the transformation efficiency. The transformation efficiency may also be presented in terms of number of transformants obtained per µg of DNA added. Around 30–50 *agaA*⁺ transformants are obtained per 10 µg of DNA.

Selecting transformants using arginase (*agaA*) as a transformation marker is simple to operate. It offers a tight yet cost effective selection for fungal transformation in all cases where arginase provides the exclusive catabolic route for arginine utilization.[12]

Acknowledgments This research was funded by the New Millennium Indian Technology Leadership Initiative of Council of Scientific and Industrial Research (NMITLI-CSIR), India. This work was also supported in part by research grant from Board of Research in Nuclear Science-Department of Atomic Energy (BRNS-DAE) and research fellowship from University Grants Commission, UGC, Council of Scientific and Industrial Research, CSIR and Department of Science and Technology Women Scientist Scheme, DST-WOS-A.

References

Bartnik E, Guzewska J, Klimczuk J, Piotrowska M, Weglenski P (1977) Regulation of arginine catabolism in *Aspergillus nidulans*. In: Smith JE, Pateman JA (eds) Genetics and physiology of *Aspergillus*. Academic, London, pp 243–254

Dave K, Punekar NS (2011) Utility of *Aspergillus niger* citrate synthase promoter for heterologous expression. J Biotechnol 155:173–177

Dave K, Ahuja M, Jayashri TN, Sirola RB, Punekar NS (2012) A novel selectable marker based on *Aspergillus niger* arginase expression. Enzyme Microb Technol 51:53–58

De Graaff L, van den Broek H, Visser J (1988) Isolation and transformation of pyruvate kinase gene of *Aspergillus nidulans*. Curr Genet 13:315–321

Durrens P, Green PM, Arst HN, Scazzocchio C (1986) Heterologous insertion of transforming DNA and generation of new deletions associated with transformation in *Aspergillus nidulans*. Mol Gen Genet 203:544–549

Goosen T, Bloemheuvel G, Gysler C, de Bie DA, van den Brock HWJ, Swart K (1987) Transformation of *Aspergillus niger* using the homologous orotidine-5'-phosphate-decarboxylase gene. Curr Genet 11:499–503

---

[12] All filamentous fungal genomes sequenced so far contain an *agaA* gene.

Jayashri TN (2006) Arginase from Aspergilli: a molecular analysis. Ph.D. thesis, Indian Institute of Technology Bombay, India

Jiang D, Zhu W, Wang Y, Sun C, Zhang K, Yang J (2013) Molecular tools for functional genomics in filamentous fungi: recent advances and new strategies. Biotechnol Adv 31:1562–1574

Lenouvel F, van de Vondervoort PJ, Visser J (2002) Disruption of the *Aspergillus niger argB* gene: a tool for transformation. Curr Genet 41:425–432

Lubertozzi D, Keasling JD (2006) Marker and promoter effects on heterologous expression in *Aspergillus nidulans*. Appl Microbiol Biotechnol 72:1014–1023

Morgan DH (1970) Selection and characterization of mutants lacking arginase in *Neurospora crassa*. Mol Gen Genet 108:291–302

Sambrook J, Russell DW (2001) Molecular Cloning: a Laboratory Manual. Cold Spring Harbour Laboratory Press, New York

Varadarajalu LP, Punekar NS (2004) Genetic transformation in Aspergilli: tools of trade. Indian J Biochem Biophys 41:205–215

Transformation of Ascomycetous Fungi Using Autonomously Replicating Vectors

13

Satoko Kanematsu and Takeo Shimizu

13.1 Introduction

Genome-integrating vectors are commonly used for transformation of fungi (Meyer 2008) because stable maintenance of integrated vectors during mitosis and meiosis is useful in experiments that require long-term cultivation. There are several issues that hamper success in fungal transformations: (1) transformation efficiency depends on the frequency of vector integration into the genome and (2) changes in genome organization at the site of vector integration may cause unexpected effects on transformants. Autonomously replicating vectors, on the other hand, are maintained independently of chromosomal replication in fungal cells. A genomic segment from *Aspergillus nidulans*, referred to as "AMA1" (autonomous maintenance in *Aspergillus*), is frequently used for the construction of autonomously replicating vectors (see Aleksenko and Clutterbuck 1997 for more detailed information). Briefly, AMA1 was discovered during the screening of elements to increase transformation efficiency in *A. nidulans*, and was found to contain two symmetrical inverted repeats and a central

spacer. Inclusion of AMA1 in vectors confers the ability for autonomous extrachromosomal replication in fungal cells, generally leading to no genome integration of AMA1-bearing vectors.

When designing experiments using AMA1-bearing vectors, the following properties should be taken into consideration compared with genome-integrating vectors: (a) the increase in transformation efficiency, (b) the localization of vectors to nuclei in fungal cells, (c) the loss of vectors from fungal cells under nonselective conditions, and (d) the potential to recover the vectors as circular plasmids from fungal transformants (Aleksenko and Clutterbuck 1997). Various advantages of AMA1-bearing vectors have been reported for *Aspergillus* and phylogenetically related species of the same class (Eurotiomycetes) in literature: increase in transformation efficiency (7–2000-fold in fungal species referred to in this chapter), improving expression of target genes and facilitating construction of genomic libraries (Fierro et al. 1996; Fierro et al. 2004; Khalaj et al. 2007; Kubodera et al. 2002; Langfelder et al. 1998; Liu et al. 2004; Ozeki et al. 1996; Shimizu et al. 2006; Shimizu et al. 2007; Storms et al. 2005; Verdoes et al. 1994). AMA1-bearing vectors are also used for transformation in other classes of fungi, including *Gibberella fujikuroi* (Sordariomycetes) (Bruckner et al. 1992), *Trichoderma* species (Sordariomycetes) (Kubodera et al. 2002), *Rosellinia necatrix* (Sordariomycetes) (Shimizu

S. Kanematsu, Ph.D. (✉) • T. Shimizu, Ph.D.
Apple Research Division, NARO, Institute of Fruit Tree Science, Shimokuriyagawa 92-24, Nabeyashiki, Morioka 020-0123, Japan
e-mail: satokok@affrc.go.jp

M.A. van den Berg and K. Maruthachalam (eds.), *Genetic Transformation Systems in Fungi, Volume 2*, Fungal Biology, DOI 10.1007/978-3-319-10503-1_13,
© Springer International Publishing Switzerland 2015

et al. 2012), and *Botrytis cinerea* (Leotiomycetes) (Rebordinos et al. 2000). These results demonstrate that the AMA1 sequence may also mediate autonomous replication in fungi other than the Eurotiomycetes. In our laboratory, we have established efficient and useful transformation systems using AMA1-bearing vectors in the plant pathogenic fungus, *R. necatrix*. For this fungus, genetic analyses were hampered because of the low transformation efficiencies obtained with genome-integrating vectors and the availability of limited other genetic tools. We have successfully improved transformation efficiencies and introduced multiple vectors simultaneously into fungal cells by co-transformation of the AMA1-bearing vectors (pAMA-H) using the transformation system of *R. necatrix* (Shimizu et al. 2012). Out of ten transformants, one had genomic integration of the AMA1-bearing vector (Shimizu et al. 2012), indicating that extrachromosomal maintenance of the AMA1-bearing vector should be confirmed by appropriate experiments. The AMA1-bearing vectors are useful genetic tools to improve transformation efficiency and to develop novel genetic approaches in fungi. In this chapter, an example using *R. necatrix* is illustrated.

13.2 General Method

AMA1-bearing vectors can be constructed by transfer of the AMA1 fragment from commercially or publically available vectors, such as pAUR316 (TaKaRa Bio, Japan), the pRG3-AMA1, and ANEp vector series (Fungal Genetic Stock Center, http://www.fgsc.net/clones.html), to the desired vector (Fig. 13.1). The AMA1-bearing vectors can be applied to any transformation system such as the protoplast-PEG/CaCl$_2$ method, electroporation, or particle bombardment (see for details the respective chapters in this book) developed for specific fungal species with genome-integrating vectors simply by alteration of the vectors. We successfully performed transformation with the AMA1-bearing vector using the protoplast-PEG/CaCl$_2$ method that was already established with a genome-integrating vector for *R. necatrix* (Kanematsu et al. 2004; Pliego et al. 2009).

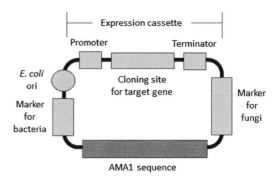

Fig. 13.1 Schematic representation of a typical AMA1-bearing vector. AMA1 confers the ability for autonomous extrachromosomal replication in fungi to the desired vector containing the ori and marker gene necessary for transformation in *E. coli*. In an expression cassette, the promoter and terminator, which are appropriate for the expression of the target gene, should be selected. A marker gene is necessary for the selection of fungal transformants with AMA1-bearing vectors

13.3 Detailed Procedure

According to the flowchart in Fig. 13.2, the transformation, validation, and application steps are detailed below.

13.3.1 Step 1: Transformation

13.3.1.1 Protoplast Production
1. *Rosellinia necatrix* W97 is cultivated for 7 days on potato dextrose agar (PDA; Difco Laboratories, Becton Dickinson, Franklin Lakes, NJ) in the dark at 25 °C.
2. Six mycelial agar discs (6 mm diameter) are excised from PDA cultures and added to 30 mL of potato dextrose broth (PDB; Difco Laboratories, Becton Dickinson, Franklin Lakes, NJ) in a 200-mL Erlenmeyer flask. The six replicate flasks are incubated statically in the dark at 25 °C for 7 days.
3. Mycelia are spun down at $5,000 \times g$ for 15 min.
4. Two aliquots of the collected mycelia are homogenized in a Waring blender with 60 mL of fresh PDB at 7,000 rpm for 30 s.
5. Ten mL of the homogenized mycelia is added to 30 mL of fresh PDB in a 200-mL Erlenmeyer flask, and 12 replicates are prepared.

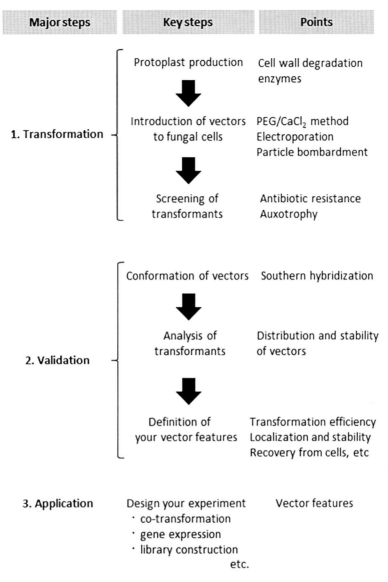

Fig. 13.2 Flowchart of procedures for the transformation system with AMA1-bearing vectors. The system consists of three major steps, each with several key steps containing important points to achieve the experimental goals

6. The homogenized mycelia are incubated statically for 2 days at 25 °C in the dark.
7. Mycelia are collected by centrifugation at 5,000×g for 15 min and suspended in 30 mL of an osmotic solution (MM: 0.6 M mannitol, 10 mM 3-morpholinopropanesulfonic acid [MOPS], pH 7.0).
8. Mycelia are spun down at 5,000×g for 15 min; the supernatant is discarded.
9. The mycelia are resuspended gently in filter-sterilized enzyme mixture containing 0.4 % Zymolyase 100T (Seikagaku Co., Tokyo, Japan) and 1 % Lysing Enzymes (Sigma-Aldrich, St. Louis, MO) in MM, and incubated at 20 °C for 2 h with gentle shaking.
10. Protoplasts are separated from the suspension by filtration through two layers of gauze.
11. The filtrate containing protoplasts is passed through a stainless steel mesh with 45-μm pores to remove any remaining debris.
12. Protoplasts are collected from the filtrate by centrifugation at 3,000×g for 10 min.

13. The collected protoplasts are washed twice with MM and gently suspended in 250 μL of MMC buffer (0.6 M mannitol, 10 mM MOPS, pH 7.0, and 10 mM $CaCl_2$).
14. The concentration of the protoplasts is determined with a haemocytometer.

13.3.1.2 Introduction of Vectors into Fungal Cells

1. Protoplasts (100 μL at 0.5 to 1×10^8 protoplasts/mL) are gently mixed with 10 μg of vector in a 50-mL conical tube (e.g., Falcon 2059, Becton Dickinson, NJ, USA), and placed on ice for 30 min (see **Note 1**).
2. 500 μL PEG (of 60 % polyethylene glycol [PEG 4000], average molecular weight 3,000 [Wako, Osaka, Japan], 10 mM MOPS, pH 7.0, and 10 mM $CaCl_2$) is added to the protoplast suspension and gently mixed.
3. After incubation at 20 °C for 20 min, 700 μL of regeneration broth (PDB containing 0.5 M glucose) is added to the tube and gently mixed.
4. The tube is incubated statically in the dark at 25 °C for 7 days (see **Note 2**).
5. Protoplasts are gently spread on YCDA plates (0.1 % yeast extract, 0.1 % casein hydrolysate [enzymatic], 0.5 M glucose, and 1.5 % agar) to enable the protoplasts to regenerate.

13.3.1.3 Screening of Transformants

1. Plates are incubated at 25 °C in the dark until small colonies appear, and then overlaid with 10 mL of PDA containing hygromycin B (80 μg/mL, Roche, Mannheim, Germany) (see **Note 3**).
2. Colonies grown on the surface of the plates (Fig. 13.3) are transferred to PDA containing hygromycin B (50 μg/mL) after incubation for 3 to 7 days at 25 °C in the dark.
3. Colonies are screened again on PDA containing hygromycin B (100 μg/mL) (see **Note 4**).

13.3.2 Step 2: Validation

Among the three major steps, the validation step is the most important for success of the study using the constructed vectors. AMA1-bearing vectors are occasionally integrated into the fungal

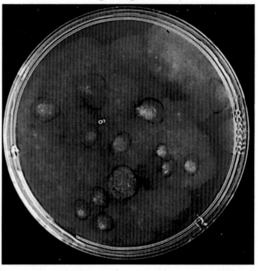

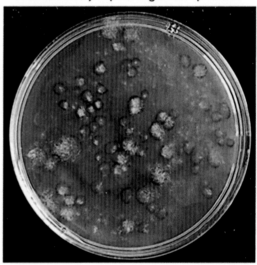

Fig. 13.3 Typical colonies grown on the surface of the overlaid PDA containing hygromycin B (80 μg/mL) (as described in "*Screening of transformants*" under Step 1). In the *upper* picture, a conventional genome-integrating vector, pCPXHY1, was transformed into *R. necatrix*. In the *lower* picture, an AMA1-bearing vector, pAMA-H, was transformed into *R. necatrix*. These transformations were performed simultaneously using aliquots from the same protoplast suspension. The transformation efficiency was increased with pAMA-H compared with pCPXHY1

genome, which can be verified through Southern blotting analysis. In Southern hybridization, the vector should be used as a control to confirm its presence in DNA samples from transformants.

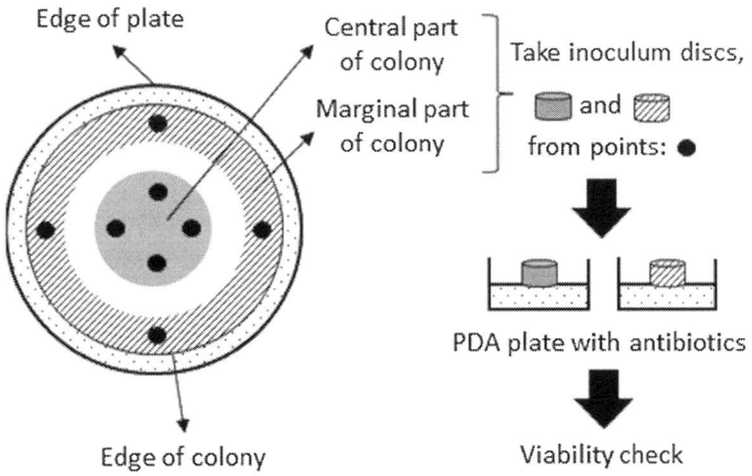

13 Transformation of Ascomycetous Fungi Using Autonomously Replicating Vectors

Fig. 13.4 Schematic representation of an assay to reveal distribution of AMA1-bearing vectors in transformant colonies. The transformants are cultured on PDA plates containing antibiotics. Inoculum discs are taken from the central and marginal parts of the colonies and transferred to PDA plates containing antibiotics, where their viability is examined

The same hybridized band pattern between the control and transformant DNA samples indicates the extrachromosomal existence of the vector in the transformant. A different pattern suggests integration of the AMA1 vector into the genome of the transformant.

13.3.2.1 Distribution of AMA1-bearing Vectors in Transformants

In *R. necatrix*, transformants with AMA1-bearing vectors show irregular or slow growth on plates with antibiotics due to the uneven distribution of AMA1-bearing vectors in the colonies, thereby affecting the phenotype of transformants. For this reason, it is necessary to investigate the distribution of AMA1-bearing vectors in the transformant colonies. The assay for vector distribution performed in *R. necatrix* is detailed below (Fig. 13.4).

1. Mycelial agar discs are excised from central parts of each colony grown on PDA plates containing antibiotics and used as primary inocula.
2. After incubation for 7 days, mycelial agar discs are excised from central and marginal parts of the growing colony and used as secondary inocula.
3. After secondary inocula are grown for 7 days on plates containing antibiotics, growth from the secondary inocula on the plates is investigated. If some of them fail to grow on the plates containing antibiotics, AMA1-bearing vectors are absent from the corresponding sections of the transformant colony, indicating uneven distribution of the vector (see **Note 5**).

13.3.2.2 Stability of the AMA1-bearing Vectors in Transformants

AMA1-bearing vectors can easily be lost from transformants under nonselective conditions. In *R. necatrix*, the stability of AMA1-bearing vectors in the transformants is investigated using the following method.

1. Transformants are cultured on PDA plates containing antibiotics for 7 days.
2. Mycelial agar discs are excised from central parts of the plates and placed on PDA plates without antibiotics for 7 days.
3. The subculture process is repeated three times under nonselective conditions, and then transformants are grown on PDA plates containing antibiotics to examine their viability. Failure to grow under selective conditions indicates that the vector has been completely lost from the transformant.

13.3.3 Step 3: Application

The efficiency of transforming with the AMA1-bearing vectors allows for the flexible co-transformation with multiple AMA1-bearing vectors for the transient and multivariate transformation system in *R. necatrix*.

13.3.3.1 Co-transformation with Multiple AMA1-bearing Vectors

1. DNA (2 µg of each vector) is gently mixed with 50 µL protoplast suspension (see **Note 6**), followed by the "*Introduction of vectors into fungal cells*" procedure mentioned in the Transformation step.
2. Transformants are selected based on appropriate antibiotic resistance.
3. Transformants are grown under selective conditions, using inoculum discs excised from the central parts of the selective culture so that stable transformants may be obtained. The stability of transformants is confirmed after two consecutive subcultures under selective conditions. Presence of vectors in the transformants needs to be confirmed by appropriate methods based on the experimental design.

13.4 Notes

1. Although 100 µL of protoplast suspension is normally used for transformation in this fungus, 50 µL of protoplast suspension is enough to perform the transformation using AMA1-bearing vectors. The amount of each buffer should be reduced by half when using the smaller protoplast suspension volume.
2. A longer (7 day) incubation period is required for *R. necatrix* as the regeneration rate from protoplasts on solid media is low.
3. Use appropriate antibiotics according to the selection markers present in the vector after determining the appropriate concentration for the target fungal species.
4. Inocula should be taken from the center of the colony until the presence of the AMA1-bearing

vector is confirmed in transformants as described under "Validation".
5. Our study demonstrated less than 50 % viability of secondary inocula from marginal parts of the colonies, although all of the secondary inocula from the central parts of the colonies could grow on plates containing antibiotics (Shimizu et al. 2012).
6. We succeeded in co-transformation with three separate vectors. Care must be taken to limit the total amount of vector added to the protoplast suspension as DNA is precipitated by adding PEG solution.

Acknowledgements We appreciate the help of Naoyuki Matsumoto and Hajime Yaegashi for their fruitful comments on the manuscript.

References

Aleksenko A, Clutterbuck AJ (1997) Autonomous plasmid replication in *Aspergillus nidulans*: AMA1 and MATE elements. Fungal Genet Biol 21:373–387

Bruckner B, Unkles SE, Weltring K, Kinghorn JR (1992) Transformation of *Gibberella fujikuroi*: effect of the *Aspergillus nidulans* AMA1 sequence on frequency and integration. Curr Genet 22:313–316

Fierro F, Kosalkova K, Gutierrez S, Martin JF (1996) Autonomously replicating plasmids carrying the AMA1 region in *Penicillium chrysogenum*. Curr Genet 29:482–489

Fierro F, Laich F, Garcia-Rico RO, Martin JF (2004) High efficiency transformation of *Penicillium nalgiovense* with integrative and autonomously replicating plasmids. Int J Food Microbiol 90:237–248

Kanematsu S, Arakawa M, Oikawa Y, Onoue M, Osaki H, Nakamura H, Ikeda K, Kuga-Uetake Y, Nitta H, Sasaki A, Suzaki K, Yoshida K, Matsumoto N (2004) A reovirus causes hypovirulence of *Rosellinia necatrix*. Phytopathology 94:561–568

Khalaj V, Eslami H, Azizi M, Rovira-Graells N, Bromley M (2007) Efficient downregulation of alb1 gene using an AMA1-based episomal expression of RNAi construct in *Aspergillus fumigatus*. FEMS Microbiol Lett 270:250–254

Kubodera T, Yamashita N, Nishimura A (2002) Transformation of *Aspergillus* sp. and *Trichoderma reesei* using the pyrithiamine resistance gene (ptrA) of *Aspergillus oryzae*. Biosci Biotechnol Biochem 66:404–406

Langfelder K, Jahn B, Gehringer H, Schmidt A, Wanner G, Brakhage AA (1998) Identification of a polyketide synthase gene (pksP) of *Aspergillus fumigatus* involved in conidial pigment biosynthesis and virulence. Med Microbiol Immunol 187:79–89

Liu W, May GS, Lionakis MS, Lewis RE, Kontoyiannis DP (2004) Extra copies of the *Aspergillus fumigatus* squalene epoxidase gene confer resistance to terbinafine: genetic approach to studying gene dose-dependent resistance to antifungals in *A. fumigatus. Antimicrob*. Agents Chemother 48:2490–2496

Meyer V (2008) Genetic engineering of filamentous fungi—progress, obstacles and future trends. Biotechnol Adv 26:177–185

Ozeki K, Kanda A, Hamachi M, Nunokawa Y (1996) Construction of a promoter probe vector autonomously maintained in *Aspergillus* and characterization of promoter regions derived from *A. niger* and *A. oryzae* genomes. Biosci Biotechnol Biochem 60:383–389

Pliego C, Kanematsu S, Ruano-Rosa D, de Vicente A, Lopez-Herrera C, Cazorla FM, Ramos C (2009) GFP sheds light on the infection process of avocado roots by *Rosellinia necatrix*. Fungal Genet Biol 46:137–145

Rebordinos L, Vallejo I, Santos M, Collado IG, Carbu M, Cantoral JM (2000) Genetic analysis and relationship to pathogenicity in *Botrytis cinerea*. Rev Iberoam Micol 17:S37–S42

Shimizu T, Kinoshita H, Nihira T (2006) Development of transformation system in *Monascus purpureus* using an autonomous replication vector with aureobasidin A resistance gene. Biotechnol Lett 28:115–120

Shimizu T, Kinoshita H, Nihira T (2007) Identification and in vivo functional analysis by gene disruption of ctnA, an activator gene involved in citrinin biosynthesis in *Monascus purpureus*. Appl Environ Microbiol 73:5097–5103

Shimizu T, Ito T, Kanematsu S (2012) Transient and multivariate system for transformation of a fungal plant pathogen, *Rosellinia necatrix*, using autonomously replicating vectors. Curr Genet 58:129–138

Storms R, Zheng Y, Li H, Sillaots S, Martinez-Perez A, Tsang A (2005) Plasmid vectors for protein production, gene expression and molecular manipulations in *Aspergillus niger*. Plasmid 53:191–204

Verdoes JC, Punt PJ, van der Berg P, Debets F, Stouthamer AH, van den Hondel CA (1994) Characterization of an efficient gene cloning strategy for *Aspergillus niger* based on an autonomously replicating plasmid: cloning of the nicB gene of *A. niger*. Gene 146:159–165

A Recyclable and Bidirectionally Selectable Marker System for Transformation of *Trichoderma*

14

Thiago M. Mello-de-Sousa, Robert L. Mach, and Astrid R. Mach-Aigner

14.1 Introduction

The described transformation strategy, which was originally developed for *Trichoderma*, combines three main advantages: (1) a high rate of homologous integration (gene replacement) events; (2) the recycling of the selection marker for unrestricted rounds of gene deletions; and (3) a bidirectional, positive selection system (Steiger et al. 2011).

Several efforts have been made in fungi in order to improve homologous integration rates by blocking the non-homologous end joining (NHEJ) mechanism that repairs double-strand breaks in DNA (reviewed by Kück and Hoff 2010). In *Neurospora* the strategy relies on the deletion of key components of NHEJ machinery, such as the *mus-51* and *mus-52* genes (homologs of the human *KU70* and *KU80*, respectively) or the *mus-53* gene (homolog of the human *LIG4*), leading to up to 100 % of transformants exhibiting integration at the homologous site (Ninomiya et al. 2004; Ishibashi et al. 2006).

T.M. Mello-de-Sousa, M.Sc. • R.L. Mach, Ph.D.
A.R. Mach-Aigner, Ph.D. (✉)
Department for Biotechnology and Microbiology,
Institute of Chemical Engineering, Vienna University
of Technology, Gumpendorfer Straße 1a,
1060 Wien, Austria
e-mail: astrid.mach-aigner@tuwien.ac.at;
tdemello@tuwien.ac.at

Besides gene targeting, the use of the Cre/loxP recombination system adapted from bacteriophage P1 has been proposed to create an efficient marker recycling system in fungi (Dennison et al. 2005; Krappmann et al. 2005; Forment et al. 2006; Florea et al. 2009; Patel et al. 2010). This approach comprises the excision of a DNA fragment that is flanked by loxP-sites by the catalytic activity of a Cre recombinase, which needs to be expressed in the fungus.

For the development of this protocol we used *T. reesei* QM6a ($\Delta tmus53\Delta pyr4$), a recombinant strain that combines both aforementioned advantages (Steiger et al. 2011). This strain was generated by transformation of the wild type strain QM6a in order to increase homologous integration events and to introduce the Cre/loxP-based marker excision system. First, the deletion of *tmus53* (the *mus-53* homolog in *Trichoderma*) yielded an NHEJ-deficient strain, in which the rates of homologous integration go up to 100 % (Steiger et al. 2011). Second, the *pyr4* gene was used as the target locus for integration of the Cre recombinase because the *pyr4* deletion causes uridine auxotrophy, which itself serves as a transformation marker (Gruber et al. 1990). The genomically integrated Cre recombinase is under control of the *xyn1* promoter and can be induced in a controlled way. This promoter is inducible by D-xylose and xylan, and shut down on glucose due to a double-lock mechanism (Mach et al. 1996; Mach-Aigner et al. 2008). For targeted deletion of a certain gene the plasmid pMS-5loxP3 is used for transformation of this strain providing an *hph/amdS* marker system. This marker offers the possibility of bidirectional, positive selection, i.e. the resistance to hygromycin B (marker insertion and gene deletion) or loss of sensitivity to fluoroacetamide (marker removal). Altogether, this transformation system allows serial targeted gene deletions with continuous marker recycling in *Trichoderma* (Steiger et al. 2011).

14.2 Materials[1]

1. Sterile, distilled water.
2. Deoxynucleoside triphosphates (dNTPs)—a mixture of dATP, dCTP, dGTP, and dTTP (10 mM each, Thermo Scientific, Waltham, MA, USA), stored at −20 °C.
3. Thermostable DNA polymerase (e.g. *Taq*, and reaction buffer supplied by manufacturer). We usually use the GoTaq Flexi DNA polymerase (Promega, Madison, WI, USA).
4. Oligonucleotide primers (custom-made, Sigma-Aldrich, St. Louis, MO, USA), resuspended to a concentration of 100 μM using 5 mM Tris pH 7.5 and stored at −20 °C.
5. PCR tubes (thin walled).
6. Thermal cycler, e.g. UNO II (Biometra GmbH, Göttingen, Germany).
7. PCR clean-up kit, e.g. QIAquick PCR Purification Kit (QIAGEN GmbH, Hilden, Germany).
8. pGEM-T vector system (Promega) or other appropriate cloning system.
9. Creator DNA cloning kit (Clontech Laboratories Inc., Mountain View, CA, USA) or other available Cre/loxP-based cloning system. A suitable alternative is to order the Cre recombinase (1,000 U/mL) separately (New England Biolabs Inc. Ipswich, MA, USA).
10. 10X Cre Recombinase Reaction Buffer: 330 mM NaCl, 500 mM Tris–HCl, 100 mM MgCl$_2$, pH 7.5, stored at −20 °C.

11. 10X BSA solution (1 mg/mL), stored at −20 °C.
12. Thermomixer or water bath.
13. Microcentrifuge.
14. Disposable polypropylene microcentrifuge tubes: 1.5 mL conical.
15. Petri dishes.
16. Hygromycin B (Calbiochem, San Diego, CA, USA).
17. 100 mg/mL ampicillin, stored at −20 °C.
18. 1 M sucrose (sterile-filtered), stored at 4 °C.
19. LB medium: 0.5 % (wt/vol) peptone from casein, 1 % (wt/vol) yeast extract, 1 % (wt/vol) NaCl, 1.5 % (wt/vol) agar.
20. *Escherichia coli* competent cells. We usually use *E. coli* supercharge EZ10 electrocompetent cells (Clontech).
21. Malt extract medium (MEX): 3 % (wt/vol) malt extract, 0.1 % (wt/vol) peptone from casein, 1.5 % (wt/vol) agar.
22. 500 mM uridine (sterile-filtered), stored at 4 °C.
23. Mandels-Andreotti (MA) agar medium, prepared according to Mandels (1985).
24. Oat spelt xylan (Sigma Aldrich).
25. Sterile cotton buds.
26. 1 M fluoroacetamide (sterile-filtrated), stored at 4 °C.
27. Igepal CA-360 (Sigma-Aldrich).[2]
28. 1 M D-xylose (sterile-filtrated), stored at 4 °C.
29. 40 % (wt/vol) D-glucose (sterile-filtrated), stored at 4 °C.
30. Incubators at 30 °C and 37 °C.

14.3 Methods

14.3.1 Strain and Cultivation Conditions

T. reesei QM6a (Δ*tmus53*Δ*pyr4*) was maintained at 30 °C on MEX plates, which were supplemented with 5 mM uridine because of the *pyr4* deletion.

---

[1] All media, solutions, and material required for protoplast preparation and transformation of *Trichoderma* has been described before by Penttilä et al. (1987).

[2] This reagent is used for growth restriction in order to obtain clearly confined colonies from streaked spores.

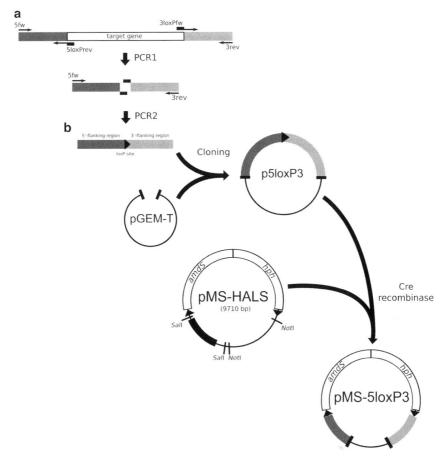

Fig. 14.1 Construction of pMS-5loxP3, the plasmid bearing the loxP-sites and the bidirectional marker system. (**a**) First, separate amplification of the 5'-flanking region (*dark grey*) and the 3'-flanking region (*light grey*) of the target gene, introducing a loxP-site (*black triangle*) between the two regions (PCR1) was performed. A second PCR (PCR2) using the two amplification products from PCR1 as templates together with the outer primers yields a fragment consisting of the 5'-flanking region, a loxP-site, and the 3'-flanking region. Thin *black arrows* (→) indicate primers; *thin black arrows* with a *black box* indicate primers introducing a loxP-site. (**b**) Cloning and Cre recombinase reaction. The obtained PCR product is subcloned into pGEM-T to obtain p5loxP3, which is used as the acceptor vector in the subsequent Cre recombinase reaction. pMS-HALS, bearing two loxP-site, which flank the *amdS* and *hph* genes, and *sacB* as a killer gene (*black segment*), is used as the donor vector. The Cre recombinase reaction yields pMS-5loxP3, bearing the 5'-flanking region of the target gene, a loxP-site, *amdS*, *hph*, a loxP-site, and the 3'-flanking region of the target gene

14.3.2 Construction of pMS-5loxP3

For the construction of the plasmid bearing the loxP-sites and the bidirectional marker system two steps are necessary (Fig. 14.1). The first step includes the construction of the p5loxP3 plasmid bearing the 5'-flanking region and the 3'-flanking region of the target gene to be deleted, and a loxP-site. In the second step, a Cre recombinase reaction is employed. The plasmid p5loxP3 is used as the acceptor vector of a fragment derived from the plasmid pMS-HALS that bears two loxP-sites flanking the *amdS* and *hph* genes, which are the marker genes for fungal transformation (Steiger et al. 2011). The resulting plasmid, pMS-5loxP3, can then be used for transformation into the fungal genome.

1. For construction of the p5loxP3 plasmid, perform a splicing-by-overlapping-extension (SOE) PCR to create a fragment bearing the

5′-flanking region (fragment 1) and the 3′-flanking region (fragment 2) of the target gene with a loxP-site in between (Fig. 14.1). In the first PCR, use primer 5fw and 5loxPrev to amplify fragment 1, and primer 3loxPfw and 3rev for amplifying fragment 2. All primers should be designed to amplify approximately 1,000-bp fragments up- and downstream of the target gene. 5fw and 3rev are regular 25–30-mer oligonucleotides. 5loxPrev and 3loxPfw include the loxP-site on their 5′-ends resulting in 58–60-mer oligonucleotides. The loxP-site consists of two 13-bp inverted repeats separated by an 8-bp spacer region that provides directionality to the recombination reaction, as follows: 5′-ATAACTTCGTATAGCATACATTATACG AAGTTATNNNNNNNNNNNNNNNNN NNNNNNNNNN-3′ (“N” represents the remaining bases of the oligonucleotides 5loxPrev and 3loxPfw dependent on the chosen target gene).

2. Purify the amplified fragments 1 and 2 using any commercial PCR clean-up kit. Combine 300 ng of fragments 1 and 2 as templates for the second PCR using the 5fw and 3rev primer. Perform the cloning of the resulting PCR product into the pGEM-T vector (Promega) using the standard protocol, resulting in plasmid p5loxP3.[3]

3. Prepare the Cre recombinase assay as follows: mix 200 ng p5loxP3 (acceptor vector), 200 ng pMS-HALS (donor vector), 2 μL 10X Cre Recombinase Reaction Buffer, 2 μL 10X BSA (1 mg/mL), and 1 μL Cre Recombinase. Add distilled H_2O to bring volume up to 20 μL. Incubate at room temperature (22–25 °C) for 15 min. Stop the reaction by incubating at 70 °C for 10 min. Transform a 0.5-μL aliquot of the Cre recombinase reaction into highly competent *E. coli* cells (e.g. supercharge EZ10 electrocompetent cells (Clontech)) and select on plates containing LB agar medium supplemented with 100 μg/mL hygromycin B, 100 μg/mL ampicillin, and 70 mg/mL sucrose. Positive clones are selected next to ampicillin on hygromycin B and sucrose because the loxP marker cassette contains the *hph* gene and the donor vector carries the *Bacillus subtilis sacB* as a killer gene, which is activated by sucrose (Gay et al. 1983). This strategy inhibits the selection of pMS-HALS, hence favouring the selection for positive clones of pMS-5loxP3.[4, 5, 6]

14.3.3 Protoplast Transformation of *Trichoderma*

The protoplasts preparation and transformations of *Trichoderma* were carried out as previously described (Penttilä et al. 1987). The amount of

---

[3] For the PCR reactions some optimization may be required according to particular primer sets. The reactions are normally carried out in 50-μL mixtures containing 1.25 U GoTaq Flexi DNA polymerase (Promega), 2 mM $MgCl_2$, 1X Green GoTaq Flexi buffer, 0.2 mM dNTPs, 0.1 μM of forward and reverse primer each, 1 μL of template DNA (10 ng plasmid DNA, 300 ng genomic DNA), and nuclease-free water. The standard PCR program consists of an initial denaturation of 3 min at 95 °C, followed by 35 cycles each comprising 30 s at 95 °C, 30 s at 60 °C, and 1 min/kb at 72 °C.

[4] For efficient gene replacement events in *Trichoderma*, it is necessary to produce cassettes with long (~1,000 bp) 5′- and 3′-regions flanking the desired integration locus (Guangtao et al. 2009). Because of this high minimum length of the flanking regions, it may be difficult to find suitable restriction enzymes for the cloning procedure of the deletion cassette. The usage of the Cre recombinase reaction is an alternative to overcome this problem. Nevertheless, classical cloning techniques may still be employed to create the desired version of pMS-5loxP3 depending on the sequences of the flanking regions of the target gene. The whole fragment bearing the two loxP-sites and the *amdS* and *hph* genes (5,375 bp) can be excised from pMS-HALS by using *Sal*I and *Not*I restriction sites (Fig. 14.1).

[5] Please note that the *pki::hph* expression cassette, which is present in pMS-HALS, is fully functional in *E. coli*. This allows the screening of positive clones from the *E. coli* transformation by selection on hygromycin B and enables the use of this plasmid as shuttle system between *E. coli* and the fungus. Consequently, the donor vector pMS-HALS is generally highly useful for the construction of any deletion cassette.

[6] If *E.coli* supercharge EZ10 electrocompetent cells are not available, other strains such DH5α or XL1-Blue can be successfully used only resulting in differences of the transformation efficiency.

DNA used is typically between 1 and 5 μg of linearized pMS-5loxP3. Selection of transformants was carried out on selection medium plates also described by Penttilä et al. (1987) supplemented with 100 μg/mL hygromycin B and 5 mM uridine.

14.3.4 Excision of the loxP Marker Cassette

The loxP marker cassette containing the *amdS* and *hph* genes can be looped out of the chromosomal DNA by the action of Cre recombinase. By cultivating the positively transformed strain on medium containing xylan the *xyn1* promoter is induced, which leads to Cre recombinase expression. After the recombination (excision) event the resulting strain contains only one copy of the loxP-site at the deleted locus and is now sensitive to hygromycin B and acetamidase-negative. A positive screening for the excision is performed on fluoroacetamide medium because fluoroacetamide has a slight toxic effect on strains still containing the *amdS* gene and reduces their growth. The resulting strain can be used as a recipient strain for further gene deletions re-applying the same system because it still contains the *cre* gene and is free of marker genes.

1. To excise the loxP marker cassette, cultivate the transformed fungal strains on MA plates containing 1 % (wt/vol) oat spelt xylan as carbon source and 5 mM uridine. Harvest some conidia after 4 days using a sterile cotton bud and streak them on a MA plate containing 0.5 μL/mL Igepal CA-360, 1.5 mg/mL fluoroacetamide, 5 mM uridine, and 1 % (wt/vol) D-xylose.
2. Check the phenotypes of the strains by plating them on a malt extract agar plate containing 100 μg/mL hygromycin B and 5 mM uridine. No growth indicates a successful excision of the loxP marker cassette. If the fungus still has the ability to grow on hygromycin B-containing medium, repeat the single-spore purification step.
3. After excision of the loxP marker cassette maintain the strains on MA agar plates containing 1 % glucose and 5 mM uridine. This is important

to avoid undesired recombination events due the expression of the Cre recombinase.

4. Characterize the transformed strains by PCR and Southern blot analysis. Once you confirmed the correct genotype, this strain can be used for a next round of gene deletion by using the recycled, bidirectional *amdS/hph* marker.[7]

Acknowledgments This work was supported by three grants from the Austrian Science Fund (FWF): [P21287, V232-B20, P24851] given to RLM and ARMA, respectively, and by a doctoral programme of Vienna University of Technology (CatMat).

References

Dennison PM, Ramsdale M, Manson CL, Brown AJ (2005) Gene disruption in *Candida albicans* using a synthetic, codon-optimised Cre-loxP system. Fungal Genet Biol 42:737–748

Florea S, Andreeva K, Machado C, Mirabito PM, Schardl CL (2009) Elimination of marker genes from transformed filamentous fungi by unselected transient transfection with a Cre-expressing plasmid. Fungal Genet Biol 46:721–730

Forment JV, Ramon D, MacCabe AP (2006) Consecutive gene deletions in *Aspergillus nidulans*: application of the Cre/loxP system. Curr Genet 50:217–224

Gay P, Lecoq D, Steinmetz M, Ferrari E, Hoch JA (1983) Cloning structural gene *sacB*, which codes for exoenzyme levansucrase of *Bacillus subtilis*: expression of the gene in *Escherichia coli*. J Bacteriol 153:1424–1431

Gruber F, Visser J, Kubicek CP, de Graaff LH (1990) Cloning of the *Trichoderma reesei pyrG* gene and its use as a homologous marker for a high-frequency transformation system. Curr Genet 18:447–451

---

[7]Performing one round of marker re-use did not result in unwanted recombination events of the present loxP-site and the new loxP-sites (Steiger et al. 2011). However, after having performed all desired rounds of gene deletions, the elimination of the Cre recombinase is strongly recommended to construct a genetically stable strain with respect to its loxP-sites. In QM6aΔ*tmus53*Δ*pyr4* the *cre* gene was directed to the *pyr4* locus. In a final transformation step the *pyr4* gene is used to replace the *cre* gene, thereby employing the *pyr4* gene as a marker. The resulting, positively transformed strain gains uridine prototrophy and does not contain the Cre recombinase anymore. This has been demonstrated by the successful *pyr4* retransformation into strain QM6aΔ*tmus53*Δ*pyr4*, which generated QM6aΔ*tmus53*. The latter is still NHEJ-deficient and can be employed for gene replacement aiming homologous integration (Steiger et al. 2011).

Guangtao Z, Hartl L, Schuster A, Polak S, Schmoll M, Wang T, Seidl V, Seiboth B (2009) Gene targeting in a nonhomologous end joining deficient *Hypocrea jecorina*. J Biotechnol 139:146–151

Ishibashi K, Suzuki K, Ando Y, Takakura C, Inoue H (2006) Non-homologous chromosomal integration of foreign DNA is completely dependent on MUS-53 (human Lig4 homolog) in Neurospora. Proc Natl Acad Sci USA 103:14871–14876

Krappmann S, Bayram O, Braus GH (2005) Deletion and allelic exchange of the *Aspergillus fumigatus veA* locus via a novel recyclable marker module. Eukaryot Cell 4:1298–1307

Kück U, Hoff B (2010) New tools for the genetic manipulation of filamentous fungi. Appl Microbiol Biotechnol 86:51–62

Mach RL, Strauss J, Zeilinger S, Schindler M, Kubicek CP (1996) Carbon catabolite repression of xylanase I (*xyn1*) gene expression in *Trichoderma reesei*. Mol Microbiol 21:1273–1281

Mach-Aigner AR, Pucher ME, Steiger MG, Bauer GE, Preis SJ, Mach RL (2008) Transcriptional regulation of *xyr1*, encoding the main regulator of the xylanolytic and cellulolytic enzyme system in *Hypocrea jecorina*. Appl Environ Microbiol 74:6554–6562

Mandels M (1985) Applications of cellulases. Biochem Soc Trans 13:414–416

Ninomiya Y, Suzuki K, Ishii C, Inoue H (2004) Highly efficient gene replacements in *Neurospora* strains deficient for nonhomologous end-joining. Proc Natl Acad Sci USA 101:12248–12253

Patel R, Lodge DJK, Baker LG (2010) Going green in *Cryptococcus neoformans*: the recycling of a selectable drug marker. Fungal Genet Biol 47:191–198

Penttilä M, Nevalainen H, Raättö M, Salminen E, Knowles J (1987) A versatile transformation system for the cellulolytic filamentous fungus *Trichoderma reesei*. Gene 61:155–164

Steiger MG, Vitikainen M, Uskonen P, Brunner K, Adam G, Pakula T, Penttilä M, Saloheimo M, Mach RL, Mach-Aigner AR (2011) Appl Environ Microbiol 77:114–121

Split-Marker-Mediated Transformation and Targeted Gene Disruption in Filamentous Fungi

15

Kuang-Ren Chung and Miin-Huey Lee

15.1 Introduction

Genetic transformation and targeted gene disruption are essential for studying and understanding gene function in filamentous fungi. Targeted gene disruption in filamentous fungi can be troublesome because of low frequencies of homologous integration (Bird and Bradshaw 1997; Chung et al. 1999; Pratt and Aramayo 2002; Segers et al. 2001; Idnurm et al. 2003). In addition, integration of foreign genes in filamentous fungi often occurs via non-homologous integration, resulting in a large number of false-positive transformants. The frequency of targeted gene disruption can be improved by a split-marker disruption strategy by fusing the target DNA fragments with truncated but overlapping within the selectable marker gene (Fairhead et al. 1996; Fu et al. 2006). Only transformants harboring a functional dominant marker gene will grow on a medium containing the selection agent. Split-marker-based transformation decreases the occurrence of multiple and tandem integrations so as to decrease the overall numbers of transformants being screened. Split-marker approach has been shown to increase the frequency of targeted gene disruption and homologous integration as high as 100 % in *Alternaria alternata* and *Cercospora* spp. (Choquer et al. 2005; You et al. 2009; Lin and Chung 2010). Two truncated, overlapping marker gene fragments are joined with a gene of interest by fusion PCR without the need for cloning and the PCR products used directly for transformation. The methods described in this article could provide better tools to analyze gene functions in filamentous fungi.

15.2 Materials and Methods

15.2.1 Solution and Medium

15.2.1.1 Wash Solution
Dissolve 58.4 g NaCl and 1.47 g $CaCl_2 \cdot H_2O$ in water, bring the final volume to 1 L, and sterilize using an autoclave.

15.2.1.2 STC Solution
Dissolve 218.6 g sorbitol in 980 mL water, add 10 mL each of 1 M Tris–HCl (pH 7.5) and 1 M $CaCl_2 \cdot H_2O$, and sterilize.

15.2.1.3 50 % PEG Solution
Dissolve 50 g polyethylene glycol (M.W. 3,350) in a pre-heated water (98 mL), add 1 mL each of 1 M Tris–HCl (pH 7.5) and 1 M $CaCl_2 \cdot H_2O$, and filter sterilize. Discard after 4 months.

K.-R. Chung (✉) • M.-H. Lee, Ph.D.
Department of Plant Pathology, National
Chung-Hsing University, 250 Kuo-Kuang Road,
Taichung 402, Taiwan
e-mail: krchung@nchu.edu.tw

M.A. van den Berg and K. Maruthachalam (eds.), *Genetic Transformation Systems in Fungi, Volume 2*, Fungal Biology, DOI 10.1007/978-3-319-10503-1_15,
© Springer International Publishing Switzerland 2015

15.2.1.4 Enzyme Solution

Mix 0.2 mL β-glucuronidase (type H2, Sigma), 0.16 g β-D-glucanase, and 6 mg lyticase with 0.5 mL of 0.4 M Na_2PO_4 (pH 5.8), 0.4 mL of 1 M $CaCl_2 \cdot H_2O$, and 1.4 g NaCl in water (final volume 20 mL), filter sterilize, and store at −20 °C (Chung et al. 2002).

15.2.1.5 Solution A

Dissolve 10 g $Ca(NO_3) \cdot 4H_2O$ in 100 mL water.

15.2.1.6 Solution B

Dissolve 2 g KH_2PO_4, 2.5 g $MgSO_4 \cdot 7H_2O$ and 1.5 g NaCl in 100 mL water (pH 5.3).

15.2.1.7 Regeneration Medium

Regeneration medium (RM) is prepared in both liquid and solid forms. In a 1-L bottle, add 10 mL each of Solution A and Solution B to 480 mL water. Add 15 g agar per liter for solid medium. In a 2-L bottle, dissolve 342.3 g sucrose and 10 g glucose in water (final volume 500 mL). Sterilize two solutions separately, mix, and dispense into sterile bottles.

15.2.2 Growth Conditions

Start culture by grinding fungal mycelium with 0.5 mL sterile water in a 1.5-mL centrifuge tube using a disposable mini pestle (Fisher Scientific) and adding the resulting suspension to 50 mL medium (potato dextrose broth or a synthetic medium). Incubate fungal culture on a rotary shaker at room temperature (*ca.* 25 °C) for 3–4 days. Blend the culture in a sterile blender cup (Fisher Scientific) for three or four 10 s pulses, add to 200 mL fresh medium, and incubate on shaker for an additional 16–18 h. Harvest fungal mycelium by low-speed centrifugation at 6,000 rpm for 10 min in an Allegra 21R centrifuge (Beckman Culter). Carefully remove supernatant using a disposable polyethylene transfer pipet. Resuspend fungal mycelium in 10 mL of wash solution, spin again, and discard supernatant.

15.2.3 Preparation of Protoplasts

Successful transformation of fungi requires competent protoplasts. Resuspend fungal mycelium in 20 mL of enzyme solution by pipetting up and down with a disposable polyethylene pipet and transfer to a 100-mL flask. Incubate the resulting suspension at 30 °C on a rotary shaker set at 100 rpm. Check protoplast release under microscope regularly. After 2 h digestion, passage the solution through Miracloth. Harvest protoplasts by low-speed centrifugation at 4,000 rpm in a F0850 Beckman rotor at 4 °C for 5 min. Discard supernatant. Wash protoplasts twice with 10 mL of STC solution. Collect protoplasts by centrifugation between washes. Discard supernatant. Gently resuspend protoplasts in 1 mL of STC and check concentration with a hemacytometer. Adjust the concentration to 10^7 protoplasts per mL in four parts of STC and one part of 50 % PEG solution (polyethylene glycol 3,350). Dispense protoplasts into a small volume (100 μL) and store them at −80 °C.

15.2.4 Generation of Split-Marker Fragments

The split-marker gene fragments flanked with a gene of interest are constructed using a fusing PCR approach (Fig. 15.1). This method completely eliminates tedious cloning procedures and allows quick generation of split-marker fragments for targeted gene disruption. Two truncated but overlapping gene fragments (WY/ and /YZ) are first amplified from a plasmid containing a suitable gene cassette. A bacterial phosphotransferase gene conferring resistance to hygromycin is often used a dominant selectable marker in filamentous fungi. Primers S1 and S2 are designed to amplify the WY/ fragment; primers S3 and S4 are used to amplify the /YZ fragment using a GoTag DNA polymerase (Promega) in a 50-μL solution using a standard PCR protocol. Two DNA fragments (0.5–1.5 kb) of a gene of interest are amplified separately by PCR with

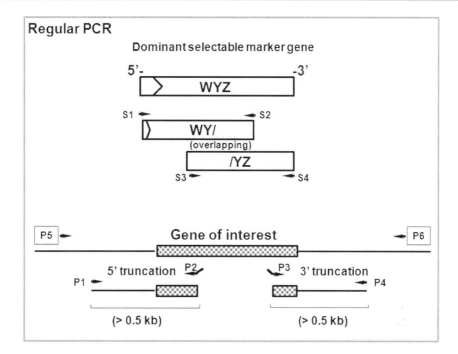

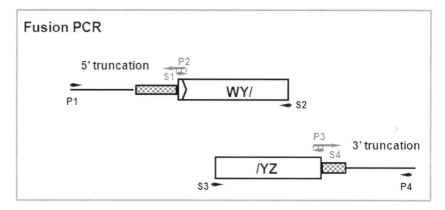

Fig. 15.1 Schematic illustration of fusion PCR for generating overlapping truncation of a dominant selectable marker gene (WYZ) fused with homologous sequence of a gene of interest. PCR is used to amplify two overlapping fragments WY/ and /YZ with the primers S1 pairing with S2 and S3 pairing with S4, respectively. Primer P2 contains a tail sequence completely complementary to the sequence of S1 and primer P3 contains partial sequence completely complementary to the sequence of S4. The 5' truncation of the target gene is amplified with the primers P1 and P2 and joined to the WY/ fragment. The 3' truncation of the target gene is amplified with the primers P3 and P4 and joined to the /YZ fragment

two gene-specific primers from fungal genomic DNA. The length of homologous sequences can be varied depending on the desired extent of deletion of target gene sequences. Using longer homologous sequences may increase the efficiency of homologous integration.

As illustrated in Fig. 15.1, primers P1 and P2 are designed to amplify the 5' region of the target gene; primers P3 and P4 are used to amplify the 3' region of the gene. The tail sequence of the P2 primer is designed to be completely complementary to the sequence of S1 and the tail sequence of

P3 is completely complementary to the sequence of S4. This is designed so that the marker gene fragment (WY/) is fused with the 5' truncation of a gene of interest with the primers P1 and S2 and the /YZ fragment fused with the 3' truncation with the primers P4 and S3 to form two chimeric DNA fragments. Note: It is not necessary to clean up PCR fragments prior to second-round amplification. The cycling profile for PCR amplification begins with a cycle of 95 °C for 3 min, immediately followed by 30 cycles of 95 °C for 30 s, 56 °C for 30 s, 72 °C for 1.5–3 min and completed by incubating at 72 °C for 10 min.

15.2.5 Transformation

PCR-generated DNA fragments are directly transformed into fungal protoplasts without any additional cleanup. Transformation of fungal protoplasts was performed using $CaCl_2$ and polyethylene glycol (Chung et al. 2002). Protoplasts frozen at −80 °C in 100 µL of STC: 50 % PEG (4:1, v/v) are placed in ice for at least 10 min. Mix split-marker DNA fragments (10 µL each) with 100 µL protoplasts in a sterile 15-mL centrifuge tube (Falcon). Leave at room temperature for 20 min. Add 1 mL of 50 % PEG gradually into the centrifuge tube, mix gently, and leave at room temperature for an additional 20 min. Add 3 mL liquid RM and place on a shaker set at 100 rpm at room temperature for 2–4 h. Add a selectable agent into each tube except the no selection control. Mix gently with molten solid RM (45 °C), pour into petri dish, and swirl gently. The plates are incubated at 28 °C. Examine daily for colony formation. Pick colonies and transfer to fresh medium. Successful disruption of a given gene in a wild-type strain can be identified quickly if the mutant strain shows any phenotypes, otherwise PCR verification is needed to confirm the disruption.

15.2.6 Homologous Integration

Because the dominant marker gene is split in separate fragments, the gene is not functional unless homologous recombination occurs between two overlapping fragments (Fig. 15.2). Fungal transformants will not grow on a medium containing the selection agent unless homologous recombination occurs between the overlapping regions of the dominant marker gene. The marker gene cassette fused with homologous flanking sequences is integrated into the target locus via double cross over recombination (Fig. 15.2). Successful integration of a marker gene fragment within a gene of interest can be validated by analytical PCR with the primers located just outside the targeted region (e.g., primers P5 and P6 in Fig. 15.1) and by Southern blot hybridization of fungal genomic DNA, digested with various endonucleases, to a gene-specific probe. For a given gene, six oligonucleotide primers are needed for generation of split-marker fragments and for verification of locus-specific integration.

Transformation of split-marker fragments in phytopathogenic fungi, *Cercospora nicotianae*, *C. beticola*, *Elsinoë fawcettii*, *Colletotrichum acutatum,* and *A. alternata*, has been shown to increase homologous integration frequency (You et al. 2007, 2009; Chen et al. 2007; Liao and Chung 2008; Weiland et al. 2010; Lin et al. 2010; Yang and Chung 2013). The split-marker approach could be useful for identifying disruptants with no obvious phenotypes because a high frequency of targeted gene disruption via homologous recombination can be achieved by screening less than 20–30 independent transformants. The target gene can be disrupted or completely replaced by the marker gene fragment, depending on the homologous DNA sequence within the gene of interest. Because the flanking fragments are generated by PCR, the deleted region can be precisely determined. The minimum flanking sequence required for efficient homologous integration varies among fungal species. However, we have observed that disruption frequency increases as the lengths of the flanking sequence on one end or both ends of the target gene increases (You et al. 2009). It is critical to have sufficient lengths of the flanking DNA sequence (>0.5 kb) when employing the split-marker approach for targeted gene disruption. The minimum overlapping sequence required for efficient recombination at

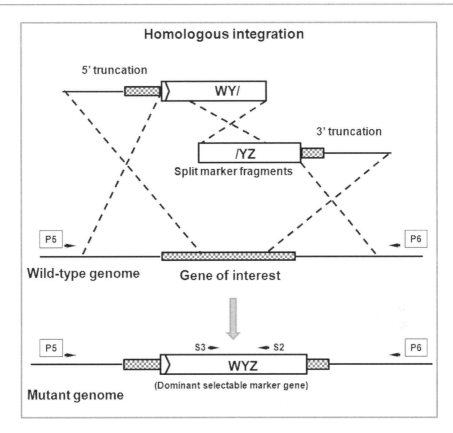

Fig. 15.2 Schematic illustration of targeted gene disruption by split-marker-based transformation. Two truncated, overlapping marker gene fragments flanked with the truncation of a target gene are directly transformed into protoplasts prepared from a wild-type fungal strain. Three cross-over events are required to generate functional marker gene and homologous integration. Only transformants containing a functional marker gene cassette will grow on a medium containing the selection agent. Employing split-marker-based gene disruption could enhance homologous recombination

the selectable marker gene remains uncertain. However, fungal disruptants have been successfully identified using two DNA fragments overlapping 200–450 bp at the selectable marker gene (Yang and Chung 2013).

15.3 Conclusion

Targeted gene disruption via homologous recombination has had a major impact on modern fungal biology. This split-marker-based transformation approach increases the frequency of recovering disruptants, presumably by increasing the frequency of homologous integration and/or by decreasing ectopic and tandem integration events in fungi. This approach was originally developed for rapid, gap repaired-mediated cloning in the budding yeast *Saccharomyces cerevisiae* (Fairhead et al. 1996). The split-marker fragments flanking by homologous sequences of target gene can be obtained by fusion PCR without the need for cloning, allowing a faster and more efficient method of generating disruption constructs. Efficient gene disruption strategies along with the other molecular techniques shall facilitate functional genomic analysis in filamentous fungi.

Acknowledgments The author would like to thank current and former Chung lab members S.L. Yang, L.H. Chen, H.C. Tsai, C.H. Lin, and B.J. You for their contributions to this work.

References

Bird D, Bradshaw R (1997) Gene targeting is locus dependent in filamentous fungus *Aspergillus nidulans*. Mol Gen Genet 255:219–225

Chen H, Lee MH, Daub ME, Chung KR (2007) Molecular analysis of the cercosporin biosynthetic gene cluster in *Cercospora nicotianae*. Mol Microbiol 64:755–770

Choquer M, Dekkers K, Ueng PP, Daub ME, Chung KR (2005) The *CTB1* gene encoding a fungal polyketide synthase is required for cercosporin biosynthesis and fungal virulence of *Cercospora nicotianae*. Mol Plant Microbe Interact 18:468–476

Chung KR, Jenns AE, Ehrenshaft M, Daub ME (1999) A novel gene required for cercosporin toxin resistance in the fungus, *Cercospora nicotianae*. Mol Gen Genet 262:382–389

Chung KR, Shilts T, Li W, Timmer LW (2002) Engineering a genetic transformation system for *Colletotrichum acutatum*, the causal fungus of lime anthracnose and postbloom fruit drop. FEMS Microbiol Lett 213:33–39

Fairhead C, Llorente B, Denis F, Soler M, Dujon B (1996) New vectors for combinatorial deletions in yeast chromosomes and for gap-repair cloning using "split-marker" recombination. Yeast 12:1439–1457

Fu J, Hettler E, Wickes BL (2006) Split marker transformation increases homologous integration frequency in *Cryptococcus neoformans*. Fungal Genet Biol 43:200–212

Idnurm A, Warnecke DC, Heinz E, Howlett BJ (2003) Characterisation of neutral trehalase and UDP-glucose:sterol glucosyltransferase genes from the plant pathogenic fungus *Leptosphaeria maculans*. Physiol Mol Plant Pathol 62:305–313

Liao HL, Chung KR (2008) Genetic dissection defines the roles of elsinochrome phytotoxin for fungal pathogenesis and conidiation of the citrus pathogen *Elsinoë fawcettii*. Mol Plant Microbe Interact 21:469–479

Lin CH, Chung KR (2010) Specialized and shared functions of the histidine kinase—and HOG1 MAP kinase–mediated signaling pathways in *Alternaria alternata*, the filamentous fungal pathogen of citrus. Fungal Genet Biol 47:818–827

Lin CH, Yang SL, Wang N, Chung KR (2010) The FUS3 MAPK signaling pathway of the citrus pathogen *Alternaria alternata* acts independently and cooperatively with the fungal redox-responsive AP1 regulator for diverse developmental, physiological and pathogenic functions. Fungal Genet Biol 47:381–391

Pratt R, Aramayo R (2002) Improving the efficiency of gene replacements in *Neurospora crassa*: a first step towards a large-scale functional genomics project. Fungal Genet Biol 37:56–71

Segers GC, Bradshaw N, Archer D, Blissett K, Oliver RP (2001) Alcohol oxidase is a novel pathogenicity factor for *Cladosporium fulvum* but aldehyde dehydrogenase is dispensable. Mol Plant Microbe Interact 14:367–377

Weiland JJ, Chung KR, Suttle JC (2010) The role of cercosporin in the virulence of *Cercospora* spp. to plant hosts. In: Lartey RT, Weiland JJ, Panella L, Crous PW, Windels CE (eds) Cercospora leaf spot of sugar beet and related species. APS Press, St. Paul, pp 109–117

Yang SL, Chung KR (2013) Similar and distinct roles of NADPH oxidases components in the tangerine pathotype of *Alternaria alternata*. Mol Plant Pathol 14:543–556

You BJ, Choquer M, Chung KR (2007) The *Colletotrichum acutatum* gene encoding a putative pH-responsive transcription regulator is a key virulence determinant during fungal pathogenesis on citrus. Mol Plant Microbe Interact 20:1149–1160

You BJ, Lee MH, Chung KR (2009) Gene-specific disruption in the filamentous fungus *Cercospora nicotianae* using a split-marker approach. Arch Microbiol 191:615–622

Part V

Tools and Applications: High Throughput Experimentation

Integrated Automation for Continuous High-Throughput Synthetic Chromosome Assembly and Transformation to Identify Improved Yeast Strains for Industrial Production of Biofuels and Bio-based Chemicals

16

Stephen R. Hughes and Steven B. Riedmuller

16.1 Overview

Production of sustainable and renewable transportation fuels and chemicals by microbial biocatalysts to supplement petroleum-based fuels and chemicals will require engineering the cell machinery of these microbes to obtain sufficiently high yields to be economical. Progress in our understanding of genomes, proteomes, and metabolomes provides new techniques to construct improved microbial catalysts for use in industrial biorefineries. Recent advances in functional genomics, synthetic biology, metabolic engineering, and systems biology have generated an increased interest in discovery, characterization, and engineering of more efficient synthetic pathways and proteins with new catalytic activities or improved native properties for optimal production of biofuels (Dellomonaco et al. 2010;

Peralta-Yahya et al. 2012; Kim et al. 2013a; Nielsen et al. 2013). Cellulosic biomass is an abundant and sustainable substrate for biofuel production (Perlack and Stokes 2011). In lignocellulosic biomass feedstocks, which include agricultural residues and wood waste, the second most abundant sugar after glucose is the pentose xylose. The inability of many microbes to metabolize the pentose sugars creates specific challenges for microbial biofuel production from cellulosic material (Ha et al. 2011; Wohlbach et al. 2011; Zhu and Zhuang 2012; Kim et al. 2013b).

The yeast *Saccharomyces cerevisiae* is currently the most widely employed industrial microbial catalyst. Native *S. cerevisiae* does not consume xylose but can be engineered for xylose consumption with a minimal set of assimilation enzymes. However, xylose fermentation remains slow and inefficient in *S. cerevisiae*, especially under anaerobic conditions. The lack of fermentative capacity in comparison to glucose, limits the economic feasibility of industrial fermentations. Consequently, improving

S.R. Hughes (✉)
Renewable Product Technology, USDA[†], ARS,
NCAUR, Room 1057, 1815 North University Street,
Peoria, IL 61604, USA
e-mail: stephen.hughes@ars.usda.gov

S.B. Riedmuller, B.A.
Hudson Robotics, Inc., Springfield, NJ, USA

[†]Mention of trade names or commercial products in this article is solely for the purpose of providing specific information and does not imply recommendation or endorsement by the United States Department of Agriculture. USDA is an equal opportunity provider and employer.

xylose utilization in industrially relevant yeasts is essential for producing economically viable biofuels from cellulosic material (Hranueli et al. 2013). A few Hemiascomycete yeasts naturally ferment pentose sugars. The best known is the xylose-fermenting yeast *Pichia* (now *Scheffersomyces*) *stipitis*, associated with wood-boring beetles that may rely on fungi to release nutrients from wood. Related yeasts cannot ferment pentoses, suggesting that xylose fermentation has evolved in this unique fungal environment. Although some details are known, much of the mechanism of xylose fermentation remains unresolved. Wohlbach and coworkers performed a comparative analysis of the genomes and transcriptomes of two xylose-fermenting species and identified several genes that, when expressed in *S. cerevisiae*, significantly improve xylose-dependent growth and xylose assimilation (Wohlbach et al. 2011).

The predominant microbially produced biofuel at the present time is ethanol mainly from starch or sugar feedstocks. However, ethanol is not an ideal fuel molecule, and lignocellulosic feedstocks are considerably more abundant than both starch and sugar. Thus, many improvements in both the feedstock and the fuel have been proposed (Nielsen et al. 2013). The principal obstacle to commercial production of fuels such as butanol, terpenoids, or higher lipids and use of feedstocks such as lignocellulosics, syngas, and atmospheric carbon dioxide is that microbial catalysts with robust yields, productivities, and titers have yet to be developed. Suitable microbial hosts for biofuel production must tolerate process stresses such as end-product toxicity and tolerance to fermentation inhibitors in order to achieve high yields and titers (Fischer et al. 2008). Implementation of engineered yeast strains in large-scale processes for production of a variety of biofuels and bio-based chemicals is ongoing. Recently, the well-characterized microorganisms *Escherichia coli* and *S. cerevisiae* were engineered to convert simple sugars into several advanced biofuels such as alcohols, fatty acid alkyl esters, alkanes, and terpenes, with high titers and yields (Zhang et al. 2011). Various yeast and Zygomycetes strains including *Rhodotorula* sp., *Yarrowia lipolytica*, *Pichia membranifaciens,* and *Thamnidium elegan* were tested for their ability to assimilate biodiesel-derived waste glycerol and convert it into value-added metabolic products to improve the economics of the biodiesel industry (Chatzifragkou et al. 2011). It remains to be seen whether the cellular machinery of *S. cerevisiae* can be re-engineered to address the process requirements for the needed products or whether non-*Saccharomyces* yeasts will become more suitable alternatives (Nielsen et al. 2013).

Classical methodologies for strain improvement typically involved a mutagenesis step, then screening of mutants for higher yield, followed by another round of mutagenesis (Hughes et al. 2012; 2013). Usually the molecular basis for increased production would cover many loci and would never be discovered. An alternative approach using recombinant DNA technology is to clone the desirable gene or gene cluster, followed by DNA sequencing and bioinformatics analysis of the sequence to identify structural and regulatory regions. Increased product yield would then be achieved by specific mutation of these regions or replacement by more efficient promoters. The field of strain improvement is moving from classical methodologies to advanced functional genomics. Functional genomics has developed as a broad new field of science that aims to characterize all the parts of a system (such as genes, proteins, and ligands) and find the interactions between the parts to discover the properties of the system being studied. This will often include engineering networks to understand and manipulate the regulatory mechanisms, and to integrate various systems (Hranueli et al. 2013).

With the completion of countless genome sequencing projects, genetic bioengineering has expanded into many applications including the integrated analysis of complex pathways, the construction of new biological parts and the redesign of native biological systems. All these areas require the well-defined and systematic assembly of multiple DNA fragments of various sizes, including chromosomes, and the optimization of gene expression levels and protein activity. Current commercial cloning products are not robust enough to support the assembly of very large or very small DNA fragments or a combination of both. In addition, current strategies are not flexible enough to allow further modifications to the original design without having to undergo complicated cloning strategies. Tsvetanova et al. (2011) have proposed a seamless, simultaneous, flexible, and highly efficient assembly of genetic material, designed for a wide size range (10s to 100,000 s base pairs), which can be performed either in vitro or within the living cells.

Recent advances in DNA synthesis technology have enabled the construction of novel genetic pathways and genomic elements, furthering our understanding of system-level phenomena. The ability to synthesize large segments of DNA allows engineering of pathways and genomes according to arbitrary sets of design principles. Dymond et al. (2011) describe the first partially synthetic eukaryotic chromosomes, *S. cerevisiae* chromosome synIXR, and semi-synVIL. When complete, the fully synthetic yeast genome will allow massive restructuring of the yeast genome, and may open the door to a new type of combinatorial genetics based entirely on variations in gene content and copy number (Dymond et al. 2011). The creation of a synthetic cell in 2010 (Gibson et al. 2010) demonstrated that it is now technically feasible to not only read whole genome sequences but also to begin synthesizing them. An outcome of this work has been a set of tools for synthesizing, assembling, engineering, and transplanting whole bacterial genomes. Although synthetic biologists are still learning how to rationally design DNA, particularly large genetic pathways or genomes, the technology is available to build them (Gibson 2014). Gibson (2012) demonstrated that the yeast *S. cerevisiae* can take up and assemble at least 38 overlapping single-stranded oligonucleotides and a linear double-stranded vector in one transformation event. These oligonucleotides can overlap by as few as 20 bp and can be as long as 200 nucleotides in length to produce kilobase-sized synthetic DNA molecules. A method for one-step assembly of DNA constructs for complete synthetic pathways in *S. cerevisiae* is described by Merryman and Gibson (2012). In a novel approach to synthetic biology, Shabi et al. (2010) compare DNA processing to word processing. Their DNA processing system provides a foundation for accomplishing complicated DNA processing tasks such as synthesis, editing, and library construction using a unified approach. The system combines a computational algorithm with biochemical protocols to plan construction of the target molecules. The plan is translated into a robotic control system that implements it to produce the target molecules while maximizing the use of existing DNA molecules and shared components. The targets can be cloned and sequenced to find a correct target molecule (Linshiz et al. 2008; Shabi et al. 2010).

Modern biotechnologies such as molecular engineering and synthetic biology are ultimately limited by their need for high-throughput measurements of biochemical reactions. In the final analysis, this will be enabled by new automated DNA technologies that can reliably convert low-cost oligonucleotides (for example, those produced on a microchip in small quantities) into accurate synthetic DNA fragments. Thousands of genome combinations can then be built and tested at an affordable price (Gibson 2014). Agrestia et al. (2010) have also developed a general ultrahigh-throughput screening platform using drop-based microfluidics that markedly increases both the scale and speed of screening. The aqueous drops dispersed in oil act as picoliter-volume reaction vessels and can be screened at rates of 1,000 per second. The system was applied to directed evolution, identifying new mutants of the enzyme horseradish peroxidase exhibiting catalytic rates more than ten times faster than the native enzyme.

The development of high-throughput screening (HTS) allows the rapid monitoring of assay parameters to detect optimized gene expression products by conducting potentially millions of biochemical or genetic tests. Different assay formats can be developed to capture multiple biological readouts from a single sample (Zanella et al. 2010; Didiot et al. 2011). The highly effective nature of HTS for identification of highly target specific compounds is attributed to its precise focus on single mechanism. This development is closely connected to changes in strategy of chemical synthesis. The vast number of compounds produced by combinatorial chemistry and the possibility of testing many compounds in a short period of time by HTS attracted the interest of researchers in many fields. With the introduction of robotics, automation, and miniaturization techniques, it became feasible to screen 50,000 compounds a day with complex work stations. High-throughput screening methods are also used to characterize metabolic and pharmacokinetic data about new drugs (Inglese et al. 2007; Martis et al. 2011).

This chapter describes a system of four robotic platforms required for continuous operation of the process to obtain improved fungal strains for production of biofuels and bio-based chemicals in an industrial biorefinery: (1) synthesis and screening of a diverse collection of systematically mutagenized gene ORFs to produce a library of optimized ORFs; (2) one-step construction of a synthetic yeast artificial chromosome (YAC) containing the optimized ORFs in a polyprotein cassette for expression of multiple genes; (3) selection of an optimal host strain that has been subjected to mutagenesis to produce a strain capable of robust growth at a biorefinery and transformation of this host strain with these collections of synthetic YACs; and (4) high-throughput screening of the transformed strains for desired industrial traits. This system is designed to produce an improved industrial microbial biocatalyst by manipulation of the host strain, assembly of an optimized synthetic chromosome without traditional cloning steps, and stable transformation of the chromosome into the engineered improved host strains for use in production of biofuels via biorefinery operations.

16.2 Pilot Biorefinery Design

In a typical pilot biorefinery, the robotic platforms are the starting point for the development of an improved industrial microbial biocatalyst, including synthesizing and screening libraries of optimized gene ORFs, one-step continuous assembly of synthetic yeast artificial chromosomes (YAC) containing the optimized ORFs, selection of an optimal host strain, transformation of this host strain with the synthetic YAC(s); and high-throughput screening of the transformed strains for desired industrial traits. A schematic of a pilot biorefinery layout showing the relationship of these operations to the other operations involved in the biorefinery, including organic and analytical chemistry laboratories, offices, bioprocessing test units, mechanical and technical support areas, and pilot plant (200–500-L scale) for fermentation at industrial conditions with appropriate feedstock, is shown in Fig. 16.1. The gene ORFs assembled and the host strain selected depend on the feedstock utilized and the potential biofuels and bio-based chemicals produced.

The robotic platforms for oligonucleotide synthesis, gene assembly, microbial strain transformation, and screening are shown at top left in Fig. 16.1. The organic and analytical chemistry areas and the bioprocessing test units are at the top right in Fig. 16.1. Each individual process, such as utilization of feedstock or growth and product formation by a transformed strain, is first tested at laboratory scale in the bioprocessing units. This synthetic biology operation allows custom feedstock bioprocessing and scale-up of engineered strains stably transformed with synthetic artificial chromosomes. After successful testing of the transformed strain in the fermentation skid under industrial conditions, the strain is transferred to the production-scale biorefinery. The plans depicted in Fig. 16.1 are for a synthetic biology research facility in place.

This pilot biorefinery allows one-step continuous construction and rapid evaluation of large libraries of recombinant microbial strains stably transformed with synthetic artificial chromosomes containing genes designed to convert the wide variety of potential feedstocks and produce the desired products (Perlack and Stokes 2011). Although many of the required genes have been identified (Dellomonaco et al. 2010; Peralta-Yahya et al. 2012; Kim et al. 2013a), it is clear additional information must be provided in the design of the artificial chromosome. Identifying the additional genetic information will require synthesis and screening of large numbers of artificial chromosome. Artificial chromosomes are able to accommodate large DNA fragments (Burke et al. 1987; Sanchez et al. 2002; Arnak et al. 2012) so they can be designed to contain movable cassettes with groups of genes with similar functions or for a given pathway that are easy to assemble, modify, and stably transform

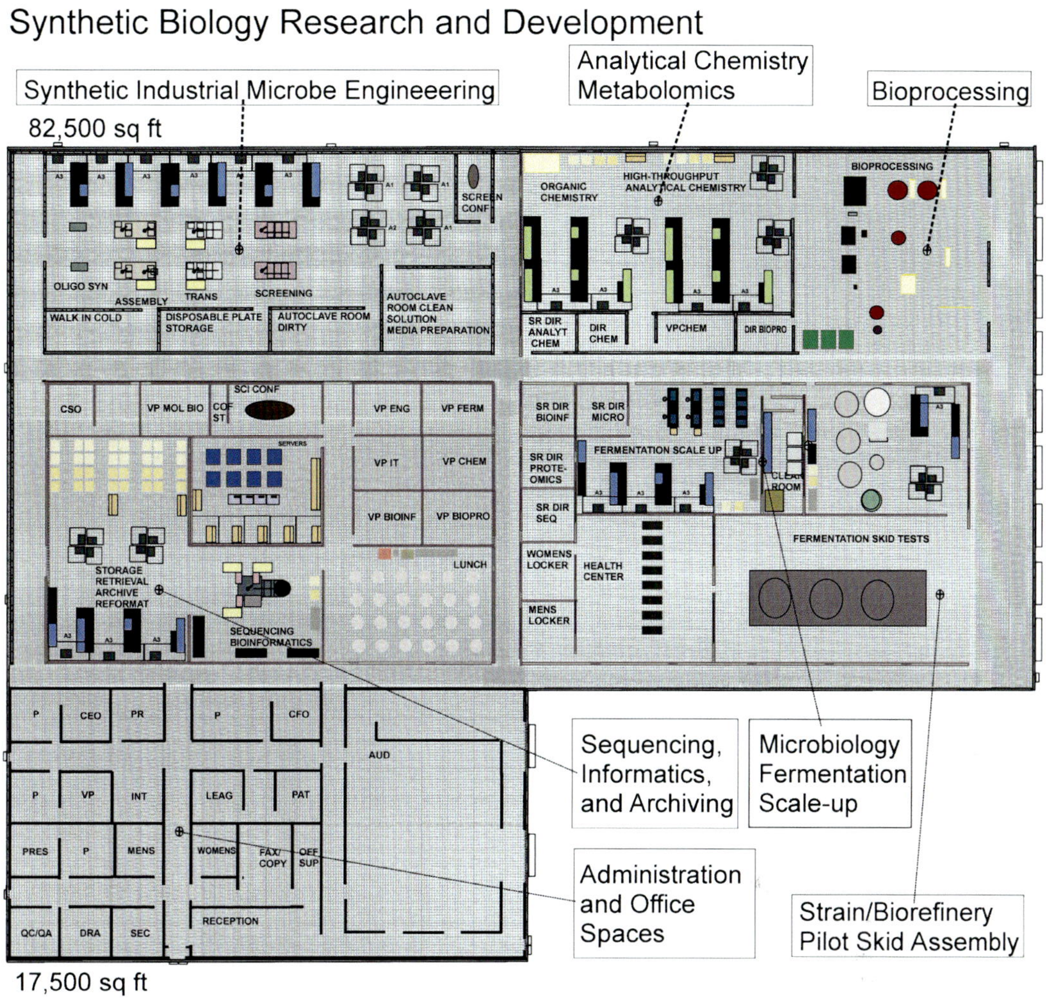

Synthetic Biology Research and Development

Fig. 16.1 Pilot biorefinery incorporating facilities for functional genomics, synthetic biology, metabolic engineering, and systems biology to support development of improved industrial microbial biocatalysts for production of biofuels and bio-based chemicals. (Scale: 1 square on background grid equals 1 foot)

in different combinations into different host microbes. A single promoter is used to express a polyprotein cassette separated by trypsin sites. Different promoters can be selected to obtain optimum expression levels. The expressed polyprotein is rapidly released by SUMO protease followed by cleavage of the polyprotein at the trypsin sites by the host yeast protease (Hughes et al. manuscript submitted). It will also be possible to insert gene ORFs for high-value coproducts such as a bioinsecticide (Hughes et al. 2007; 2008) or a noncaloric peptide sweetener (Pinkelman et al. manuscript submitted) to increase profitability of the biorefinery.

16.3 Synthetic Optimized Gene ORF Assembly

One method for assembling gene ORFs to construct an artificial chromosome is amino acid scanning mutagenesis (AASM). The AASM algorithm enables synthesis of a complete set of mutations across a gene ORF, replacing each amino acid in the encoded polypeptide or protein with all other possible amino acids (Hughes et al. 2007; 2008). Mutagenized ORFs of any gene encoding a polypeptide or protein of interest are produced using PCR to separately replace the original triplet codons for each amino acid at each codon position. This produces a library of separate mutant gene products encoding each of the possible amino acids at each position in the polypeptide chain. The assembled clones may be screened for desired phenotypic properties such as biochemical activity and/or binding site recognition and then assembled into an artificial chromosome.

To initiate mutation of a gene ORF, a first set of overlapping oligonucleotides is constructed in a brickwise manner as shown in Fig. 16.2. The gene ORF sequence is divided into a first group of adjacent oligonucleotide segments (preferably of equal length with the possible exception of the last segment at the 3' or 5' end). The oligonucleotide segments will typically span substantially the length of the sequence, except for a short portion of the 5' and 3' ends, about 10 bases long. These adjacent oligonucleotide segments will serve as a template for the design of mutant oligonucleotide sequences or primers which may then be used to synthesize mutant nucleotide sequences (mutant ORFs) with altered codons. The length of the oligonucleotides is not critical and may vary. The other strand is similarly divided into a set of adjacent oligonucleotides, with the provision that the oligonucleotides from the two strands overlap so that their ends are not coincident. These oligonucleotide primers are illustrated in Fig. 16.2, Set 3, as arrows directly above and below the gene ORFs.

Within each oligonucleotide segment on the selected strand, a first block of four (or more) adjacent triplet codons are selected for modification. The location of the first codon blocks is not critical; although they are typically located near the 5'end of the oligonucleotide. Each triplet codon within the selected block is randomly changed to substitute codons encoding all other amino acids, generating a first set of mutant oligonucleotides. Each codon block selected in the first mutation step (i.e., set 3) is labelled with the numeral 2 in Fig. 16.2. Following their construction, the first mutant oligonucleotides are used to synthesize a first library of mutagenized ORFs of the gene of interest by PCR. The mutagenized ORFs independently encode all of the possible

permutations of amino acids within each of the first block of codons.

The mutagenesis steps used for constructing the first library of mutagenized ORFs are repeated, using PCR with the PCR primers shifted toward the 3' end of the mutagenized ORFs from the first set. As in the first set, one strand is divided into a second group of adjacent oligonucleotide segments. However, the oligonucleotide segments of this second group begin at approximately the center to the 3' end codon of the first block of codons. The oligonucleotide segments of this second group are shifted toward the 3' end of the mutant nucleotide sequence by approximately 2–4 codons as shown in Fig. 16.2, set 4. Within each oligonucleotide segment on the selected strand, a second block of four (or more) adjacent codons are selected for further modification. The 5' end of the second block of codons is adjacent to the 3' end of the first block (in the first group of oligonucleotides above). Each of the second block of codons in the second group of oligonucleotide sequences is labelled 3 in Fig. 16.2. Again, within each selected block, each of the codons is randomly changed to substitute codons encoding all other amino acids, generating a second set of mutagenized ORFs. The process is repeated for sets 5 and 6 in Fig. 16.2.

Using an average gene length of 1,000 bp, the number of clones for screening produced by a 4-codon algorithmic substitution can be estimated. A set of overlapping oligonucleotides of approximately 50 bp in length are synthesized to assemble a clone of this gene. A second set of oligonucleotides is produced to assemble a clone with an identical sequence, but the overlap is offset by approximately 25 bp. Once these two clone sets are produced there is no section of the clone that is not covered by an overlap when introduction of a 4-codon set (12 bp) of randomization codons is shifted down along both clone set sequences leaving at least 10 bp overlap at the 3' end of each oligonucleotide. This 4-codon randomization set substituted into each of the two identical assembled clones will give rise to $20^4 = 160,000$ possible versions for all 20 amino acids at each of the positions in the mutant peptides corresponding to these

16 Integrated Automation for Continuous High-Throughput Synthetic Chromosome...

Automated Gene Assembly and Amino Acid Scanning Routine

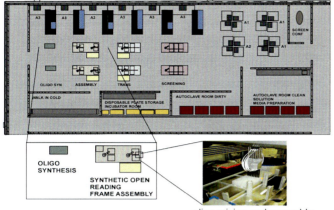

- Oligonucleotide synthesis
- PCR assembly
- Resulting in assembled clonal open reading frame collections
- Performed without DNAsI or other costly molecular biology enzymes and reagents
- Seamlessly incorporated into synthetic chromosome assembly routines
- Possible to use very large numbers of resulting assembled clones in seaming to increase diversity lost in traditional cloning

ORF = open reading frame

Gene Source Identification
- From Published Source (assemble)
- From a gene collection (JCVI, HIP, JGI., ATCC, .etc.)
- Isolate from genomic DNA and clone out

Oligo Design — Overlapping brick construction of oligo sets
- SET 1 WILD TYPE GENE ASSEMBLY
- SET 2 CODON OPTIMIZED FOR CELL BETTER TRANSCRIPTION AND TRANSLATION
- SET 3 FIRST MUTATIONAL SET IN BLOCKS OF FOUR CODONS
- SET 4 SECOND MUTATIONAL SET FOR SHIFTED FOUR CODONS COVER REST
- SET 5 THIRD MUTATIONAL SET FOR SHIFTED FOUR CODONS COVER REST
- SET 6 FOURTH FINAL MUTATIONAL SET FOR SHIFTED FOUR CODONS COVER REST
- FINAL AASM OPTIMIZED ORF

Fig. 16.2 Amino acid scanning mutagenesis (AASM) algorithm showing the stepwise progression of gene ORF mutagenesis using overlapping oligonucleotides and PCR assembly to produce a library of mutagenized gene ORFs by systematically replacing each codon across the length of the original gene ORF until all of the codons have been substituted

four codons. If this were a 1,000 bp gene, shifting this randomization through the whole clone would give approximately 50 different 4-codon randomization substitutions, each generating 160,000 clones to screen after expression of the protein encoded by the gene. The result would be a set of 18 million $(50 \times 160,000)$ optimized clones. This level of screening can occur on most commercial liquid handler-based functional proteomic robotic platforms (Hughes et al. 2005; 2006; Patent 2011).

The advantage of the AASM strategy for synthesizing optimized gene ORFs is that the contribution of every codon is considered. The screening for optimized function in the gene products may identify a mutagenized gene or group of genes that have not yet been identified as potentially providing the functions or pathways needed for conversion of biomass feedstocks to biofuels. Mutagenizing the entire gene ORF at all codons allows production of genes optimized for codon usage, expression levels, solubility, and functionality. It is almost impossible to predict which codon change might increase properties that will optimize the final functionality of the gene product (Angov 2011).

16.4 One-Step Assembly of Artificial Chromosome by Seaming Optimized Gene ORFS

Yeast artificial chromosomes (YACs) have been employed for the cloning and manipulation of large deoxyribonucleic acid (DNA) inserts (up to 3 Mb pairs) in yeast. The capacity of YACs to accommodate large DNA fragments can be used to clone clusters of genes surrounded by their native DNA context, where regulatory elements are located (Burke et al. 1987; Sanchez et al. 2002; Arnak et al. 2012). This is important for biotechnology when YACs are used for engineering genetic determinants of new biochemical pathways for production of secondary metabolites, and for heterologous protein expression. YACs contain a yeast autonomously replicating sequence (ARS1) necessary for replication, with its associated centromere (CEN4) DNA sequence

for segregation at cell division, and two telomere-like DNA sequences (TEL) derived from Tetrahymena thermophila (Kuhn and Ludwig 1994). The centromere and telomere sequences allow maintenance of large cloned DNA as stable, single-copy yeast linear chromosomes (Bruschi and Gjuracic 2002). They function like naturally existing chromosomes showing comparable stability (Dymond et al. 2011; Arnak et al. 2012).

The one-step production of an artificial chromosome using high fidelity polymerase chain reaction (HF-PCR) assembly and seaming allows production of polyprotein expression cassettes for entire metabolic pathways and for combinations of whole genome libraries (Hughes et al. 2009) to obtain the optimal set of ORFs optimized by AASM. It also allows the placement of custom synthetic promoters for adjusting expression levels. For increased profitability of biorefineries, it also is possible to seam gene ORFs into the YAC sequence for expression of high-value coproducts. As an indicator of the level of polyprotein expression, the gene for green fluorescent protein (GFP) can be inserted for expression at the carboxy terminus of the polyprotein.

A schematic of the automated one-step PCR assembly of a yeast artificial chromosome (YAC) library by seaming metabolic pathway gene ORFs for enhanced utilization of xylose from cellulosic biomass and mutagenized optimized gene ORFs for expression of valuable coproducts driven by a selected promoter sequence and expressing the green fluorescent protein (GFP) sequence as a marker for stable transformation of host strain and high expression levels of polyprotein from the synthetic YAC is presented in Fig. 16.3. The XI and XKS gene open reading frames (ORFs) to express enzymes from the metabolic pathway for xylose were obtained by HF-PCR amplification using plasmids from Hughes et al. (2009) as templates. The yeast promoter and SUMO expression tag were obtained by HF-PCR from the pCR8 plasmid. The gene ORF for the high-value natural sweetener (brazzein) coproduct selected from the optimized ORFs produced by AASM, indicated as a collection of mutagenized brazzein sequences at the right on Fig. 16.3, is seamed into the cassette.

Automated Assembly of Expression Synthetic Chromosome with Polyprotein Open Reading Frame and Transformation

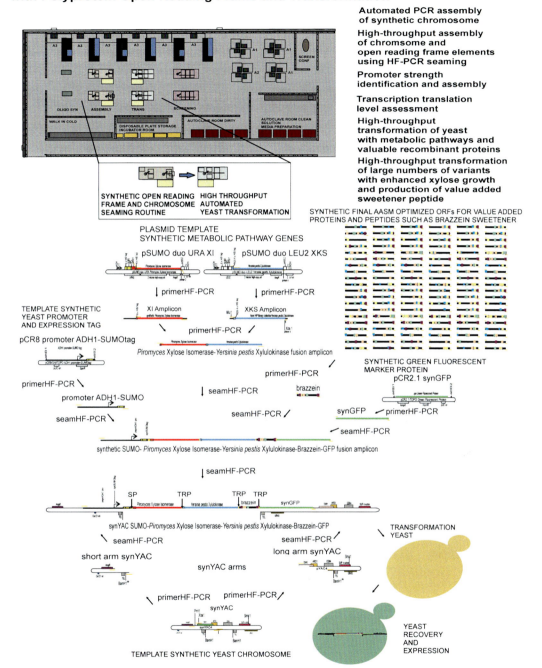

Fig. 16.3 Automated one-step PCR assembly of yeast artificial chromosome (YAC) library by seaming metabolic pathway gene ORFs and mutagenized optimized gene ORFs for high-value coproduct (brazzein) with selected promoter sequence and SUMO expression tag plus green fluorescent protein (GFP) sequence as a marker for stable transformation of host strain and high expression levels of polyprotein from synthetic YAC

The GFP sequence obtained by HF-PCR from the pCR2.1 plasmid is added to the cassette as a marker for transformation into the yeast strain. The plasmid pYAC4 is digested with the restriction enzymes and the two arms from the pYAC4 digest seamed in to produce the artificial expression chromosome YAC SUMO-XI-XKS-Brazz-GFP that is transformed directly into *S. cerevisiae* for replication and expression (Pinkelman et al. manuscript submitted).

A similar process could be followed to assemble a polyprotein expression cassette containing in-frame synthetic gene ORFs for expression of other enzymes and proteins to enhance xylose utilization (Ha et al. 2011; Kim et al. 2013a) or for insertion of gene ORFs expressing other high-value peptide products to improve cost-effectiveness of cellulosic biofuel production (Hughes et al. 2008) behind an optimized promoter with custom expression fusion tags selected for desired expression levels and extra- or intracellular protein localization. Yeast artificial chromosomes are ideal for multigene cassette insertion because they allow for stable incorporation of DNA fragments larger than 100 kb (Arnak et al. 2012). This one-step HF-PCR amplification assembly and seaming allows complete assembly of large numbers of chromosomes much more rapidly than traditional cloning and molecular biology procedures to prepare plasmids, transform into bacteria, repeat plasmid preparation, and then restrict to produce the linearized YAC. Several different procedures are available for transformation of yeast cells including spheroplast generation, electroporation, alkali cation, or polyethylene glycol (PEG) treatment (Ito et al. 1983; Lin-Cereghino et al. 2005). The transformation of *S. cerevisiae* cells with the YAC constructed as shown in Fig. 16.3 was performed using the alkali–cation procedure.

16.5 Mutagenized Improved Host Strains for Use with Synthetic YAC Libraries

Suitable microbial hosts for biofuel production must tolerate process stresses such as end-product toxicity and tolerance to fermentation inhibitors in order to achieve high yields and titers (Fischer et al. 2008). Recently, increasing attention has been directed toward developing microbial catalysts for ethanol production at elevated temperatures (Abdel-Banat et al. 2010). Fermentation processes conducted at elevated temperatures will significantly reduce cooling costs, improve efficiency of simultaneous saccharification and fermentation, allow continuous ethanol removal by evaporation under reduced pressure, and reduce risk of contamination. The temperatures suitable for conventional strains of *S. cerevisiae* are relatively low (25–30 °C). The yeast *Kluyveromyces marxianus* has been reported to grow at 47 °C and above and to produce ethanol at temperatures above 40 °C (Abdel-Banat et al. 2010; Nonklang et al. 2008).

Furthermore, *K. marxianus* has the ability to grow on a wide variety of substrates not utilized by *S. cerevisiae* such as xylose, xylitol, cellobiose, lactose, arabinose, and glycerol (Rodrussamee et al. 2011). Because of these advantages, *K. marxianus* is currently being developed as a viable alternative to *S. cerevisiae* for ethanol production (Rodrussamee et al. 2011).

To produce *K. marxianus* strains with improved thermotolerance and enhanced ability to produce ethanol under microaerophilic industrial conditions utilizing both xylose and glucose, wild-type *K. marxianus* NRRL Y-1109 cultures were irradiated with UV-C using automated protocols on a robotic platform for picking and spreading irradiated cultures and for processing the resulting plates (Hughes et al. 2013). UV-C irradiation is a standard technique (James and Kilbey 1977; Pang et al. 2010; Hughes et al. 2012) for inducing mutations in yeast. It produces large numbers of random mutations broadly and uniformly over the whole genome to generate unique strains. The irradiated plates were incubated under anaerobic conditions on xylose or glucose for 5 months at 46 °C. Two *K. marxianus* mutant strains survived and were isolated from the glucose plates. Both mutant strains, but not wild-type, grew aerobically on glucose at 47 °C. All strains grew anaerobically at 46 °C on glucose, galactose, galacturonic acid, and pectin; however, only one mutant strain grew anaerobically on xylose at 46 °C. It also produced ethanol with glucose or galacturonic acid as substrate (Hughes et al. 2013). This mutant strain has potential application as a host strain for transformation with YACs for biofuel production at elevated temperature from constituents of starch, sucrose, pectin, and cellulosic biomass. Other microbial strains that are naturally capable of converting renewable feedstocks to biofuels and producing high-value bio-based chemicals (Dellomonaco et al. 2010; Young et al. 2010; Peralta-Yahya et al. 2012) may also be used as host strains, subjected to mutagenesis and selection for optimized characteristics, and then transformed with artificial chromosomes designed for these microbial strains.

An example of the process for high-throughput mutagenesis of microbial strains via UV-C irradiation to produce optimized host strains for production of cellulosic biofuels is diagrammed in Fig. 16.4 (Hughes et al. 2012). A culture of wild-type *Scheffersomyces stipitis* NRRL Y-7124 was placed into a Marsh RR-0014 deep trough plate with baffled bottom. The plates were placed 14 cm below a source of 234 nm UV-C radiation and irradiated for 1 min. Using an automated protocol on the robotic workcell, samples were spread onto xylose plates. The spread plates were wrapped, sealed, and placed into a Mitsubishi anaerobic chamber at 28 °C for 5 months to select for strains that could survive anaerobically on xylose for an extended period of time unlike the wild-type strain. Two colonies, designated 14 and 22, were found still growing when the spread plates were unwrapped. Duplicate samples were picked from these colonies, spread onto glucose or xylose plates, and incubated at 28 °C for 2 weeks anaerobically to check that these isolates were still capable of growth on glucose and to eliminate background. Five surviving colonies were picked from the re-spread anaerobic xylose plates onto xylose plates and incubated aerobically at 28 °C for 3 days to verify growth capability on xylose was still present and to provide starter cultures for the second round of irradiation. Cultures from isolates 14 and 22 were placed into Marsh RR-0014 deep trough plates 14 cm below a source of UV-C radiation and irradiated at 234 nm for 4 h. Spread plates were prepared on the robotic workcell and placed into a Mitsubishi anaerobic chamber at 28 °C for 5 months. One sample from each of the 42 surviving colonies on the anaerobic xylose plates was spread onto xylose or glucose plates (one sample per plate) and incubated at 28 °C aerobically to check that these strains were still capable of growth on glucose and xylose and to obtain single isolates. After 3 days of aerobic growth on xylose, 4 colonies were considerably larger than those on any of the other plates. These samples, designated 22-1-1, 22-1-11, and 22-1-12, 14-2-6, were screened for ability to utilize sugars from hydrolysates of agricultural waste (Hughes et al. 2012).

16.6 High-Throughput Automated Screening for Improved Strains

High-throughput screening (HTS) typically refers to a process in which large numbers of molecules are screened to identify biologically active molecules as candidates for further validation in additional experiments. This may involve screening thousands to millions of molecules. HTS assays can be biochemical or cell-based. Biochemical assays are target-based with the objective of identifying molecules with a desired function and include such in vitro assays as assessment of enzymatic activity. The screening of specific molecular targets is the most straightforward method for identifying small molecules with a specific, desired function. An effective HTS assay must clearly define the response to be measured and the parameter or parameters used to measure the response (Inglese et al. 2007; Thorne et al. 2010; Martis et al. 2011). The goal is to develop a robust, functional assay that is easily automated and provides good signal-to-background ratio and Z factor scores (Inglese et al. 2007; An and Tolliday 2010). This increases objectivity and reduces subjectivity in data evaluation, providing easier identification of true hits (Martis et al. 2011).

Cell-based assays monitor the response of the cell to stimuli and include assays such as reporter gene assays or phenotypic assays

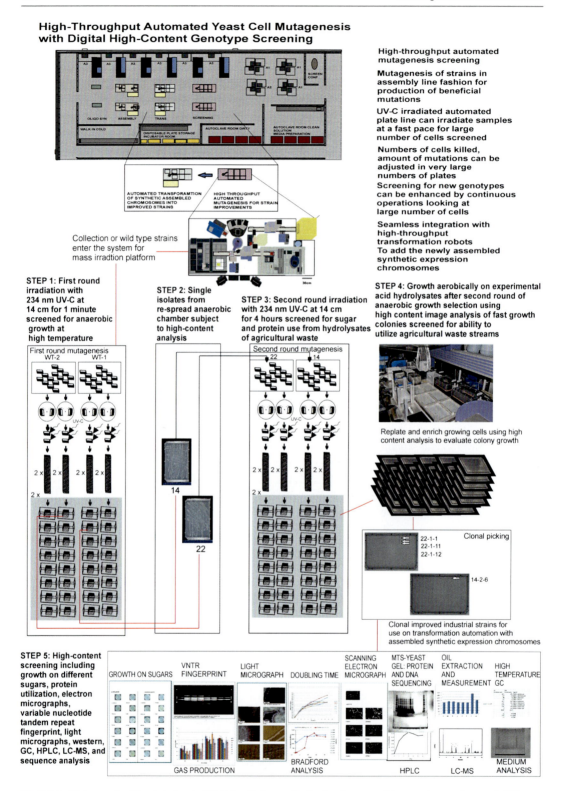

Fig. 16.4 Mutagenesis of host strain via UV-C irradiation and selection for anaerobic growth on xylose and ability to utilize agricultural waste

(Didiot et al. 2011; Martis et al. 2011). Combining HTS with cellular imaging to collect quantitative data from complex biological systems is called high-content screening (HCS) (Zanella et al. 2010). HCS was introduced to meet the need for automation of cellular assays and has emerged as a powerful approach for evaluating protein and gene function in cell culture.

In the example shown in Fig. 16.4, the optimized microbial strains were selected by high-content screening for anaerobic growth on xylose and ability to utilize agricultural waste as substrate. Selection for growth at high temperature has also been used for identifying strains that grow at elevated temperatures (Hughes et al. 2013), a desirable trait for an industrial biocatalyst. Selection for production of oil for biodiesel by an oleaginous yeast using Sudan black or for the production of ammonia as a high-value coproduct using pH measurement have also been used to screen strains for biofuel production (Hughes unpublished data). Examples of the results for these plate screening assays are depicted at the top right of Fig. 16.5. The plate on the left shows the higher the level of oil (lipids) production for biodiesel by the *Yarrowia* strains, the more intense the color exhibited with Sudan black. The plate on the right demonstrates that the level of ammonia (for fertilizer) produced by the *Yarrowia* strains (raising the pH) can be determined by the intensity of the bromothymol blue color. Additional high-content screening assays are shown at the bottom of Fig. 16.4. These include aerobic and anerobic growth of strains on selected sugar substrates, variable nucleotide tandem repeat (VNTR) analysis, gas production, light micrographs, doubling time, growth on protein, scanning electron micrographs, protein and DNA sequencing, HPLC, oil production, LC-MS, GC, and gel electrophoresis.

The robotic platform diagrammed in Fig. 16.4 can typically screen 100 plates per day per line. The platform has 2 lines so that if the unit runs 7 days a week, it can screen and process 1,400 plates a week. The number of strains per plate depends on the process and the number of wells per plate. The number of cells per well usually ranges from 10^7 to 10^9 and the number of wells per plate is typically 96 or 384.

An example of a biorefinery producing renewable biofuels and bio-based chemicals using processes catalyzed by microbial biocatalysts including *K. marxianus*, *Y. lipolytica*, *Rhodotorula glutinis*, *Scheffersomyces* stipitis, and *S. cerevisiae* is depicted in Fig. 16.5. The initial bioconversion stage uses a mutant *K. marxianus* yeast strain to produce bioethanol from sugars. The resulting sugar-depleted solids (mostly protein) can be used in a second stage by the oleaginous yeast *Y. lipolytica* to produce bio-based ammonia for fertilizer and are further degraded by *Y. lipolytica* proteases to peptides and free amino acids for animal feed. The lignocellulosic fraction can be ground and treated to release sugars for fermentation in a third stage by a recombinant cellulosic *S. cerevisiae*, which can also be engineered to express valuable peptide coproducts. The residual protein and lignin solids can be jet cooked and passed to a fourth-stage fermenter where *R. glutinis* converts methane into isoprenoid intermediates. The residues can be combined and transferred into pyrocracking and hydroformylation reactions to convert ammonia, protein, isoprenes, lignins, and oils into renewable gas. Any remaining waste can be thermoconverted to biochar as a humus soil enhancer. The integration of multiple technologies for utilization of sugarcane and agave, and sugarcane waste has the potential to contribute to economic and environmental sustainability.

16.7 Perspective

Tremendous advances in gene and genome engineering have been made in the past decade. Each advance brings the goal of sustainable energy closer. Pursuing this goal using biocatalytic processes with microbial biocatalysts will most likely require the construction of synthetic chromosomes containing the necessary cell machinery to enable the microbes to efficiently convert biomass into biofuels and other bio-based products. The tools available from synthetic biology, metabolic engineering, systems biology, and

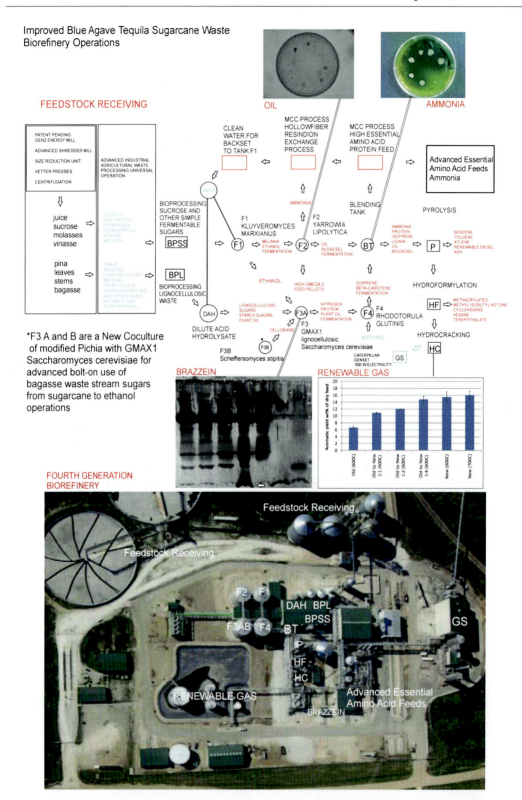

Fig. 16.5 Photograph and schematic diagram of a biorefinery using a process developed for microbial conversion of sugarcane and coffee waste in Colombia to biofuels and bio-based products and adapted for utilization of sugarcane, sugarcane waste, and agave waste from tequila production in Mexico

functional genomics provide the means to synthesize these artificial chromosomes. However, although many of the required genes have been identified, it is clear, additional information must be provided in the design of the artificial chromosome. Identifying that information will require synthesis and screening of large numbers of synthetic genes and chromosomes. One-step continuous automated assembly of synthetic artificial chromosomes from optimized gene ORFs and collections of gene ORFs is an important tool for accomplishing this research. In identifying the requisite genes and groups of genes it will be necessary to organize the process by grouping genes with similar functions, producing cassettes that can be combined and interchanged, simplifying screening assays, selecting regulatory elements for better expression, and, ultimately, obtaining a core cassette that will function as a basis for an artificial chromosome in numerous optimized host cells to efficiently convert biomass into biofuels and other bio-based products.

References

Abdel-Banat BM, Hoshida H, Ano A, Nonklang S, Akada R (2010) High-temperature fermentation: how can processes for ethanol production at high temperatures become superior to the traditional process using mesophilic yeast? Appl Microbiol Biotechnol 85(4): 861–867

Agrestia JJ, Antipov E, Abate AR, Ahn K, Rowat AC, Baret J-C, Marquez M, Klibanov AM, Griffiths AD, Weitz DA (2010) Ultrahigh-throughput screening in drop-based microfluidics for directed evolution. Proc Natl Acad Sci U S A 107(9):4004–4009

An FW, Tolliday N (2010) Cell-based assays for high-throughput screening. Mol Biotechnol 45:180–186

Angov E (2011) Codon usage: nature's roadmap to expression and folding of proteins. Biotechnol J 6:650–659

Arnak R, Bruschi CV, Tosato V (2012) Yeast Artificial Chromosomes. In: Encyclopedia of Life Sciences (eLS). John Wiley & Sons Ltd, Chichester; Wiley online Library. http://www.els.net. doi:10.1002/9780470 015902.a0000379.pub3

Bruschi CV, Gjuracic K (2002) Yeast Artificial Chromosomes. In Encyclopedia of Life Sciences. John Wiley & Sons, Wiley online Library. doi:10.1038/npg.els.0000379

Burke DT, Carle GF, Olson MV (1987) Cloning of large segments of exogenous DNA into yeast by means of artificial chromosome vectors. Science 236:806–812

Chatzifragkou A, Makri A, Belka A, Bellou S, Mavrou M, Mastoridou M, Mystrioti P, Onjaro G, Aggelis G, Papanikolaou S (2011) Biotechnological conversions of biodiesel derived waste glycerol by yeast and fungal species. Energy 36(2):1097–1108

Dellomonaco C, Fava F, Gonzalez R (2010) The path to next generation biofuels: successes and challenges in the era of synthetic biology. Microb Cell Fact 9:3

Didiot M-C, Serafini S, Pfeifer MJ, King FJ, Parker CN (2011) Multiplexed reporter gene assays: monitoring the cell viability and the compound kinetics on luciferase activity. J Biomol Screen 16:786–793

Dymond JS, Richardson SM, Coombes CE, Babatz T, Müller H, Annaluru N, Blake WJ, Schwerzmann JW, Junbiao D, Lindstrom DL, Boeke AC, Gottschling DE, Chandrasegaran S, Bader JS, Boeke JD (2011) Synthetic chromosome arms function in yeast and generate phenotypic diversity by design. Nature 477(7365):471–476

Fischer CR, Klein-Marcuschamer D, Stephanopoulos G (2008) Selection and optimization of microbial hosts for biofuels production. Metab Eng 10(6):295–304

Gibson DG (2012) Oligonucleotide assembly in yeast to produce synthetic DNA fragments. Methods Mol Biol 852:11–21. doi:10.1007/978-1-61779-564-0_2

Gibson DG (2014) Programming biological operating systems: genome design, assembly and activation. Nat Methods 11(5):521–526

Gibson DG, Glass JI, Lartigue C, Noskov VN, Chuang RY, Algire MA, Benders GA, Montague MG, Ma L, Moodie MM, Merryman C, Vashee S, Krishnakumar R, Assad-Garcia N, Andrews-Pfannkoch C, Denisova EA, Young L, Qi ZQ, Segall-Shapiro TH, Calvey CH, Parmar PP, Hutchison CA 3rd, Smith HO, Venter JC (2010) Creation of a bacterial cell controlled by a chemically synthesized genome. Science 329(5987):52–56

Ha S-J, Galazka JM, Kim SR, Choi J-H, Yang X, Seo J-H, Glass NL, Cate JHD, Jin Y-S (2011) Engineered *Saccharomyces cerevisiae* capable of simultaneous cellobiose and xylose fermentation. Proc Natl Acad Sci U S A 108(2):504–509

Hranueli D, Starcevic A, Zucko J, Rojas JD, Diminic J, Baranasic D, Gacesa R, Padilla G, Long PF, Cullum J (2013) Synthetic biology: a novel approach for the construction of industrial microorganisms. Food Technol Biotechnol 51(1):3–11

Hughes SR, Riedmuller SB, Mertens JA, Li X-L, Bischoff KM, Cotta MA, Farrelly PJ (2005) Development of a liquid handler component for a plasmid-based functional proteomic robotic workcell. J Assoc Lab Autom 10:287–300

Hughes SR, Riedmuller SB, Mertens JA, Li X-L, Bischoff KM, Qureshi N, Cotta MA, Farrelly PJ (2006) High-throughput screening of cellulase F mutants from multiplexed plasmid sets using an automated plate assay on a functional proteomic robotic workcell. Proteome Sci 4:10

Hughes SR, Dowd PF, Hector RE, Riedmuller SB, Bartolett S, Mertens JA, Qureshi N, Liu S, Bischoff KM, Li X, Jackson JS Jr, Sterner D, Panavas T, Cotta

MA, Farrelly PJ, Butt T (2007) Cost-effective high-throughput fully automated construction of a multiplex library of mutagenized open reading frames for an insecticidal peptide using a plasmid-based functional proteomic robotic workcell with improved vacuum system. J Assoc Lab Autom 12(4):202–212

Hughes SR, Dowd PF, Hector RE, Panavas T, Sterner DE, Qureshi N, Bischoff KM, Bang SS, Mertens JA, Johnson ET, Li XL, Jackson JS, Caughey RJ, Riedmuller SB, Bartolett S, Liu S, Rich JO, Farrelly PJ, Butt TR, Labaer J, Cotta MA (2008) Lycotoxin-1 insecticidal peptide optimized by amino acid scanning mutagenesis and expressed as a coproduct in an ethanologenic *Saccharomyces cerevisiae* strain. J Pept Sci 14(9):1039–1050

Hughes SR, Hector RE, Rich JO, Qureshi N, Bischoff KM, Dien BS, Saha BC, Liu S, Cox EJ, Jackson JS Jr, Sterner DE, Butt TR, Labaer J, Cotta MA (2009) Automated yeast mating protocol using open reading frames from *Saccharomyces cerevisiae* genome to improve yeast strains for cellulosic ethanol production. J Assoc Lab Autom 14:190–199

Hughes SR, Gibbons WR, Bang SS, Pinkelman R, Bischoff KM, Slininger PJ, Qureshi N, Kurtzman CP, Liu S, Saha BC, Jackson JS, Cotta MA, Rich JO, Javers JE (2012) Random UV-C mutagenesis of *Scheffersomyces* (formerly *Pichia*) *stipitis* NRRL Y-7124 to improve anaerobic growth on lignocellulosic sugars. J Ind Microbiol Biotechnol 39(1):163–173. doi:10.1007/s10295-011-1012-x. Epub 2011 Jul 12

Hughes SR, Bang SS, Cox EJ, Schoepke A, Ochwat K, Pinkelman R, Nelson D, Qureshi N, Gibbons WR, Kurtzman CP, Bischoff KM, Liu S, Cote GL, Rich JO, Jones MA, Cedeño D, Doran-Peterson J, Riaño-Herrera NM, Rodríguez-Valencia N, López-Núñez JC (2013) Automated UV-C mutagenesis of *Kluyveromyces marxianus* NRRL Y-1109 and selection for microaerophilic growth and ethanol production at elevated temperature on biomass sugars. J Lab Autom 18(4):276–290. doi:10.1177/2211068213480037. Epub 2013 Mar 29

Inglese J, Johnson RL, Simeonov A, Xia M, Zheng W, Austin CP, Auld DS (2007) High-throughput screening assays for the identification of chemical probes. Nat Chem Biol 3(8):466–479

Ito H, Fukuda Y, Murata K, Kimura A (1983) Transformation of intact yeast cells treated with alkali cations. J Bacteriol 153(1):163

James AP, Kilbey BJ (1977) The timing of UV mutagenesis in yeast: a pedigree analysis of induced recessive mutation. Genetics 87:237–248

Kim B, Du J, Eriksen DT, Zhao H (2013a) Combinatorial design of a highly efficient xylose-utilizing pathway in *Saccharomyces cerevisiae* for the production of cellulosic biofuels. Appl Environ Microbiol 79(3):931–941

Kim SR, Skerker JM, Kang W, Lesmana A, Wei N, Arkin AP, Jin Y-S (2013b) Rational and evolutionary engineering approaches uncover a small set of genetic changes efficient for rapid xylose fermentation in *Saccharomyces cerevisiae*. PLoS One 8(2):e57048. doi:10.1371/journal.pone.0057048

Kuhn RM, Ludwig RA (1994) Complete sequence of the yeast artificial chromosome cloning vector pYAC4. Gene 141:125–127

Lin-Cereghino J, Wong WW, Xiong S, Giang W, Luong LT, Vu J, Johnson SD, Lin-Cereghino GP (2005) Condensed protocol for competent cell preparation and transformation of the methylotrophic yeast *Pichia pastoris*. Biotechniques 38(1):44–48

Linshiz G, Yehezkel TB, Kaplan S, Gronau I, Ravid S, Adar R, Shapiro E (2008) Recursive construction of perfect DNA molecules from imperfect oligonucleotides. Mol Syst Biol 4:191

Martis EA, Radhakrishnan R, Badve RR (2011) High-throughput screening: the hits and leads of drug discovery: an overview. J Appl Pharm Sci 01(01):02–10

Merryman C, Gibson DG (2012) Methods and applications for assembling large DNA constructs. Metab Eng 14:196–204

Nielsen J, Larsson C, van Maris A, Pronk J (2013) Metabolic engineering of yeast for production of fuels. Curr Opin Biotechnol 24:398–404

Nonklang S, Abdel-Banat BMA, Cha-aim K, Moonjai N, Hoshida H, Limtong S, Yamada M, Akada R (2008) High-temperature ethanol fermentation and transformation with linear DNA in the thermotolerant yeast *Kluyveromyces marxianus* DMKU3-1042. Appl Environ Microbiol 74(24):7514–7521

Pang ZW, Liang JJ, Qin XJ, Wang JR, Feng JX, Huang RB (2010) Multiple induced mutagenesis for improvement of ethanol production by *Kluyveromyces*. Biotechnol Lett 32:1847–1851

Patent Application, Docket No. 0150.06 10—Stephen R. Hughes, Serial No. 13/246,096 "Amino Acid Scanning Mutagenesis Process for Automated Production of a Library of Synthetic Mutagenized Gene Sequences" filed 9-27-2011 LOG #275214

Peralta-Yahya PP, Zhang F, del Cardayre SB, Keasling JD (2012) Microbial engineering for the production of advanced biofuels. Nature 488:320–328

Perlack RD, Stokes BJ (Study Leads) (2011) U.S. billion-ton update: biomass supply for a bioenergy and bioproducts industry. United States Department of Energy ORNL/TM-2011/224. Oak Ridge National Laboratory, Oak Ridge, TN. 227 p. DOI10.2172/1023318. http://www1.eere.energy.gov/bioenergy/pdfs/billion_ton_update.pdf (accessed May 24, 2014).

Rodrussamee N, Lertwattanasakul N, Hirata K, Suprayogi LS, Kosaka T, Yamada M (2011) Growth and ethanol fermentation ability on hexose and pentose sugars and glucose effect under various conditions in thermotolerant yeast *Kluyveromyces marxianus*. Appl Microbiol Biotechnol 90:1573–1586

Sanchez CP, Preuss M, Lanzer M (2002) Construction and screening of YAC libraries. In: Donlan DL (ed) Malaria methods and protocols, methods in molecular medicine™, vol 72. Humana Press, New York, pp 291–304

Shabi U, Kaplan S, Linshiz G, Yehezkel TB, Buaron H, Mazor Y, Shapiro E (2010) Processing DNA molecules as text. Syst Synth Biol 4:227–236

Thorne N, Auld DS, Inglese J (2010) Apparent activity in high-throughput screening: origins of compound-dependent assay interference. Curr Opin Chem Biol 14(3):315–324. doi:10.1016/j.cbpa.2010.03.020. Epub 2010 Apr 22

Tsvetanova B, Peng L, Liang X, Li K, Yang JP, Ho T, Shirley J, Xu L, Potter J, Kudlicki W, Peterson T, Katzen F (2011) Genetic assembly tools for synthetic biology. Methods Enzymol 498:327–348. doi:10.1016/B978-0-12-385120-8.00014-0

Wohlbach DJ, Kuo A, Sato TK, Potts KM, Salamov AA, LaButti KM, Sun H, Alicia Clum A, Pangilinan JL, Lindquist EA, Lucas S, Lapidus A, Jin M, Gunawan C, Balan V, Dale BE, Jeffries TW, Zinkel R, Barry KW, Grigoriev IV, Gasch AP (2011) Comparative genomics of xylose-fermenting fungi for enhanced biofuel production. Proc Natl Acad Sci U S A 108(32):13212–13217

Young E, Lee S-M, Alper H (2010) Optimizing pentose utilization in yeast: the need for novel tools and approaches. Biotechnol Biofuels 3:24

Zanella F, Lorens JB, Link W (2010) High content screening: seeing is believing. Trends Microbiol 28(5):237–245

Zhang F, Rodriguez S, Keasling JD (2011) Metabolic engineering of microbial pathways for advanced biofuels production. Curr Opin Biotechnol 22(6):775–783

Zhu JY, Zhuang XS (2012) Conceptual net energy output for biofuel production from lignocellulosic biomass through biorefining. Prog Energ Combust 38:583–598. doi:10.1016/j.pecs.2012.03.00

Imaging Flow Cytometry and High-Throughput Microscopy for Automated Macroscopic Morphological Analysis of Filamentous Fungi

17

Aydin Golabgir, Daniela Ehgartner,
Lukas Neutsch, Andreas E. Posch,
Peter Sagmeister, and Christoph Herwig

17.1 Introduction

Fungi exhibit cellular and macroscopic morphological variations in response to genetic as well as environmental influences. Morphological variations have been shown to be an indicator for assessing properties such as pathogenicity (Hnisz et al. 2011), efficiency of industrial bioprocesses (Posch et al. 2013), as well as ecological impacts (Pomati and Nizzetto 2013). Morphological characterization of fungi is highly dependent on the application. Whereas assessing pathogenicity might depend on the choice of color and cell size as indicators of underlying genetic variations, bioprocess applications have considered macro-morphological attributes, such as affinity for pellet growth, rate of branching, and septation frequency as quality attributes for production processes (Krull et al. 2013; Posch et al. 2012; Krabben and Nielsen 1998). Choosing a collection of suitable methods for morphological characterization, consisting of image acquisition/processing and data analysis, depend on the variables to be measured. However, for all cases, the description of quantitative morphological features with sufficient statistical power is only possible by methods that allow for high-throughput analysis of multiple cells or cell populations in a given sample. Hence, to be of practical use, applied measurement methodologies should be robust, fast, and not subject to observer bias.

The presented acquisition methods, (1) imaging flow cytometry and (2) whole-slide microscopy are chosen based on the versatility of applications and statistical rigor respectively. Flow cytometry devices with imaging capability present a stand-alone, standardized technological platform for high-throughput sampling with diverse applicability. Combination of the flow cytometer signals, such as fluorescence intensity, with "in-flow" images, opens diverse research possibilities for assessment of the molecular and genetic basis for morphological variations. On the other hand, whole-slide microscopy in combination with automated image analysis has been shown capable of achieving statistically

---

A. Golabgir, D.I. M.Sc. (✉)
Research Division Biochemical Engineering, Vienna University of Technology, Gumpendorferstraße 1a/166-4, Vienna 1060, Austria
e-mail: aydin.golabgir@tuwien.ac.at

D. Ehgartner, D.I. M.Sc. • A.E. Posch, Ph.D.
C. Herwig, Ph.D.
CD Laboratory for Mechanistic and Physiological Methods for Improved Bioprocesses, Vienna, Austria

L. Neutsch, Ph.D., M.Pharm.Sci.
Department of Biochemical Engineering, Vienna University of Technology, Vienna, Austria

P. Sagmeister
Vienna University of Technology, Graz, Styria, Austria

M.A. van den Berg and K. Maruthachalam (eds.), *Genetic Transformation Systems in Fungi, Volume 2*, Fungal Biology, DOI 10.1007/978-3-319-10503-1_17,
© Springer International Publishing Switzerland 2015

verified morphological characterization (Posch et al. 2012; Cox et al. 1998).

Despite the availability of several automatic image acquisition systems, required image processing and data analysis steps are often a limiting factor due to the absence of one-fit-all software packages. Here, in addition to presenting two automated image acquisition methods, we provide a step-by-step guide for implementation of simple Matlab scripts capable of performing common data analysis tasks such as computation of morphological variables as a first step towards implementation of more advanced data analysis algorithms such as regression, classification, and clustering. The ability to setup fast and reliable methods for automatic and high-throughput morphological analysis as well as being able to implement custom algorithms for efficient data analysis is not only a useful tool for mycological research, but will also improve our ability to design improved biotechnological processes.

17.2 Materials

Materials are grouped according to the choice of the image acquisition platform.

17.2.1 Microscopy

1. 1-mL-pipette-tips (cut tips to avoid size exclusion).
2. Lactophenol blue (Art. No.: 3097.1, Carl Roth, Germany).
3. Microscope slide (25×75 mm).
4. High Precision Microscope Cover Glasses (24×60 mm) (Art. No.: LH26.1, Carl Roth, Germany).
5. Brightfield microscope (Leitz, Germany) equipped with a 6.3 magnifying lense, 5 mega-pixel microscopy CCD color camera (DP25, Olympus, Germany), and a fully automated x-y-z stage (Märzhäuser, Austria).

6. Microscope control program analysis 5 (Olympus, Germany).
7. MATLAB 2013a including Image Processing Toolbox.

17.2.2 Imaging Flow Cytometry

1. Milli-Q water.
2. 1-mL-pipette-tips (cut tips to avoid size exclusion).
3. Flow cytometer, e.g., CytoSense (CytoBuoy, Netherland) equipped with a PixeLINK PL-B741 1.3MP monochrome camera.
4. Flow cytometer software, e.g., CytoClus (CytoBuoy, Netherlands).
5. MATLAB 2013a including Image Processing Toolbox (MATLAB, Image Processing Toolbox Release 2013a).

17.3 Methods

The workflow for morphological analysis consists of the choice of an image acquisition platform and implementation of subsequent image and data processing routines. Figure 17.1 provides an overview of the presented methodologies. All of the presented methods have been developed for the analysis of *Penicillium chrysogenum* grown in submerged cultures, but could be extended to other similar organisms.

17.3.1 Sample Preparation

The importance of sound protocols for sample workup and conditioning prior to the actual process of image acquisition and evaluation is often underrated. Since factors such as pH-value and osmolarity are known to affect morphology, it has to be assured that sample preparation does not interfere with results from morphological analysis. Staining steps may be included in order to focus the image-based analysis on selected

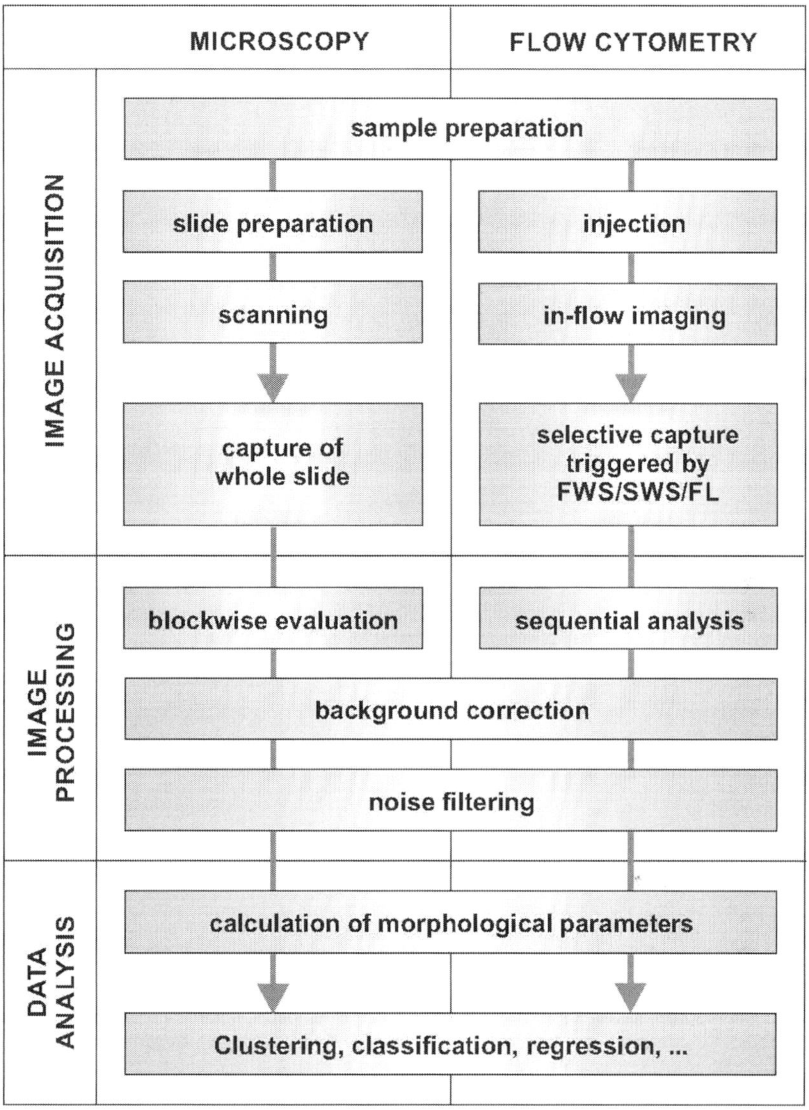

17 Imaging Flow Cytometry and High-Throughput Microscopy for Automated...

Fig. 17.1 Overview of the steps involved in presented methods

elements in the sample, which can be distinguished by specific structural or functional parameters of the fungal cell. Individual sample preparation procedures for the presented image acquisition methods are presented in Sects. 17.3.2.1 and 17.3.2.2.

17.3.2 Image Acquisition

The choice of the most suitable method for image acquisition depends on the nature of the sample and the analysis criteria with regard to throughput, accuracy, and investigated morphological features. Selection of a specific method will also be influenced by the available equipment in most laboratories.

The maximum number of images acquired per run, which influences the statistical robustness of a method, is one of the main differentiating factors of the two presented image acquisition systems. The flow cytometry system is capable of taking up to 150 images per run, each containing a single hyphal element, whereas scanning a microscope slide (at 63× magnification) can yield up to 900 images, each containing up to ten individual macromorphological fungal objects (such as pellets, mycelial clumps, and branched hyphae). Therefore, the microscopy-based system is at an advantage when it comes to recording a large number of images.

In both systems, discriminatory parameters (such as size and fluorescence intensity) may be utilized for confining the analysis to particles with given properties. However, in microscopy-based systems these parameters are not easily accessible, and the sorting process is carried out *post-hoc* (by analyzing the image and including or excluding certain structures). Flow cytometric analysis allows for in-situ control of image acquisition (by triggering acquisition based on the optical parameters of the particle in the flow cell). The signals required for selective acquisition are more readily available in-flow cytometry systems. Sample handling and mounting on the device is another important factor. In microscopy systems, it usually involves spreading the sample on glass slides, which is more tedious and less standardized compared to fully automated systems of sample injection.

The maximum size of particles being analyzed has traditionally been a limiting factor for flow cytometric analyses of fungal samples. The system presented here has been specifically designed to allow for analysis of biomass particles up to a size range of 1.5 mm, which is still in line with the requirements of many applications. However, a potential shortcoming of imaging flow cytometry is that the maximum possible exposure time is determined by the relatively fast transit of the fungal biomass element through the flow cell.

This may interfere with analysis of certain image features or stains that require prolonged signal integration times.

17.3.2.1 Microscopy

1. Dilute the sample with to about 1 g/Lbiomass dry cell weight. For pipetting, use cut pipette tips to avoid size exclusion effects.
2. Add 100 µL/mL lactophenol blue solution to stain the sample.
3. Transfer 50 µL of the stained sample to a microscope slide. Ensure homogenous sample distribution.
4. Place a high precision cover slip on the slide using tweezers. Avoid agglomeration of biomass elements and inclusion of air bubbles, dust, or debris. The use of a high precision slip ensures that the images remain in focus throughout scanning of the entire slide, eliminating the need for manually adjusting focus at each step.
5. Transfer the microscope slide to the automated microscope stage.
6. Adjust acquisition parameters (i.e. contrast, brightness, exposure time) based on illumination settings and desired image features.
7. Scan the whole slide. For analysis of *P. chrysogenum*, at 63× magnification, typically 600–800 images are taken (adjacently arranged in a grid pattern without overlap as shown in Fig. 17.2).
8. Measure multiple slide replicates (in our case 3, amounting to approx. 2,000 images) of a sample to account for variations in sampling and sample preparation steps.

17.3.2.2 Imaging Flow Cytometry

1. Sample is diluted with MQ water to a final concentration of approximately 10^4 cells/mL and stained if required. 5–10 mL of diluted sample is needed for one measurement.
2. Sample measurement is performed by the flow cytometer via the standardized routine.
3. Depending on the particular selection criteria for taking pictures, the in-flow images are recorded.

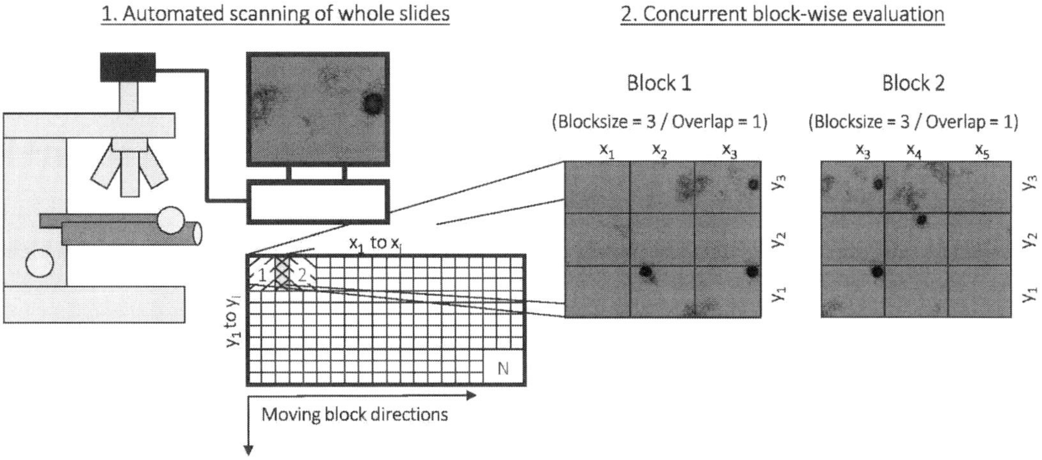

17 Imaging Flow Cytometry and High-Throughput Microscopy for Automated...

Fig. 17.2 Implemented workflow for automated morphological analysis. Step 1 is controlled by the image recording software Analysis5; step 2 by the evaluation routine implemented in Matlab. (From Posch AE, Spadiut O, Herwig C. A novel method for fast and statistically verified morphological characterization of filamentous fungi. Fungal Genet Biol. 2012 Jul;49(7):499–510 with permission.)

17.3.3 Image Processing and Data Analysis

Following the acquisition of a series of images, either via whole-slide microscopy or imaging flow cytometry, images are analyzed using custom-built Matlab scripts. In the following sections, the introduced programming functions (such as *imread* and *im2double*) either refer to inbuilt functions of MATLAB or are part of the MATLAB Image Processing Toolbox (MATLAB, Image Processing Toolbox Release 2013a).

17.3.3.1 Image Processing for Whole-Slide Microscopy

In contrast to Imaging Flow Cytometry where only one hyphal biomass element is captured on an image, in whole-slide microscopy, each image often contains more than one biomass element. It is therefore required to either combine all images prior to evaluation or evaluate a set of multiple images sequentially because if only one image were to be evaluated at a time, the biomass elements spanning an image border would not be evaluated correctly. An iterative evaluation method of overlapping composite image blocks allows for the evaluation of all biomass elements by using a combination of 4 or 9 microscope images at each iterative evaluation step.

1. Export/store all images of a slide in a folder
2. Implement the following Matlab functionalities in a script (m-file):
   (a) Create an index of block numbers and be able to choose a set of n images (4 or 9) that are adjacent to each other. Implement a loop functionality that traverses through all images in a blockwise fashion as illustrated in Fig. 17.2.
   (b) Open the set of 4 or 9 microscope images corresponding to a block using the *imread* function (MATLAB, Image Processing Toolbox Release 2013a) and combine them into one image .
   (c) Convert the images from RGB format to grayscale using the *rgb2gray* function (MATLAB, Image Processing Toolbox Release 2013a). Further convert the image values to double precision using the *im2double* function (MATLAB, Image Processing Toolbox Release 2013a).
   (d) Convert the images to binary (black/white) using the *im2bw* function (MATLAB, Image Processing Toolbox Release 2013a). Adjust the threshold value based on particular light conditions.

(e) Apply mean and median filters to enhance image quality.

(f) Convert to binary (black/white) using the *im2bw* function. Use a threshold value according to the camera settings of the microscope.

(g) Identify connected objects using the *bwconncomp* function (MATLAB, Image Processing Toolbox Release 2013a) and append to list of objects (excluding border-touching elements and eliminating duplicates stemming from overlapping composite image blocks)

(h) Go back to step b and evaluate the next block (repeat until all images have been analyzed)

(i) Save the list of objects as a MAT file for subsequent data analysis steps.

3. Find the length of a pixel (magnification) in the camera software. This will depend on the resolution of the installed camera. Square this value to get the area [μm^2] corresponding to each pixel.

17.3.3.2 Image Processing for Imaging Flow Cytometry

Image acquisition via imaging flow cytometry eliminates the need for blockwise evaluation because each image captures only one hyphal biomass element. In rare cases where the biomass elements is not captured wholly and touches the image border, the image can be discarded automatically.

1. Export images from the Cytosense software (using the *export all images* function) to a common folder

2. Perform following activities in a Matlab script (m-file). Programming functions (such as *rg2gray* and *im2double*) either refer to inbuilt functions of MATLAB or are part of the MATLAB Image Processing Toolbox (MATLAB, Image Processing Toolbox Release 2013a).

(a) Open an image and a background image.

(b) Convert both images from RGB format to grayscale using *rgb2gray* function. Also convert the image data format to double precision using the *im2double* function.

(c) Subtract the background image from the actual image being analyzed.

(d) Convert to binary (black/white) using the *im2bw* function. Use a threshold value according to the camera settings of the flow cytometer. For our settings values between 0.04 and 0.1 work best.

(e) Apply mean and median filters to enhance the image quality and remove noise.

(f) Convert to binary (black/white) again. Use a threshold value according to the camera settings of the flow cytometer. For our settings values between 0.4 and 0.6 work best.

(g) Detect all connected objects using the *bwconncomp* function. Ideally, if the right pre-processing parameters are chosen, each image should result in one connected object only. If additional small elements are found, i.e. due to noise, the largest element should be chosen automatically using the *bwconncomp* function to output the area of the connected objects.

(h) Append the largest connected object to a list for later processing and iterate steps a–g until all images of a flow cytometer run have been analyzed.

(i) Save the list of objects as a MAT file for subsequent data analysis steps.

3. Compute the length and area of each pixel from the scale bar on one of the output images. The resolution (number of pixels) of each exported image should be the same for each analyzed sample.

(a) Count the number of pixels in the scale bar of an image by cropping the image, loading it in Matlab, and then using the *length* function. For instance: Length of scale bar = 670 pixels = 450 μm.

(b) Length of a pixel = 450/670 = 0.67 μm/pixel.

(c) Area of a pixel = 0.67 × 0.67 = 0.45 μm^2/pixel.

17.3.3.3 Data Analysis and Calculation of Morphological Variables

The output of either method (microscopy or imaging flow cytometry) will be the same type of Matlab structure, namely a collection of binary images of individual hyphal elements. Subsequent data analysis steps will be the same for both image acquisition methods. Here, we describe the procedure for deriving a set of commonly-used morphological variables using easy-to-implement Matlab routines. For statistical verification of calculated morphological data, the reader is referred to Posch et al. (2012).

Area

Area of the 2d slice through the hyphal element is proportional to the size of the hyphal element. This property can be calculated according to the following procedure.

1. Use the *regionprops* function to get a list of all region properties.
   properties = regionprops(input_bw, 'all');
2. In the resulting structure, access the "Area" matrix using "props.Area."
3. Multiply the pixel number by the "area of each pixel [μm^2/pixel]" (calculated in previous steps) to get the area of the hyphal element in units of μm^2.
4. The properties structure contains a collection of variables such as circularity and roughness which can be used for a variety of tasks, such as classification of hyphal elements.

Equivalent Diameter and Major Axis Length

The "equivalent diameter" parameter is a scalar that specifies the diameter of a circle with the same area as the region. Similarly, the "major axis length" specifies the length of the major axis of the ellipse that has the same normalized second central moment as the region. Both parameters provide an appropriate measure of the size of the biomass elements and are calculated by the *regionprops* function, which returns these values in units of pixels. Multiplication by pixel length (calculated in previous steps) returns the *EquivDiameter* and *MajorAxisLength* in units of μm.

Classification of Hyphal Elements

Each fungal object can be classified into distinct morphological classes, such as pellets, large and small clumps, branched and unbranched hyphae.

The decision for such classification is based on a combination of morphological parameter and simple if/else rules. Combination of classification and area can be used to derive the area fraction of each class for a sample, such as area fraction of pellets or large clumps (Fig. 17.3).

Hyphal Growth Unit

The hyphal growth unit (HGU), defined as the average length of a hyphae supporting a growing tip, is used for studying the growth kinetics and morphology of filamentous organisms. It can be given by the equation:

$$\mathrm{HGU} = \frac{L_\mathrm{t}}{N_\mathrm{t}}$$

where L_t is total mycelial length, and N_t is the total number of tips. The procedure for calculating the total mycelial length and the total number of tips is as follows:

1. Starting from the binary image of the previous steps, apply the *bwmorph* function with the "skel" method. The operation "skel" removes pixels on the boundaries of objects but does not allow objects to break apart. The pixels remaining make up the image skeleton. Use "inf" as the n (input) parameter, which causes the skeleton to have a width of 1 pixel.
   T1 = bwmorph(input_bw,'skel',inf);
2. Apply the **bwmorph** function with the "shrink" method in order to eliminate tips which are smaller than a specified length. Use the minimum tip size [pixels] as the n (input) parameter.
   T2 = bwmorph(T1,'shrink',2/pxl_length);
3. The total hyphal length is calculated by considering the length of a rectangle having the same area and perimeter as the hyphal object (Cox et al. 1998).

$$\mathrm{Total\ length} = \frac{\mathrm{Perimeter} + \sqrt{\mathrm{Perimeter}^2 - 16 \times \mathrm{Area}}}{4}$$

4. Apply the **bwmorph** function with the "endpoint" method in order to identify all of the endpoints.
   T3 = bwmorph(T2,'endpoints');

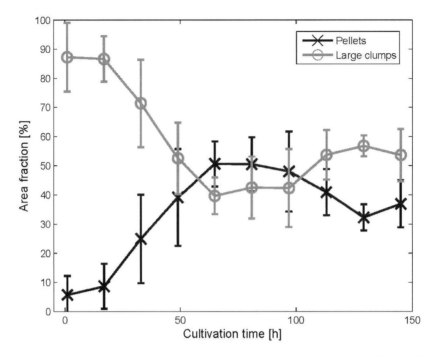

Fig. 17.3 Time course of area fraction of pellets and large clumps over a P. chrysogenum fed-batch cultivation using the whole-slide microscopy method

5. Appy the **bwconncomp** function to find all connected objects in the resulting image. The NumObjects property of the resulting structure is equivalent to the number of tips.
6. Dividing total length by number of tips results in the hyphal growth unit parameter.

17.4 Illustrative Examples

17.4.1 Imaging Flow Cytometry

The advantage of the imaging flow cytometry method lies in its ability to combine classical morphological analysis with flow cytometer detector signals. Equipped with appropriate staining methods, it would be possible to differentiate hyphal elements on a functional level (i.e. pathogenicity, live/dead), and these subpopulations can then be analyzed with respect to macroscopic morphological differences. In comparison to whole-slide microscopy, an advantage of this method lies in the fact that overlapping and touching elements are eliminated since the flow cytometer adjusts the flow such that every image contains only one biomass element.

As an example for calculation of common size parameters, Fig. 17.4 depicts a hyphal element and the corresponding calculated values. Figure 17.5 shows an example of skeletonization used for calculation of total hyphal length, number of tips, and HGU. Figure 17.6 shows an image of a pellet and the corresponding detector signals of the flow cytometer which can be used as a criteria for capturing images at specific signal levels. Lack of detailed focus for larger particles is a clear drawback of the in-flow image acquisition method as seen by the example of Fig. 17.6. Further optimization of the method may result in images with improved quality.

17.4.2 Whole-Slide Microscopy

Application of the described whole-slide microscopy method has been previously shown to significantly enhance characterization of bioprocesses (Posch et al. 2013). As an example, classification

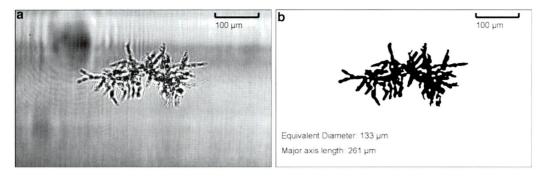

Fig. 17.4 (a) Image of a branched hyphae captured by the in-flow camera. (b) Calculation of the equivalent diameter and major axis length of a biomass element

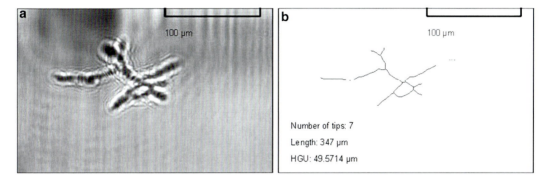

Fig. 17.5 (a) Image of a branched hyphae captured by the in-flow camera. (b) Calculation of the number of tips, total length, and hyphal growth unit from an image taken in the flow cell of the flow cytometer is performed via skeletonization of the image

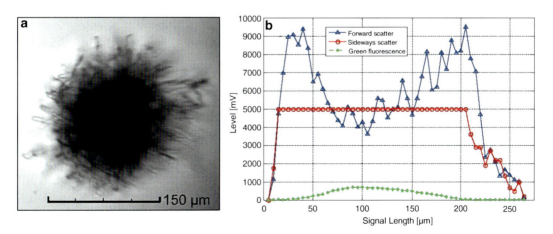

Fig. 17.6 (a) Image of a pellet captured by the in-flow camera. (b) detector signals of the flow cytometer which can be used as a criteria for capturing images at specific signal levels. The sideways scatter detector has become saturated due to the limitations of the device. The size of the pellet can be easily calculated using the forward scatter signal

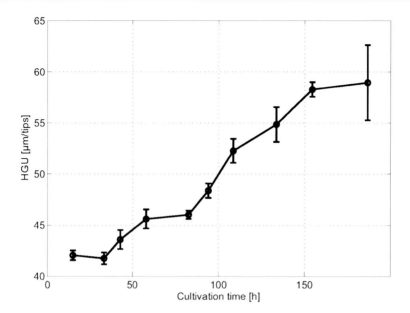

Fig. 17.7 Time course of hyphal growth unit (HGU) over a P. chrysogenum fed-batch cultivation using the whole-slide microscopy method

of biomass into distinct morphological classes according to a rule-based method can provide the time course of morphological variations over a typical industrial cultivation process of *P. chrysogenum* (Fig. 17.3). The dynamics of pellet growth and subsequent breakage provides insights for optimization of rheological properties of the culture. The time course of the hyphal growth unit over process time (Fig. 17.7) can serve for assessing the effects of process conditions, such as agitation speed, on the branching behavior of the organisms.

Acknowledgements The authors would like to thank Sandoz GmbH for providing the strains and guidance. Financial support was provided by the Austrian research funding association (FFG) under the scope of the COMET program within the research network "Process Analytical Chemistry (PAC)" (contract # 825340).

References

Cox P, Paul G, Thomas C (1998) Image analysis of the morphology of filamentous micro-organisms. Microbiology 144(Pt 4):817–827

Hnisz D, Tscherner M, Kuchler K (2011) Morphological and molecular genetic analysis of epigenetic switching of the human fungal pathogen Candida albicans. Methods Mol Biol 734:303–315

Krabben P, Nielsen J. (1998). Modeling the mycelium morphology of Penicillium species in submerged cultures. In: Schügerl K, ed. Relation between morphology and process performances [Internet]. Springer: Berlin. [cited 2013 Apr 25]. p. 125–52. Available from: http://link.springer.com/chapter/10.1007/BFb0102281

Krull R, Wucherpfennig T, Esfandabadi ME, Walisko R, Melzer G, Hempel DC et al (2013) Characterization and control of fungal morphology for improved production performance in biotechnology. J Biotechnol 163(2):112–123

MATLAB and Image Processing Toolbox Release. (2013a). Natick: The MathWorks, Inc.

Pomati F, Nizzetto L (2013) Assessing triclosan-induced ecological and trans-generational effects in natural phytoplankton communities: a trait-based field method. Ecotoxicology 22(5):779–794

Posch AE, Spadiut O, Herwig C (2012) A novel method for fast and statistically verified morphological characterization of filamentous fungi. Fungal Genet Biol 49(7):499–510

Posch AE, Herwig C, Spadiut O (2013) Science-based bioprocess design for filamentous fungi. Trends Biotechnol 31(1):37–44

Yeast Cell Electroporation in Droplet-Based Microfluidic Chip

18

Qiuxian Cai and Chunxiong Luo

18.1 Introduction

Electroporation has been widely used in cell lysis and gene delivery. When applying an external electric pulse in the vicinity of a cell, nanopores could form reversibly or irreversibly on the plasma membrane. High intensity and long duration of the electric field generates permanent nanopores so that intercellular contents will be released for further studies; Low intensity and short duration results in transient nanopores, allowing for temporal permeabilization to foreign molecules.

In a traditional molecular biology lab, electroporation is done with electroporators. With the advent of more and more integrated microfluidic technology, electroporation has been successfully demonstrated in a variety of microfluidic devices (Luo et al. 2006; Lin et al. 2004; Lu et al. 2005; Khine et al. 2005). Compared with electroporators, electroporation in microfluidics has a single-cell resolution with less cell sample and reagent, lower applied voltage, and the most important thing—

higher cell viability and electroporation efficiency. Besides, the microfluidic electroporation device is compatible with microscopy so that a simultaneous microscopic observation is possible.

In this protocol, we suggest a droplet-based microfluidic device in the application of electroporation. The advantage of using droplets is that the solution in each droplet can be kept at the same velocity in the microfluidic channel, thereby the electroporation for each cell can be precisely controlled.

18.2 Materials and Equipments

18.2.1 Photomasks

1. A photomask design software, like L-edit, Auto CAD, Ai, etc. The software is used both to draw the masks and define the geometries of the microfluidic channels precisely.

18.2.2 Reagents and Equipment for Photolithography

1. Silicon wafers. The Silicon wafers (3–4", 500 μm thick, Single Side Polished, Orientation <111>) were purchased from Luoyang Single Crystal Silicon Co., Ltd. It serves as the substrate for the structures formed by photoresist.

Q. Cai • C. Luo, Ph.D. (✉)
The State Key Laboratory for Artificial
Microstructures and Mesoscopic Physics, School of
Physics; Center for Qualitative Biology, Academy for
Advanced Interdisciplinary Studies, Peking
University, Beijing 100871, China
e-mail: caiqiuxian@gmail.com; pkuluocx@pku.edu.cn

M.A. van den Berg and K. Maruthachalam (eds.), *Genetic Transformation Systems in Fungi, Volume 2*, Fungal Biology, DOI 10.1007/978-3-319-10503-1_18,
© Springer International Publishing Switzerland 2015

2. AZ ® 50XT, a positive photoresist. Positive resist becomes soluble to developer when exposed to UV light.
3. AZ ® photoresist developer, AZ ® 400 K or AZ ® 421 K. Photoresist developer is used to remove the AZ photoresist exposed to UV light. All the AZ materials in this protocol were bought from Clariant.
4. Spin coater. The wafer is spin coated with AZ photoresist by using the WS-400BZ-6VPP/LITE spin coater (Laurell Technologies Corporation). The thickness of the AZ layer can be specified by controlling the spinning speed.
5. Hot plate. A hot plate is needed to harden the AZ after spinning coating.
6. Mask aligner with UV lamp (e.g. Karl Suss MA6 Mask Aligner). Mask aligner is used to transfer the feature from mask to the wafer with AZ photoresist.
7. Petri dish. A petri dish is used as a container to develop the masters.
8. Wafer tweezers. The surface of the wafer should be carefully protected.

18.2.3 Microelectrodes and PDMS Microfluidic Channels Fabrication

1. Microscope slides. The glass slide serves as the substrate for the microelectrodes and PDMS microfluidic channels.
2. Sputter coater. In this protocol, we used the JGP450A Magnetron sputtering equipment (Sky Technology Development Co. Ltd., Chinese Academy of Sciences) to magnetron sputter the glass slides with structures formed by AZ photoresist.3. Photoresist remover. Acetone solvent is used as remover in this protocol.
3. PDMS (Polydimethylsiloxane). PDMS RTV 615 and PMDS Sylgard 184 are two commonly used silicon-based organic polymers for making microfluidic devices. PDMS RTV 615 is utilized in this protocol.

4. Stir bar. This is used to mix the silicone rubber compounds and the curing agents in matched kits.
5. Plastic gloves.[1] Because of the toxicity of photoresist, developer as well as non-curing PDMS, one should wear gloves when handling them.
6. Aluminum foil. Cover the wafer to protect it from dust.
7. Digital scale. A scale is used to weigh PDMS and curing agent.
8. Desiccator and vacuum pump. A vacuum pump connected to a desiccator is used to remove the air bubbles from PDMS.
9. Oven. High temperature helps PDMS to cure faster.
10. Puncher. The inlet and outlet holes are punched with a puncher.
11. Plasma surface treatment machine. Before binding the PDMS to the slide, PDMS and slide surfaces are oxidized in oxygen plasma to strengthen the bonding.

18.2.4 Yeast Cell Preparation

1. Yeast cells are diluted to 3×10^7 cells/mL in PBS solution.
2. Fluorescein is the target molecule to be transformed into yeast cells.

18.2.5 Yeast Cell Electroporation

1. Function generator. A function generator controls the voltage and frequency provided to the microelectrodes.
2. Digital storage oscilloscope. A digital storage oscilloscope records the resulting voltage when aqueous droplets are in contact with both microelectrodes.
3. Injection pump. A constant fluid flow in the microfluidic channels is generated and controlled by injection pump.

---

[1]Latex hampers PDMS curing, so you'd better not wear latex gloves when handling non-cured PDMS.

4. Inverted Fluorescent Microscope. The fluid flow in microfluidic channels is monitored by a microscope.
5. 1 mL and 25 µL syringes. 1 mL syringes are used to inject soybean oil. 25 µL syringes are used to inject yeast cell solution.
6. The common non-conductive soybean oil ($\rho=0.92$ g/cm^3) is the oil phase in this protocol.

18.3 Methods

18.3.1 Photomasks Design

1. Two types of photomasks are used for this protocol[2] (Fig. 18.1). Mask I and II are respectively used for the microfabrication of microfluidic channels and microelectrodes. The width of microfluidic channels is 300 µm for both water and oil channels. The junction of the aqueous and oil phase is 30 µm (Fig. 18.1I). The width of microelectrodes is 20 µm and the spacing is 20 µm between the two electrodes (Fig. 18.1II).

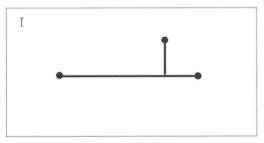
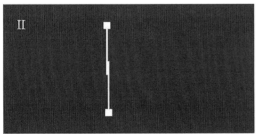

Fig. 18.1

---

[2] When using a negative photoresist such as SU-8, the dark and light region of photomask should be designed in an opposite way.

18.3.2 Photolithography

Master fabrication should be completed in a clean room where atmospheric dust is minimized. Before entering the clean room, full body cover, goggles, shoe covers, gloves and hair cap should be put on. Procedures of photolithography for microelectrodes (Fig. 18.2a) and channels (Fig. 18.2b) are detailed below.

1. Before spin coating the wafer, the hot plate and UV lamp are turned on to warm up. The hot plate is set to 110 °C. The UV lamp needs 30 min to warm up.
2. Cover the inner well of the spin coater with aluminum foil to collect photoresist thrown off during spinning.
3. Program the spin coater according to the spin curve of AZ 50XT. Usually, a two-step spin coating is applied. For the first step, set at 500 rpm for 10 s to make the photoresist spread over the wafer. To fabricate a 20 µm high photoresist layer, the second step is to be set at 1,600 rpm for 60 s.
4. Carefully take out the wafer with wafer tweezers. Clean the wafer using a nitrogen gun. Place the wafer in the spin coater, and make sure that the shiny side points up. Center the wafer on the coater chuck.
5. Pour 2–3 mL AZ 50XT photoresist onto the shiny side of the wafer. Run the coater immediately.
6. Soft bake the well-coated wafer on the preheated 110 °C hot plate for 2 min. Cover the wafer with aluminum foil or Petri dish to protect it from dust.
7. Insert the mask into the mask aligner and put the photoresist coated wafer onto the chuck. Align the mask and wafer, and make sure all the features appear on the wafer. Then expose the wafer to UV light through the mask I for 65s. Adjust the exposure time according to the strength of the UV radiation.
8. Submerge the UV exposed wafer in AZ 400 K developer for 2 min to develop the master.
9. Rinse the wafer with water for 10 s to remove the developer completely.
10. Use a microscope slide as substrate to repeat the photolithography with mask II. If necessary, use a smaller spin coater chuck to fix the microscope slide.

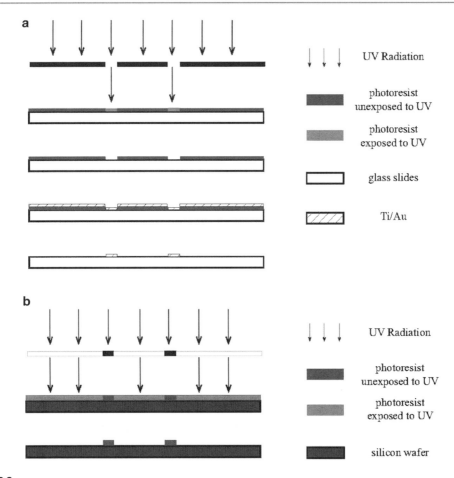

Fig. 18.2

18.3.3 Microelectrodes Fabrication

1. Put the slide from step 3.2.10 into the sputter coater. Coat the slide with 20 nm titanium and 200 nm gold.
2. Remove AZ photoresist by rinsing the sputter coated slide with acetone, so that the titanium and gold coated on the slide form the microelectrodes.

18.3.4 PDMS Microfluidic Channel Fabrication

1. Pour PDMS and its curing agent into a non-latex cup. The mass ratio between PDMS and curing agent should be 7:1–10:1. Make sure the PDMS is over 5 mm.[3]
2. Use a stir bar to mix PDMS and the curing agent.
3. Fix the master in a clean Petri dish. Pour the well-mixed PDMS mixture into the Petri dish. Cover the Petri dish.
4. Vacuum the Petri dish from step 3.4.3 in a desiccator until all air bubbles are out of the PDMS. Normally this step takes at least 30 min, but it depends on the vacuum pressure used.

---

[3]For a 4-in. wafer 50 g PDMS with 5 g curing agent should be enough. If your wafer size is different, adjust the amount of PDMS accordingly.

5. Bake the well-vacuumed PDMS in an oven at 80 °C for 30 min.
6. Peel off the PDMS from the master.
7. Cut the PDMS from step 3.4.6 to fit the slide with the microelectrode from step 3.3.2.
8. Punch through the inlet and outlet holes with a puncher.[4]
9. Place both the PDMS and the slide into the plasma surface treatment machine. Oxidize PDMS and the glass surfaces for 1 min.
10. Align the microfluidic channels with the microelectrodes; make sure the microelectrodes stride over the channel.[5] Once the alignment is finished, drop the PDMS onto the slide. Don't bend or twist the PDMS.
11. Bake the device in the oven at 80 °C for at least 3 h to render the hydrophobicity of PDMS surface.

18.3.5 Yeast Cell Preparation

1. Grow yeast to 1×10^8 cells/mL (OD_{660} is 0.6).
2. Harvest the yeast cells by centrifugation and prepare a 3×10^7 cells/mL solution in PBS with 40 μM Fluorescein.[6]

18.3.6 Devices Setup and Yeast Cell Electroporation

Before loading the cells into the chip, the devices including the fluorescent microscope connected to the computer, the injection Pump with the syringe, the function generator, as well as the fabricated chip should be prepared and set up for electroporation (Fig. 18.3).
1. Fill a 25 μL syringe with yeast cells to be treated and the 1 mL syringe with soybean oil.
2. Connect the syringes with the injection pump.

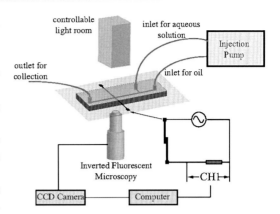

Fig. 18.3

3. Connect the microfluidic device with the injection pump.
4. Connect the microelectrodes with the function generator and the digital storage oscilloscope.
5. Set the outlet of the function generator to a square signal with a peak-to-peak voltage of 18 V_{pp} and a frequency of 1 kHz.
6. Fix the microfluidic device on the microscope stage.
7. Start the oil phase flow at a speed of 120 nL/s equivalent to 20 mm/s.
8. Start the aqueous phase (yeast) flow at a speed of 4 nL/s.
9. Observe the encapsulation of single yeast cells into aqueous droplets. When droplet formation is stabilized, start the function generator and the digital storage oscilloscope.
10. Connect the outlet of the microfluidic device with the 0.5 mL tube and collect the treated yeast cells.
11. Leave the emulsion from step 3.6.10 for 1 h until the mixture has been separated in to an oil and aqueous phase.
12. Carefully remove the upper oil layer.
13. Using the PBS buffer to dilute the solution remained from 3.6.12 to 1/500 of its original concentration.
14. Image the yeast cells under the inverted fluorescent microscope (Fig. 18.4). Compared with original cell solution (Fig. 18.4a, Fig. 18.4b), the fluorescein has

---

[4] To prevent liquid leakage from the inlet and outlet, the diameter of the puncher should be a little smaller than that of the tubing.

[5] Since plasma bonding is irreversible, don't attach the PDMS to the slide until they are well aligned.

[6] For transformation and transfection process, fluorescein can be changed to customized target plasmids or DNA sequences.

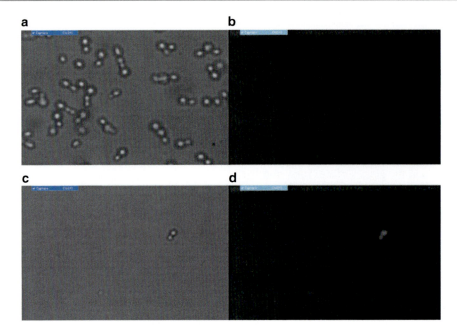

Fig. 18.4

been introduced into the yeast cells, while the yeast cells remained intact (Fig. 18.4c, Fig. 18.4d).

18.4 Concluding Remarks

Yeast cell encapsulation into picoliter-volume aqueous droplets provides a controllable way for electroporation. This method also has potential for integration with other microfluidic chips of aqueous droplet-based application and provides a new system for cell electroporation.

References

Khine M, Lau A, Ionescu-Zanetti C, Seo J, Lee LP (2005) A single cell electroporation chip. Lab Chip 5:38–43

Lin YC, Li M, Wu CC (2004) Simulation and experimental demonstration of the electric field assisted electroporation microchip for in vitro gene delivery enhancement. Lab Chip 4:104–108

Lu H, Schmidt MA, Jensn KF (2005) A microfluidic electroporation device for cell lysis. Lab Chip 5:23–29

Luo C, Yang X, Fu Q, Sun M, Ouyang Q, Chen Y, Ji H (2006) Picoliter-volume aqueous droplets in oil: Electrochemical detection and yeast cell electroporation. Electrophoresis 27:1977–1983

Identification of T-DNA Integration Sites: TAIL-PCR and Sequence Analysis

19

Jaehyuk Choi, Junhyun Jeon, and Yong-Hwan Lee

19.1 Introduction

Agrobacterium tumefaciens is a bacterial plant pathogen causes crown gall disease. *A. tumefaciens* can transfer a DNA segment from its tumor-inducing plasmid (Drummond et al. 1977). The transferred DNA (called T-DNA) is enclosed in-between the so-called left and right borders (LB and RB). Its capability of gene transfer has been widely used as a genetic engineering tool in plants (Tinland 1996). *Agrobacterium*-mediated transformation (AMT) has also been introduced into other organisms such as bacteria, fungi, algae, mammals (Piers et al. 1996; de Groot et al. 1998; Cheney et al. 2001; Kunik et al. 2001; Kelly and Kado 2002). Especially, AMT was quickly applied to more than 100 fungal species due to highly efficient transformation (Michielse et al. 2005). One of the fungal species for which AMT was intensively employed was *Magna-porthe oryzae*, which

J. Choi (✉)
Division of Life Sciences, College of Life Sciences and Bioengineering, Incheon National University, Incheon, Republic of Korea
e-mail: jaehyukc@incheon.ac.kr

J. Jeon, Ph.D.
Department of Agricultural Biotechnology, Seoul National University, Seoul, Republic of Korea

Y.-H. Lee, Ph.D.
Center for Fungal Genetic Resources, Seoul National University, Seoul, Republic of Korea
e-mail: yonglee@snu.ac.kr

is a causal agent of rice blast disease (Rho et al. 2001). Thus far, more than 200,000 AMT transformants have been generated by the *Magnaporthe* community (Xu et al. 2006). A mutant library consisting of 21,070 transformants was generated by our colleagues and the high-throughput screening system was employed to yield more than 180,000 data points for both genotypes and phenotypes of the transformants (Jeon et al. 2007). Analyses using Southern blot, thermal asymmetric interlaced (TAIL)-PCR, and sequencing revealed that more than 70 % of the tested transformants had single copy of T-DNA integration and 1,110 T-DNA insertion sites were identified in the genome of *M. oryzae* (Choi et al. 2007).

Here, we provide a detailed protocol for analyzing T-DNA insertion patterns. The isolation of T-DNA and its flanking sequences is a key step in identification of T-DNA insertion sites. TAIL- and inverse-PCRs were mainly used for the purpose (Liu and Whittier 1995; Ochman et al. 1988). TAIL-PCR is a hemi-specific PCR method where specific (SP) and random primers were used in combination (Fig. 19.1a). This PCR method directly isolates the target region without any pre- or posttreatment (Liu and Whittier 1995). Thus, TAIL-PCR is suitable for amplifying flanking regions from a large number of samples in parallel. One issue with TAIL-PCR is that random arbitrary degenerate primers can generate nonspecific products. Multi-rounds of PCRs are required to increase specific products. Unlike TAIL-PCR, inverse-PCR protocols involve enzyme digestion

M.A. van den Berg and K. Maruthachalam (eds.), *Genetic Transformation Systems in Fungi, Volume 2*, Fungal Biology, DOI 10.1007/978-3-319-10503-1_19,
© Springer International Publishing Switzerland 2015

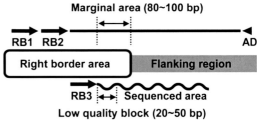

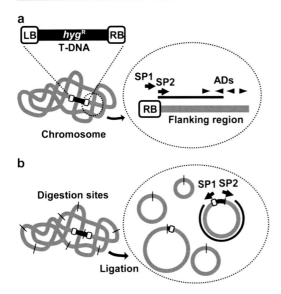

Fig. 19.1 Principles of PCRs to isolate T-DNA and its flanking sequences. Schematic diagram of TAIL-PCR (**a**) and inverse-PCR (**b**). *LB* left border, *RB* right border, *SP* specific primer, *AD* arbitrary degenerate primer, and hyg^R hygromycin resistance gene cassette

of transformant's DNA and ligation steps, which require more time and labor (Ochman et al. 1988). However, it specifically finds targets because only specific primers are used for the rescue of flanking regions (Fig. 19.1b). Taken together, TAIL-PCR is suitable for large-scale isolation of flanking regions and inverse PCR is good for improving specificity.

19.2 Materials

1. Extracted genomic DNA.
2. Specific primers (SPs): Both right and left borders (RB and LB, respectively) can be used when SPs are designed. At least three sets of SPs are required for one side border. The first specific primer (e.g., RB1) is located inside the border area and amplification goes toward end of the border. We called the area from the third specific primer (e.g., RB3) to the end of border as "marginal area" (Fig. 19.2) and at least 70 bp of marginal area are recommended. If the marginal area is too short, no border sequence is detected when isolated flanking sequences are BLAST searched with NCBI sequence database. Moreover, longer marginal areas are needed for the LB because truncation (i.e., loss of border sequence) happens frequently during T-DNA integration and LB is more susceptible for truncation than RB (Tinland 1996).

Fig. 19.2 Location of SPs and marginal area in RB. *Arrows* indicate SPs and the *arrowhead* the AD primer. The *line* from RB2 to AD is the isolated PCR product by TAIL-PCR. *RB* right border, *AD* arbitrary degenerate primer

3. Arbitrary degenerate primers (ADs): ADs are mixed primers consisting of different specific primers. For example, the degenerate primer sequence "NGTCGASWGANA" consist of 64 specific primers ($4(N) \times 2(S) \times 2(W) \times 4(N) = 64$). We use "64" as degeneracy of this primer. ADs with higher degeneracy lead to more products but also the number of unwanted products will be increased. Different composition of nucleotides indicates different melting temperatures.
4. 96-well ELISA plate: This plate is used for storage of diluted genomic DNA.
5. 96-well reaction plate: Also called 96-well PCR plates.

19.3 Methods

19.3.1 Large-Scale TAIL-PCR

1. Dilute genomic DNA in the 96-well ELISA plate to a final concentration as 10–30 ng/µL in the PCR. Add 3 µL of the diluted genomic DNA to the 96-well reaction plate (i.e., PCR tubes).
2. Prepare the PCR mixture (Table 19.1). Except template genomic DNA, all the other reagents are mixed together in a 2 mL microcentrifuge

19 Identification of T-DNA Integration Sites: TAIL-PCR and Sequence Analysis

Table 19.1 Composition of the TAIL-PCR mixture

Components (units)	Final concentration	Original concentration	×1 (μL)	×100 (μL)
Template DNA (ng/μL)	10–30	70–200	3	
dNTP mixture (mM)	0.20 each	2.5 each	1.6	160
10× Reaction buffer	1×	10×	2	200
$MgCl_2$ (mM)	0.12	1.5	1.6	160
SP (μM)	0.2	10	0.4	40
AD (μM)	3	10	6	600
Sterilized distilled water	–	–	5.2	520
Taq polymerase (U)	0.01	1	0.2	20
Total			20	2,000

tube and distributed in a 96-well reaction plate. The final concentration of SP is adjusted to 0.2 μM in primary reactions, respectively. AD is used at 3–4 μM according to its degeneracy. We tested different commercial Taq DNA polymerases and found no difference in the efficiency of rescuing T-DNA flanks. Concentration of dNTP or $MgCl_2$ can differ according to polymerases.

3. First round PCR. We used the following PCR cycling parameters: (1) 94 °C for 3 min; (2) 5 cycles of 94 °C for 10 s, 62 °C for 1 min, and 72 °C for 1 min; (3) 2 cycles of 94 °C for 30 s, 25 °C for 3 min (ramping to 72 °C for 3 min), and 72 °C for 2.5 min; (4) 15 cycles of 94 °C for 30 s, 65 °C for 1 min, 72 °C for 2.5 min, 94 °C for 30 s, 65 °C for 1 min, 72 °C for 2.5 min, 94 °C for 30 s, 44 °C for 1 min, and 72 °C for 2.5 min; and (5) 72 °C for 7 min. The "ramping" option increases temperature slowly to help AD primers to anneal as much as possible. If a PCR machine doesn't have this option, one can alternatively add three steps of different temperatures (e.g., 40 °C for 1 min, 50 °C for 1 min, and 60 °C for 1 min). The annealing temperature for the SP (i.e., 65 °C here) should be the actual temperature of your SPs. They are different according to the composition and length of SPs.

4. Load 3 μL of the primary PCR products on a 1.5 % agarose gel. If the multiple bands with same sizes were shown in all lanes (Fig. 19.3a), the PCR worked well. At this stage, the targeted specific products are less

than nonspecific products, which cause multiple bands.

5. Dilute the primary PCR products threefold. For example, if 17 μL of the primary PCR product is left, add 34 μL of water to the PCR product. Prepare a new 96-well ELISA plate and add 147 μL of water to each well. Transfer 3 μL of diluted primary PCR products to the wells (i.e., 50-fold dilution). Mix well by pipetting.

6. The 150-fold diluted PCR products are used as template DNA for the secondary PCR. Because 3 μL of template DNA is used in a 20 μL reaction volume, the primary PCR product is diluted 1,000-fold finally $(51/17 \times 150/3 \times 20/3)$. All the conditions are the same as for the primary PCR except the increased concentration of second SPs from 0.2 to 0.4 μM.

7. Second round PCR. We used the following PCR cycling parameters: (1) 94 °C for 3 min; (2) 5 cycles of 94 °C for 10 s, 62 °C for 1 min, and 72 °C for 1 min; (3) 15 cycles of 94 °C for 30 s, 66 °C for 1 min, 72 °C for 2.5 min, 94 °C for 30 s, 65 °C for 1 min, 72 °C for 2.5 min, 94 °C for 30 s, 44 °C for 1 min, and 72 °C for 2.5 min; and (5) 72 °C for 7 min.

8. Load 3 μL of the secondary PCR products on agarose gel. Single or multiple bands can be shown but size of the bands should be different in all lanes (Fig. 19.3b). If similar pattern of bands is observed in all lanes like the primary PCR products, we may not obtain positive results from these PCR products. Then it may be better to try another PCR

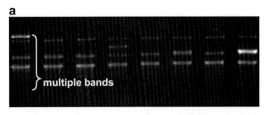
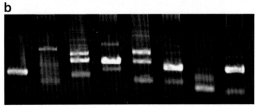

Fig. 19.3 TAIL-PCR products loaded on the agarose gels. First (**a**) and second (**b**) round of PCR products

from the beginning using different AD primers or lowering the annealing temperature.
9. Perform purification of PCR products for better sequencing results. Purification is optional but improves sequencing quality significantly. Lots of primers (SPs and ADs) remain after the second round PCR and can interrupt sequencing reaction. To save time and effort, enzyme purification using ExoSAP-IT® (USB, Cleveland, OH, USA) is suggested. Add 1 μL of the ExoSAP-IT® enzyme to each well of 96-well reaction plate and incubate at 37 °C for 20 min. Inactivate the enzyme at 80 °C for 15 min before sequencing.
10. Purified PCR products are sequenced starting from the third SPs (e.g., RB3), which also improves selectivity of the correct targets.

19.3.2 Identification of T-DNA Integration

1. Following BLAST search with downloaded sequences, three major types can be found: (1) border + flanking area, (2) flanking only, and (3) border only or with vector backbone. The portions of three types were 68 %, 28 %, and 4 %, respectively, when we screened ~2,000 sequences (Choi et al. 2007). "Type 1" sequences can be used for determination of insertion positions. "Type 2" sequences often are obtained when the "marginal area" is too short or border truncation is severe. "Type 3" sequences arise from irregular integration of T-DNA such as tandemly repeated T-DNA and read-through of vector backbone. This irregular integration could happen frequently (10–20 %) (Meng et al. 2007).
2. Determine T-DNA insertion site. Ideally, "type 1" sequences contain border and flanking area without any unmatched region (named "gap") between them. This type is called "complete junction" (Choi et al. 2007). The boundary between border and flanking area is named as "T-DNA insertion site." However, some "type 1" sequences have a gap between border and flanking area (called "incomplete junction"). Low sequencing quality, irregular integration of T-DNA, and multiple T-DNA integration might cause this gap. In this case, we replace the gapped area with a virtual flanking region and determine the new boundary as the T-DNA insertion site. However, such prediction with a virtual flanking region sometimes introduces an error in determining of the insertion site, depending on the length of the virtual region. Even in a mutant having complete junctions at both borders, the mutant may have deletion, addition, or translocation of genomic DNA. Such genomic change allows one inserted T-DNA in a mutant had more than two values of the insertion positions. For example, when determined positions at both borders have 10-bp difference in a mutant, the mutant is regarded as having genomic deletion during T-DNA integration. However, the computer program counts it as two different mutants with 10-bp difference in T-DNA insertion position. Such genomic deletion was frequently (78 %) observed in our analysis for T-DNA insertion sites (Choi et al. 2007). Thus, a new concept for T-DNA insertion site should be employed in counting the number of T-DNA mutants not to overestimate it. We defined "T-DNA tagged

location" as position information with a buffering range (Choi et al. 2007). The buffering range is set as 35 bp because most errors (>98 %) were limited to that length. That is, when one T-DNA insertion site is determined, the next one should be apart at least 35 bp from the first one.

19.3.3 Patterns of T-DNA Integration in Fungi

1. *More truncation in LB*. Integrated T-DNAs are often observed with partial loss of its border sequences (Mayerhofer et al. 1991). The phenomenon is called "truncation" (Tinland 1996). It is a well-known feature which happens during T-DNA integration in plants (Forsbach et al. 2003). As shown in plants, LBs are truncated more often than RBs in *M. oryzae*: more than 60 % of LBs were truncated while less than 10 % of RBs were truncated (Choi et al. 2007; Meng et al. 2007; Li et al. 2007). Conservation of RB is explained by protection of the bacterial virulence protein, VirD2. When T-DNA is transported to the host, virulence proteins, VirD2 and VirE2, guide the T-DNA to the nucleus (Tinland 1996). Because VirD2 binds covalently to the 5' end of T-DNA, right borders are protected against nucleolytic degradation. Therefore, due to more truncation of LB, a longer marginal area is suggested for LB sequences in primer design of the TAIL-PCR.

2. *Actual microhomology exists only in LB*. Microhomology means shared nucleotides between border and flanking sequences. In general, this is more frequently observed in LB than RB border (Tinland 1996; Mayerhofer et al. 1991; Forsbach et al. 2003). However, because two random sequences can make microhomology at a certain ratio ($1/4^n$, where n means the number of shared nucleotides), the portion ($T/4^n$, where T means total number of the analyzed sequence samples) should be subtracted from the observed frequency to calculate actual microhomology. In our analysis, microhomology was found in 85 % of LB and 31 % of RB samples (Choi et al. 2007). However, actual microhomology existed only in LB and ranged from 2 to 6 bp (Choi et al. 2007). Short homology in LB might be an anchoring region for T-DNA integration via illegitimate recombination model (Tinland 1996).

3. *Deletion, duplication, addition, and rearrangement can occur in the genome of a host*. The host genome sequence is also changed during T-DNA integration. Deletion of target sites is prevalent in fungi (~80 %) and plants (~90 %) (Choi et al. 2007; Forsbach et al. 2003). The deleted lengths are mostly short (less than 30 bp) but it can be up to ~2 kb. Duplication (addition of the same sequence with the other flanking region), addition (filler DNA), and rearrangement of the genomic sequences is observed at a low frequency. In general, the frequency is lower in fungi than plants.

4. *T-DNA insertion is frequently observed in the physically bended area of the genome*. The insertion positions of T-DNA integration might be favorable at highly bendable areas (Choi et al. 2007; Zhang et al. 2007). Peaked bendability was observed within 100 bp distance from the insertion positions (both sides), suggesting that the regions are structurally bendable or flexible for T-DNA integration (Choi et al. 2007). Bended structures of DNAs are usually found in promoter areas where transcription factors recognize the position for transcription initiation (Perez-Martin et al. 1994). This might explain why T-DNA insertions were often found in the promoter regions (Choi et al. 2007; Forsbach et al. 2003).

Acknowledgements This work was supported by the National Research Foundation of Korea grant funded by the Korea government (2008-0061897, 2013-003196 (to Y.H.L.), and 2013R1A1A1010928 (to J.C)), and the Next-Generation BioGreen 21 Program of Rural Development Administration in Korea (PJ00821201 (to Y.H.L)).

References

Cheney D, Metz B, Stiller J (2001) *Agrobacterium-mediated* genetic transformation in the macroscopic marine red alga *Porphyra yezoensis*. J Phycol 37:11

Choi J, Park J, Jeon J, Chi MH, Goh J, Yoo SY, Park J, Jung K, Kim H, Park SY, Rho HS, Kim S, Kim BR, Han SS, Kang S, Lee YH (2007) Genome-wide analysis of T-DNA integration into the chromosomes of *Magnaporthe oryzae*. Mol Microbiol 66:371–382

de Groot MJA, Bundock P, Hooykaas PJJ, Beijersbergen AGM (1998) *Agrobacterium tumefaciens*-mediated transformation of filamentous fungi. Nat Biotechnol 16:839–842

Drummond MH, Gordon MP, Nester EW, Chilton MD (1977) Foreign DNA of bacterial plasmid origin is transcribed in crown gall tumors. Nature 269:535–536

Forsbach A, Schubert D, Lechtenberg B, Gils M, Schmidt R (2003) A comprehensive characterization of single-copy T-DNA insertions in the *Arabidopsis thaliana* genome. Plant Mol Biol 52:161–176

Jeon J, Park S-Y, Chi M-H, Choi J, Park J, Rho H-S, Kim S, Goh J, Yoo S, Choi J, Park J-Y, Yi M, Yang S, Kwon M-J, Han S-S, Kim BR, Khang CH, Park B, Lim S-E, Jung K, Kong S, Karunakaran M, Oh H-S, Kim H, Kim S, Park J, Kang S, Choi W-B, Kang S, Lee Y-H (2007) Genome-wide functional analysis of pathogenicity genes in the rice blast fungus. Nat Genet 39:561–565

Kelly BA, Kado CI (2002) *Agrobacterium*-mediated T-DNA transfer and integration into the chromosome of *Streptomyces lividans*. Mol Plant Pathol 3:125–134

Kunik T, Tzfira T, Kapulnik Y, Gafni Y, Dingwall C, Citovsky V (2001) Genetic transformation of HeLa cells by *Agrobacterium*. Proc Natl Acad Sci U S A 98: 1871–1876

Li GH, Zhou ZZ, Liu GF, Zheng FC, He CZ (2007) Characterization of T-DNA insertion patterns in the genome of rice blast fungus *Magnaporthe oryzae*. Curr Genet 51:233–243

Liu YG, Whittier RF (1995) Thermal asymmetric interlaced PCR: automatable amplification and sequencing of insert end fragments from P1 and YAC clones for chromosome walking. Genomics 25:674–681

Mayerhofer R, Koncz-Kalman Z, Nawrath C, Bakkeren G, Crameri A, Angelis K, Redei GP, Schell J, Hohn B, Koncz C (1991) T-DNA integration: a mode of illegitimate recombination in plants. EMBO J 10:697–704

Meng Y, Patel G, Heist M, Betts MF, Tucker SL, Galadima N, Donofrio NM, Brown D, Mitchell TK, Li L, Xu JR, Orbach M, Thon M, Dean RA, Farman ML (2007) A systematic analysis of T-DNA insertion events in *Magnaporthe oryzae*. Fungal Genet Biol 44:1050–1064

Michielse CB, Hooykaas PJJ, van den Hondel CA, Ram AFJ (2005) *Agrobacterium*-mediated transformation as a tool for functional genomics in fungi. Curr Genet 48:1–17

Ochman H, Gerber AS, Hartl DL (1988) Genetic applications of an inverse polymerase chain reaction. Genetics 120:621–623

Perez-Martin J, Rojo F, de Lorenzo V (1994) Promoters responsive to DNA bending: a common theme in prokaryotic gene expression. Microbiol Mol Biol Rev 58:268–290

Piers KL, Heath JD, Liang X, Stephens KM, Nester EW (1996) *Agrobacterium tumefaciens*-mediated transformation of yeast. Proc Natl Acad Sci U S A 93: 1613–1618

Rho HS, Kang S, Lee YH (2001) *Agrobacterium tumefaciens-mediated* transformation of the plant pathogenic fungus, *Magnaporthe grisea*. Mol Cells 12:407–411

Tinland B (1996) The integration of T-DNA into plant genomes. Trends Plant Sci 1:178–184

Xu J-R, Peng Y-L, Dickman MB, Sharon A (2006) The dawn of fungal pathogen genomics. Annu Rev Phytopathol 44:337–366

Zhang J, Guo D, Chang YX, You CJ, Li XW, Dai XX, Weng QJ, Zhang JW, Chen GX, Li XH, Liu HF, Han B, Zhang QF, Wu CY (2007) Non-random distribution of T-DNA insertions at various levels of the genome hierarchy as revealed by analyzing 13 804 T-DNA flanking sequences from an enhancer-trap mutant library. Plant J 49:947–959

Part VI

Tools and Applications: Comprehensive Approaches in Selected Fungi

Genetic and Genomic Manipulations in *Aspergillus niger*

20

Adrian Tsang, Annie Bellemare, Corinne Darmond, and Janny Bakhuis

20.1 Introduction

The filamentous fungi in the genus *Aspergillus* display tremendous nutritional flexibility and metabolic capacity. They are known for their propensity to produce high levels of extracellular proteins and metabolites. *Aspergillus niger* is a particularly important species for the industrial production of homologous and heterologous enzymes as well as organic acids. Present globally and commonly found in decaying plant debris, it is proficient in decomposing complex carbohydrates and other macromolecules in plant-derived biomass. This organism has a long history of safe use in the food industry (Schuster et al. 2002). Citric acid and extracellular enzymes produced in *A. niger* have received "generally regarded as safe," or GRAS, status from the United States Food and Drug Administration, and have been used safely in food and feed applications. It is an important industrial workhorse in the production of citric acid (Magnuson and Lasure 2004) and many polysaccharide-degrading enzymes (Schuster et al. 2002).

Gene transformation is a powerful tool for numerous applications including protein production, regulation of primary and secondary metabolites, and studies on gene regulation, gene function, and development. Gene transformation in *A. niger* was achieved nearly 30 years ago (Buxton et al. 1985; Kelly and Hynes 1985). Since then many transformation protocols and selectable markers have been developed. For improving production of active proteins, genetic engineering efforts have been directed toward controlling post-translational modifications (Li et al. 2013), limiting the production of extracellular proteases to reduce proteolysis (Punt et al. 2008), and enhancing protein secretion (Krijgsheld et al. 2013b). Genes and pathways have also been manipulated through transformation techniques to modulate the production of primary and secondary metabolites (Li et al. 2012; Zabala et al. 2012). In the first part of this review, we summarize the methods, promoters, and selectable markers that have been used for genetic transformation in *A. niger*. The use of *A. niger* in the production of homologous and heterologous proteins will also be described.

The genome sequence of *A. niger* strain CBS 513.88, which has been used for protein production in industry, was published in 2007 (Pel et al. 2007). This was followed by the publication in 2011 of the genome sequence of the wild-type parent of the citric-acid producer strain ATCC 1015 (Andersen et al. 2011). Together with gene transformation techniques, the availability of

A. Tsang (✉) • A. Bellemare • C. Darmond
Centre for Structural and Functional Genomics,
Concordia University, 7141 Sherbrooke Street West,
Montreal, QC, Canada H4B1R6
e-mail: adrian.tsang@concordia.ca

J. Bakhuis
DSM Biotechnology Center, Delft, The Netherlands

M.A. van den Berg and K. Maruthachalam (eds.), *Genetic Transformation Systems in Fungi, Volume 2*, Fungal Biology, DOI 10.1007/978-3-319-10503-1_20,
© Springer International Publishing Switzerland 2015

high-quality genome sequences has initiated many genome-based studies. In the second part of this article, we review the salient features of the sequenced *A. niger* genomes and the use of genomic information in studies of gene function, gene regulation, and protein production.

20.2 Gene Transformation

20.2.1 Host Strains

Despite the lack of public availability of *A. niger* strains used in industry owing to their proprietary nature, several other genetically marked *A. niger* strains are available to the research community, many of them maintained in public repositories including the Agricultural Research Service culture collection (NRRL) of the U.S. Department of Agriculture (http://nrrl.ncaur.usda.gov), CBS-KNAW Fungal Biodiversity Centre (CBS) of the Royal Netherlands Academy of Arts and Sciences (http://www.cbs.knaw.nl), Fungal Genetics Stock Center (FGSC) (http://www.fgsc.net), American Type Culture Collection (ATCC) (http://www.atcc.org), and National Collection of Industrial Microorganisms (NCIM) (http://www.ncl-india.org).

The most commonly used laboratory strains of *A. niger* are derived from strain NRRL3 (ATCC 9029, CBS 120.49, NCIM 545). The University of Wageningen obtained CBS 120.49 from the Fungal Biodiversity Centre and renamed it N400 (FGSC A1143). Strain N402 (ATCC 64974, FGSC A733) is a derivative of N400 that displays short conidiophores. Strain N593 (ATCC 64973) carries a mutation in the orotidine-5′-phosphate-decarboxylase, *pyrG*, gene, and was selected from 5-fluoro-orotic acid-resistant mutants following UV mutagenesis (Goosen et al. 1987). Strains N402 and N593 are the parents of numerous genetically marked strains, some of which are maintained by the Fungal Genetic Stock Center. The industrial enzyme producer, strain CBS 513.88, is a derivative of NRRL 3122 (ATCC 22343, CBS 115989) which was isolated following mutagenesis and screening for increased glucoamylase production. Strain ATCC 1015 (NRRL 328, CBS 113.46,

FGSC A1144, NCIM 588) and its spontaneous mutant ATCC 11414 (NRRL 2270) have been used for the production of citric acid. These strains and NCIM565 have been used extensively to study gene expression, gene function, protein production, and genome manipulations.

20.2.2 Methods of Transformation

Several methods for transformation of *A. niger* have been described. Advantages and limitations of the different methods have been reviewed by Meyer (2008). Protoplast-mediated transformation (Buxton et al. 1985), modified from a protocol originally developed for yeast, is the most widely used method. It provides good transformation efficiency of up to 1,000 transformants per µg DNA for integrative vectors (Buxton et al. 1985) to 10,000 transformants per µg DNA for episomal vectors (Verdoes et al. 1994b). Recently, this method has been adapted for high-throughput platform with transformation and regeneration conducted in microtiterplate (Gielesen and van den Berg 2013). Transformation by electroporation of either protoplasts (Ward et al. 1989) or germinating spores yields fewer transformants, 10–500 per µg DNA for integrative and episomal vectors, respectively (Ozeki et al. 1994). Agrobacterium-mediated transformation has also been attempted, but it is rather inefficient with up to five transformants per µg of DNA (de Groot et al. 1998; Michielse et al. 2005). All transformation techniques need optimization for the specific strain of interest.

20.2.3 Selectable Markers

For the isolation of *A. niger* transformants, a wide range of selectable markers is available. Both dominant selectable markers that do not require creating an appropriate mutant host strain and prototrophic nutritional markers that require the construction of complementary auxotrophic recipient hosts have been used. The *amd*S (Kelly and Hynes 1985) and *pyrG* (Hartingsveldt et al. 1987) selection markers can be used as bidirectional-

20 Genetic and Genomic Manipulations in *Aspergillus niger*

Table 20.1 Selectable markers used for transformation

Marker	Encoded function	Method of selection	Reference
argB	Ornithine cabamoyl transferase	Arginine prototrophy	(Buxton et al. 1985; Lenouvel et al. 2001)
amdS	Acetamidase	Acetamide utilization	(Kelly and Hynes 1985)
pyrG	Orotidine 5-Monophosphate decarboxylase	Uridine prototrophy	(Hartingsveldt et al. 1987)
Hph	Hygromycin phosphotransferase	Hygromycin B resistance	(Punt et al. 1987)
oliC	Subunit 9 of mitochondrial ATP synthetase	Oligomycin resistance	(Ward et al. 1988)
bleR	Bleomycin/phleomycin inactivating protein	Phleomycin resistance	(Mattern et al. 1988)
niaD	Nitrate reductase	Nitrate utilization	(Unkles et al. 1989)
sC	Adenosine triphosphate sulfurylase	Selenate resistance	(Buxton et al. 1989)
Bar	Phosphinothricin acetyl transferase	L-Phosphinothricin resistance	(Ahuja and Punekar 2008)
agaA	Arginase	Arginine prototrophy	(Dave et al. 2012)

dominant markers: either presence or absence of the marker can be selected. Selectable markers that have been used for *A. niger* transformation are summarized in Table 20.1.

20.2.4 Expression Vectors

No autonomously replicating sequences have been isolated from *A. niger*. As in other fungal species, therefore, transformation results from genomic integration of transforming DNA at non-homologous sites as well as at the homologous region (Hynes 1996). However, with the isolation of the AMA1 sequence from *A. nidulans* (Gems et al. 1991), autonomously replicating vectors for use in *A. niger* have been constructed (Storms et al. 2005; Verdoes et al. 1994b). Advantages of such vectors are increased transformation efficiencies (see previous section) and easy plasmid rescue. The downside of the AMA1-containing vectors is that they are mitotically unstable. For stable protein production and for genetic modifications of the host strains (e.g., targeted gene disruptions), vectors that integrate into the host genome are preferred.

The functional elements of *A. niger* expression vectors are fundamentally the same as those used in other eukaryotic expression systems. Different genetic manipulations (e.g., protein production, gene regulation studies, or genome manipulations) require plasmid vectors for *A. niger* containing different elements. A set of plasmids suitable for

each of these purposes has been described by Storms et al. (2005).

All *A. niger* transformation vectors are shuttle vectors, containing sequences required for amplification and selection in *E. coli*. A selectable marker (Table 20.1), necessary to select *A. niger* transformants, may be included on the shuttle vector or on a separate vector since co-transformation works very efficiently. For the production of homologous and heterologous proteins the typical expression cassette includes, next to the ORF coding for the target protein, promoter and transcriptional regulatory sequences and sequences required for termination and processing of the RNA transcript. Table 20.2 summarizes the promoter-transcriptional regulatory sequences useful in expression cassettes for *A. niger*. Both constitutive and inducible promoters have been described. The choice of promoter depends largely on the conditions under which the recombinant protein is to be produced.

If the recombinant protein is to be secreted into the medium, a region encoding a secretion signal has to be included, either from the expressed gene itself or from another eukaryote since they are functionally similar (Salovuori et al. 1987). For producing heterologous proteins, amino-terminal fusion of an efficiently secreted endogenous protein, e.g. fusion of the native glucoamylase gene (Ward et al. 1990), to the protein of interest can result in higher levels of secreted protein. The fusion partner may serve as a carrier to facilitate the translocation of the

Table 20.2 Promoters used for gene expression

Promoter	Protein	Regulation	Reference
amdS	Acetamidase	Inducible	(Turnbull et al. 1990)
glaA	Glucoamylase	Inducible	(Fowler et al. 1990)
gpdA	Glyceraldehyde-3-phosphate dehydrogenase	Constitutive	(Archer et al. 1990)
pkiA	Pyruvate kinase	Constitutive	(de Graaff et al. 1992)
aphA	Acid phosphatase	Inducible	(Piddington et al. 1993)
adhA	Alcohol dehydrogenase	Constitutive	(Davies 1994)
xlnA	Xylanase	Inducible	(de Graaff et al. 1994)
abnA	Arabinase	Inducible	(Flipphi et al. 1994)
alcA/alcR	Alcohol dehydrogenase I	Inducible	(Nikolaev et al. 2002)
amy (*taka*)	Taka amylase	Inducible	(Prathumpai et al. 2004)
hERα-ERE	Human estrogen receptor	Inducible	(Pachlinger et al. 2005)
sucA	Beta-fructofuranosidase	Inducible	(Roth and Dersch 2010)
gdhA	Glutamate dehydrogenase	Constitutive	(Fleissner and Dersch 2010)
citA	Citrate synthetase	Constitutive	(Dave and Punekar 2011)
mbfA	Multiprotein bridging factor	Constitutive	(Blumhoff et al. 2013)
coxA	Subunit IV of cytochrome c oxidase	Constitutive	(Blumhoff et al. 2013)
srpB	Nucleolar protein Srp40	Constitutive	(Blumhoff et al. 2013)
tvdA	Transport vesicle docking protein	Constitutive	(Blumhoff et al. 2013)
mdhA	Malate dehydrogenase precursor	Constitutive	(Blumhoff et al. 2013)
manB	Filamentous growth protein Dfg5	Constitutive	(Blumhoff et al. 2013)

foreign protein in the secretory pathway and protects the heterologous portion from degradation (Gouka et al. 1997).

A novel, alternative route for producing secreted proteins in *A. niger* called "peroxicretion" has been described (Sagt et al. 2009) and could be an option for the secretion of intracellular proteins. The method involves import of the protein into peroxisomes followed by fusion of the peroxisomes with the plasma membrane to release their protein content into the extracellular medium. To achieve this, a peroxisomal import signal (SKL tag) has to be fused C-terminally to the protein of interest and the peroxisomes of the *A. niger* production host strain need to be "decorated" with v-SNARE protein.

To promote homologous integration, the expression vector needs to contain sequences homologous to the *A.niger* genomic region where integration is directed. The frequency of homologous recombination can be dramatically improved if the host strain is defective in non-homologous end joining; e.g., by deletion of *kusA* (homologue of yeast *ku70*) which strongly reduces random integration (Meyer et al. 2007); see also chapter 10.2.

20.3 Production of Extracellular Proteins

As a protein production platform, *A. niger* is mainly used to produce extracellular proteins. Many endogenous, extracellular proteins have been overproduced. Most of them encode enzymes for breaking down starch, cellulose, hemicelluloses, or pectin. The production and characterization of these enzymes have been curated (Murphy et al. 2011) and are periodically updated in the mycoCLAP database (https://mycoclap.fungalgenomics.ca).

Production and secretion of heterologous proteins appear to be protein-dependent; some proteins, although driven by the same promoter and secretion signal, are secreted at much lower levels than others. The many factors that can play a role in the proper production of proteins of interest have been reviewed previously (Fleissner and Dersch 2010; Gouka et al. 1997). Genes encoding extracellular proteins from a wide range of organisms have been expressed in *A. niger*. Table 20.3 provides an overview of heterologous proteins produced and secreted by *A. niger*.

20 Genetic and Genomic Manipulations in *Aspergillus niger*

Table 20.3 Production and secretion of heterologous proteins

Organism	Protein function	Reference
Vertebrates		
Human	Cytokines	(Broekhuijsen et al. 1993; Krasevec et al. 2000b)
	Serine protease	(Wiebe et al. 2001)
	Protease inhibitors	(Karnaukhova et al. 2007; Mikosch et al. 1996)
	Lysozymes	(Spencer et al. 1999)
	Antibodies	(Ward et al. 2004)
	Antigens	(James et al. 2012; Pluddemann and Van Zyl 2003)
Cattle	Pancreatic trypsin inhibitor	(MacKenzie et al. 1998)
	Enterokinase	(Svetina et al. 2000)
Camel	Chymosin	(Kappeler et al. 2006)
Horse	Lysozymes	(Spencer et al. 1999)
Hen	Lysozyme	(Archer et al. 1990; Jeenes et al. 1994)
Pig	Phospholipase	(Roberts et al. 1992)
Invertebrate		
Cattle tick	Cell surface glycoprotein	(Turnbull et al. 1990)
Plants		
Barley	Alpha amylase	(Juge et al. 1998)
Potato	Alpha-glucan phosphorylase	(Koda et al. 2005)
Fungi		
Agaricus meleangris	Pyranose dehydrogenase	(Pisanelli et al. 2010)
Aspergillus aculeatus	Beta-mannanase	(Sandgren et al. 2003)
	Mannanase	(van Zyl et al. 2009)
Aspergillus kawachii	Endoglucanase	(Goedegebuur et al. 2002)
Aspergillus terreus	Cis-aconitate decarboxylase	(Li et al. 2011a; Blumhoff et al. 2013)
Aspergillus tubingensis	Endo-polygalacturonase	(Bussink et al. 1991)
	Arabinoxylan Arabinofuranohydrolase	(Gielkens et al. 1997)
Aureobasidium pullulans	Endoglucanase	(Tambor et al. 2012)
Bionectria ochroleuca	Endoglucanase	(Goedegebuur et al. 2002)
Caldariomyces fumago	Chloroperoxidase	(Conesa et al. 2001)
Chaetomium brasiliense	Endoglucanase	(Goedegebuur et al. 2002)
Coprinopsis cinerea	Cellobiose dehydrogenase	(Turbe-Doan et al. 2013)
Emericella desertorum	Endoglucanase	(Goedegebuur et al. 2002)
Fusarium equiseti	Endoglucanase	(Goedegebuur et al. 2002)
Fusarium solani subsp. cucurbitae	Endoglucanase	(Goedegebuur et al. 2002)
Gloeophyllum trabeum	Endoglucanase	(Tambor et al. 2012)
Humicola grisea	Endoglucanase	(Goedegebuur et al. 2002; Sandgren et al. 2003)
Hypocrea schweinitzii	Endoglucanase	(Goedegebuur et al. 2002)
Morchella costata	α-glucan lyases	(Bojsen et al. 1999)
Myceliophthora thermophila	Endoglucanase	(Tambor et al. 2012)
Penicillium chrysogenum	Penicillin biosynthetic gene cluster	(Smith et al. 1990)
Phanerochaete chrysosporium	Beta-mannanase	(Benech et al. 2007)
	Xylanase	(Decelle et al. 2004)
	Manganese peroxidase	(Cortes-Espinosa et al. 2011; Conesa et al. 2002)

(continued)

Table 20.3 (continued)

Organism	Protein function	Reference
Phanerochaete flavido-alba	Laccase	(Benghazi et al. 2014)
Pleurotus eryngii	Peroxidase	(Eibes et al. 2009)
	Laccase	(Rodríguez et al. 2008)
Podospora anserina	Cellobiose dehydrogenase	(Turbe-Doan et al. 2013)
Pycnoporus cinnabarinus	Laccase	(Camarero et al. 2012; Record et al. 2002)
Pycnoporus sanguineus	Tyrosinase	(Halaouli et al. 2006)
Rhizopus oryzae	Fumarase	(de Jongh and Nielsen 2008)
Stachybotrys echinata	Endoglucanase	(Goedegebuur et al. 2002)
Talaromyces emersonii	Glucoamylase	(Nielsen et al. 2002)
Trametes versicolor	laccase	(Bohlin et al. 2006)
Trichoderma koningii	Endoglucanase	(Goedegebuur et al. 2002)
Trichoderma reesei	Endoglucanase	(Sandgren et al. 2003; Rose and van Zyl 2002)
	Xylanase	(Rose and van Zyl 2002)
Trichoderma viride	Endoglucanase	(Goedegebuur et al. 2002)
Bacteria		
Erwinia carotovora	Pectin lyase	(Bartling et al. 1996)
Escherichia coli	Aconitases	(Blumhoff et al. 2013)
Escherichia coli	Glucuronidase	(Roberts et al. 1989)
Thermotoga Maritima	Xylanase	(Zhang et al. 2008)
Virus		
Infectious bursal disease virus	Viral capsid protein	(Azizi et al. 2013)

A strategy frequently employed in heterologous gene expression is the fusion of the foreign gene sequence to native fungal genes such as the highly expressed glucoamylase gene (Broekhuijsen et al. 1993; Conesa et al. 2000; Cortes-Espinosa et al. 2011; Halaouli et al. 2006; James et al. 2012; MacKenzie et al. 1998; Mikosch et al. 1996; Roberts et al. 1992; Spencer et al. 1999). Attempts have also been made to use only the catalytic domain of this gene (Wiebe et al. 2001). In some cases a linker containing the KEX2 processing site is placed between the native gene and the foreign gene to facilitate post-translational cleavage of the polypeptide (Broekhuijsen et al. 1993; James et al. 2012; Karnaukhova et al. 2007; Krasevec et al. 2000a, b; MacKenzie et al. 1998; Mikosch et al. 1996; Svetina et al. 2000; Wiebe et al. 2001). Svetina et al. (2000) designed expression vectors both with and without the KEX2 linker and found that the fused protein without the linker was produced ten times higher than the KEX2-cleaved recombinant.

Other strategies for improving protein production and secretion include codon optimization and manipulating the unfolded protein response. Koda et al. (2005) successfully increased the production of the potato alpha-glucan phosphorylase by optimizing the codons to conform to the preferred *A. niger* codons. The production of a manganese peroxidase from *Phanerochaete chrysosporium* was improved by over-expression of two ER chaperone genes, the calnexin *clxA* and the binding protein *bipA* genes (Conesa et al. 2002).

20.4 Genomes of *Aspergillus niger* Strains

20.4.1 Organization and Statistics of Sequenced Genomes

The genome sequence of two *A. niger* strains and an extensive library of cDNA sequences (Semova et al. 2006) have been publicly available for several

Table 20.4 Overall statistics of three *A. niger* genomes

	CBS 513.88	ATCC 1015	NRRL3
Genome size (million base pairs)	33.98	34.85	35.25
Number of contigs	468	24	15
Number of gaps in the assembly	449	15	7
Number of telomeres retrieved	Not determined	10	16
Number of predicted open-reading frames	14,165	11,200	12,684

years. The industrial enzyme producer, strain CBS 513.88, was sequenced by DSM of the Netherlands (Pel et al. 2007). The wild-type citric acid producer ATCC 1015 (NRRL 328, CBS 113.46) was sequenced by the Joint Genome Institute of the U.S. Department of Energy (Andersen et al. 2011). We at Concordia University have recently sequenced NRRL3 (ATCC 9029, CBS 120.49, N400), a strain commonly used for genetic and molecular studies in research laboratories (unpublished data). Table 20.4 shows the overall statistics of these three genome sequences.

Linkage group analysis (Debets et al. 1993) and electrophoretic karyotyping (Verdoes et al. 1994a) revealed that the *A. niger* genome is arranged into eight chromosomes. The high-quality genome sequence of strain ATCC 1015 genome has been assembled into eight chromosomes with 15 gaps, eight of which correspond to the centromeres (Andersen et al. 2011). The CBS513.88 genome was assembled into 19 supercontigs (Pel et al. 2007) and displays extensive synteny with the genome of ATCC 1015 (Andersen et al. 2011). Based on its synteny to the ATCC 1015 genome, 186 out of 449 contig gaps in the CBS 513.88 genome were filled (Andersen et al. 2011). The NRRL3 genome is close to complete with only seven gaps corresponding to seven of the eight centromeres. While displaying extensive synteny, the CBS 513.88 and ATCC 1015 genomes exhibit a high degree of polymorphisms. The average density of single nucleotide polymorphism (SNP) between the two strains is 7.8 SNPs/kb. On the other hand, the NRRL3 and ATCC 1015 genomes are nearly identical with only 34 SNPs, or 1 SNP/Mb, between them (unpublished data). The near identity prompts speculation that NRRL3 and ATCC 1015 were originally derived from the same isolate.

The polymorphisms between strains bring into question whether the tools developed based on strain CBS 513.88 can be used for strain ATCC 1015/NRRL3 or vice versa. We have used extensively the same expression cassette to successfully produce recombinant proteins in derivatives of both strains CBS 513.88 and NRRL3. Whether cassettes used for gene replacement, which often contain 1.5–3.0 kb of genomic sequence to facilitate homologous recombination, can interchangeably be used between derivatives of CBS 513.88 and ATCC 1015/NRRL3 may depend on the genomic regions in question because the polymorphisms between the strains are not randomly distributed (Andersen et al. 2011).

20.4.2 Large Gene Families of Relevance to Industry

Different computational methods were used to predict the number of genes in the three genomes. This has led to the prediction of different numbers of open-reading frames for the three strains despite the close similarity of their genome sequences. Detailed comparison of the called genes revealed that the number of genes in ATCC 1015 is under-predicted while there is an over-prediction of genes in CBS 513.88. The gene complement for the two strains is very similar. Notable exceptions are two additional genes encoding alpha-amylases which appear to have been acquired by CBS 513.88 from *Aspergillus oryzae* through lateral gene transfer and three polyketide synthase genes, including the predicted ochratoxin A cluster, that are specific to CBS 513.88 (Andersen et al. 2011).

Aspergillus niger can grow on a wide variety of substrates. This is reflected by the large repertoire of genes encoding enzymes for biomass degradation (>170 glycoside hydrolases, polysaccharide lyases, and carbohydrate esterases) and solute transporters (~860). Fungal-specific transcription regulators represent another family of proteins that has expanded substantially in the Aspergilli, especially in *A. niger*. For example, there are 226 proteins containing the fungal-specific transcription factor domain in *A. niger* compared to 182 in *A. oryzae*, 61 in *Neurospora crassa*, and 26 in *Saccharomyces cerevisiae* (Pel et al. 2007). Taken together, these expanded protein families suggest that the regulation of nutrient metabolism and transport is finely controlled to respond to different environmental conditions or that there is considerable redundancy in the regulation. Mutational studies suggest that the regulation of nutrient metabolism is tightly controlled. For example, mutations in the regulator *xlnR* gene result in defective growth on xylan and a dramatic reduction in the transcription of genes encoding xylan-degrading enzymes (van Peij et al. 1998b).

20.5 Genome-Wide and Genome-Based Analyses

20.5.1 Transcriptomes and Proteomes

The genome sequences offer an important reference for genome-wide gene expression studies. In addition to providing insights into gene expression and gene function, examination of transcriptomes and proteomes can be used to improve structural and functional annotation of genes (Andersen et al. 2011; Tsang et al. 2009; Wright et al. 2009). Three basic technologies have been used to profile whole transcriptomes of *A. niger*. Affymetrix GeneChips, constructed with the annotated CBS 513.88 gene set, have been used to examine the regulation of genes encoding pectinolytic enzymes (Martens-Uzunova and Schaap 2009; Martens-Uzunova et al. 2006), secretion stress response (Guillemette et al. 2007), response to changes in ambient pH (Andersen et al. 2009), the role of the transcription factor HacA[CA] in the response to secretion stress (Carvalho et al. 2012), and conidial germination (Novodvorska et al. 2013; van Leeuwen et al. 2013). The Affymetrix GeneChips have also been used to analyze transcriptomes of single hyphal tips (de Bekker et al. 2011).

Oligonucleotides microarrays, based on gene models predicted in both CBS 153.88 and ATCC 1015 genomes, have also been manufactured. These DNA microarrays have been used to examine the transcriptomes of *A. niger* cultivated in sugarcane bagasse (de Souza et al. 2011) and to reveal the downstream targets of several gene regulators including the regulator of xylanolytic and cellulolytic enzymes XlnR and the regulator of arabinolytic enzymes AraR (de Souza et al. 2013), the regulator of the D-galactose oxidoreductive pathway GalX (Gruben et al. 2012), and the oxalic acid repression factor OafA (Poulsen et al. 2012).

Both the gene chips and DNA microarrays rely on nucleic acid hybridization to produce measurable signals. Background signals caused by nonspecific hybridization mean that a substantial population of transcripts expressed at low levels cannot be quantified. With the introduction of massively parallel sequencing technology, whole transcriptome sequencing as a method of profiling transcriptomes has become popular. Sequencing is more quantitative than hybridization. The near absence of background produced by RNA sequencing allows the detection and quantification of low-abundance transcripts. This approach has been used to evaluate the complex enzyme mixtures deployed by *A. niger* to break down biomass derived from plant cell wall (Delmas et al. 2012).

Mass spectrometry can be used to identify and quantify the proteins present in the proteomes. Investigation into the proteomes of *A. niger* has focused on the exo-proteomes, or the collection of extracellular proteins, under different growth conditions (Adav et al. 2010; Braaksma et al. 2010; Lu et al. 2010; Tsang et al. 2009). The microsomal proteome following xylose induction has also been examined by shot-gun proteomics (Ferreira de Oliveira et al. 2010). Attempts have been made to use proteomic data to correct gene models (Wright et al. 2009). By employing an integrated proteomic and transcriptomic approach, Jacobs et al. (2009) identified the differentially regulated proteins in enzyme-overproducing strains and used the knowledge for strains improvement. Nitsche et al. (2012) used integrated transcriptomics and exo-proteomics to examine the carbon starvation response in order to generate mutations and to produce proteins for downstream characterization.

20.5.2 Genome-Based Examination of Biochemical Pathways

Bioinformatic analysis of the *A. niger* genome sequence in conjunction with experimental investigations has been used to reveal the complexity of biochemical pathways and genetic networks. Based on published data, Andersen et al. (2008) used the genome sequence to reconstruct the metabolic network. Yuan et al. (2006, 2008b) used a combined bioinformatics and experimental approach to examine inulin and starch utilization. In the degradation of inulin they showed that, in addition to the previously known extracellular inulin-modifying enzymes, there are two invertase-like proteins that lack a secretory signal peptide. They also revealed that inulinolytic enzymes are under the control of the catabolite repressor CreA and that sucrose or a derivative of sucrose, but not fructose, is the inducer of inulinolytic genes (Yuan et al. 2006). *Aspergillus niger* is well known for its ability to hydrolyze starch. The major enzymes activities in starch hydrolysis are alpha-amylase (GH13), glucoamylase (GH15), and alpha-glucosidase (GH31). Mining the CBS 513.88 genome revealed 17 previously unknown proteins from the GH13, GH15, and GH31 families. Furthermore, the regulator of starch-degrading enzymes transcription AmyR controls only a subset of these enzymes, suggesting that most of these uncharacterized GH13, GH15, and GH31 proteins are involved in other cellular processes (Yuan et al. 2008b). Srivastava et al. (2010) used pathway reconstruction methods to identify the components of the eugenol to vanillin bioconversion pathway in the genome.

Comparative genomic analysis was used to define the differences between *A. niger* and other Eurotiales species in the regulation of pentose metabolism (Battaglia et al. 2011a, b). Kwon et al. (2011) identified six genes encoding small monomeric GTPases in the genome. By characterizing the phenotypes of deletion mutants, they were able to define the roles of the Rho GTPases in polarity establishment, cell wall integrity, and septum formation. Following the identification of genes predicted to encode the components in heme biosynthesis (Franken et al. 2011), Franken et al. (2012) deleted the *hemA* gene, encoding 5′-aminolevulinic acid synthase, to investigate heme metabolism and demonstrated differences in heme biosynthesis between *A. niger* and *Saccharomyces cerevisiae*.

20.5.3 Functional Characterization of Genes in the Post-genomic Era

Sequences of known enzymes or protein domains have been used to identify orthologues in *A. niger* by searching genome and cDNA sequences. A search of the *A. niger* genome revealed 21 genes for GH28 proteins, over half of which have not been characterized previously. Expression and characterization of the newly uncovered GH28 genes showed that seven of them encode exo-acting pectinases (Martens-Uzunova et al. 2006). The *xeg12A* gene encoding GH12 xyloglucanase was initially identified as an *A. aculeatus* orthologue (Master et al. 2008). The *eroA* and *ervA* genes were identified as functional orthologues of the *S. cerevisiae* ERO1 and ERV2 thiol oxidases involved in protein folding in the endoplasmic reticulum (Harvey et al. 2010). A GH26 mannanase was initially identified by domain analysis of the CBS 513.88 proteome (Zhao et al. 2011). Ten of the thirteen laccase-like multicopper oxidases identified in the genome were cloned and expressed. The recombinant multicopper oxidases exhibit different substrate specificities towards aromatic compounds (Tamayo-Ramos et al. 2011, 2012). A novel, extracellular GH31 alpha-xylosidase AxlA was purified from a commercial enzyme preparation and identified by mass spectrometry. The gene encoding AxlA was identified by matching the mass spectral data to the ATCC 1015 proteome (Scott-Craig et al. 2011). Among the 14 genes encoding GH18 proteins in the *A. niger* genome, the product of *cfc1* was singled out for characterization because it and its close orthologues in other *Aspergillus* species occupy a distinct clade in the phylogenetic tree of GH18 chitinases. Biochemical analysis of Cfc1 revealed that it is an exochitinase that releases monomers of *N*-acetylglucosamine from the reducing end of chitin oligosaccharides (van Munster et al. 2012). The *vmaD* gene was identified by sequence similarity to the subunit of the eukaryotic vacuolar-H(+)-ATPase. The ability of *vmaD* to complement cell wall mutants suggests that the ATP-driven transport of protons and acidification of the vacuole is important for the integrity of the fungal cell wall (Schachtschabel et al. 2012). Sorbitol dehydrogenase SdhA and galactitol dehydrogenase LadB of the oxidoreductive D-galactose catabolic pathway were initially identified by sequence similarity to related enzymes and their identity subsequently confirmed by gene deletion studies and by complementation of yeast mutants (Koivistoinen et al. 2012; Mojzita et al. 2012).

20.5.4 Polyketide Synthase Genes

The discovery in the *A. niger* genome of many genes predicted to be involved in the production of secondary metabolites has stimulated several studies into the role of polyketide synthases (PKS) in secondary metabolites biosynthesis. By genetic complementation of an albino mutant and by characterizing the phenotype of a gene disruption mutant, the PKS gene *alba/fwnA* has been shown to be responsible for producing the spore pigment precursor and a family of naphtha-gamma-pyrones (Chiang et al. 2011; Jørgensen et al. 2011). A type III PKS (Accession number XP_001390871) was identified from the CBS 513.88 genome and expressed in *E. coli*, and its recombinant product was shown to catalyze synthesis of alkyl pyrones (Li et al. 2011b). The kotanin biosynthetic cluster was identified by disrupting the closely linked genes encoding

O-methyltransferase, P450 monooxygenase, and polyketide synthase (*ktnB, ktnC, ktnS*) (Gil Girol et al. 2012).

Bioinformatics analysis revealed 34 PKS genes in strain CBS 513.88 and 31 PKS genes in strain ATCC 1015. The sequences of the three extra PKS genes found in CBS 513.88 (An01g01130, An11g05940, and An15g07920) were used to design PCR primers to examine by PCR amplification the distribution of these three PKS genes in 119 *A. niger* strains. These results show that the distribution of these three PKS genes is strain-specific. Furthermore, the presence of the An15g07920 PKS positively correlated with the production of ochratoxin in the *A. niger* strains tested (Ferracin et al. 2012). The genome-based analysis of genes encoding secondary metabolites will possibly lead to the discovery of compounds that are useful as well as harmful to other organisms.

20.5.5 Transcription Regulators

The *A. niger* genome harbors over 200 genes encoding predicted fungal-specific transcription regulators. Years before the genome was sequenced, transcription regulators were identified in *A. niger* and related species (de Vries, 2003). The availability of the genome sequence not only accelerates the discovery of new regulators, it supports the detailed transcriptomic and proteomic analyses of the downstream targets of the transcription regulators. Table 20.5 summarizes the role of transcription regulators characterized in *A. niger*.

The regulation of the expression of cellulase and hemicellulase genes by XlnR is by far the most intensely studied regulatory system in *A. niger* (Battaglia et al. 2011a, b; de Vries et al. 1999; Hasper 2004; Hasper et al. 2000; Mach-Aigner et al. 2012; Omony et al. 2011; van Peij et al. 1998a, b). XlnR is known to regulate expression of 20–30 enzymes, transporters, and possibly other transcription factors (Stricker et al. 2008). One bottleneck in the development of *A. niger* as a protein production platform is the degradation of the proteins of interest by extracellular proteases. Given that the *A. niger* genome

20 Genetic and Genomic Manipulations in *Aspergillus niger*

Table 20.5 Summary of characterized transcription factors

Gene name	Protein function	Inducer	Reference
acuB	Regulator of acetate metabolism	Acetate	(Meijer et al. 2009; Papadopoulou and Sealy-Lewis 1999; Sealy-Lewis and Fairhurst 1998)
amyR	Regulator of starch-degrading enzymes	Maltose, starch	(vanKuyk et al. 2012; Yuan et al. 2008b)
araR	Regulator of arabinolytic enzymes	Arabinose, arabitol, arabinan	(Battaglia et al. 2011b; de Souza et al. 2013)
area	Regulator of nitrogen utilization	Ammonium	(Lenouvel et al. 2001)
azaR	Regulator of azaphilones biosynthesis	Not known	(Zabala et al. 2012)
brlA	Regulator of conidiophore development	Activated by FlbA	(Krijgsheld et al. 2013a, b; Lee and Adams 1994, 1996)
creA	Repressor of carbon catabolism	Glucose	(Delmas et al. 2012)
flbA	Regulator of G-protein signaling protein, inhibition of vegetative growth enabling asexual development	Activated by FluG protein	(Krijgsheld et al. 2013a, b; Lee and Adams 1994, 1996)
galX	Regulator of the D-galactose oxido-reductive pathway	Galactose	(Gruben et al. 2012)
hacA	Regulator of the unfolded protein response	Environmental stress	(Carvalho et al. 2010; Guillemette et al. 2011; Mulder et al. 2004)
inuR	Regulator of inulinolytic and sucrolytic enzymes	Sucrose, inulin	(Yuan et al. 2008a)
oafA	Repressor of oxalic acid production	Gluconic acid	(Poulsen et al. 2012)
pacC	pH regulator	Extracellular pH	(Andersen et al. 2009; Denison 2000; Van Den Hombergh et al. 1996)
prtT	Regulator of secreted proteases	Not known	(Punt et al. 2008)
xlnR	Regulator of cellulolytic and hemicellulolytic enzymes	Xylose	(Battaglia et al. 2011b; Stricker et al. 2008)

contains close to 200 genes coding for proteases (Pel et al. 2007), effective strategies to reduce the levels of extracellular proteases would be highly useful. Punt et al. (2008) showed that deletion of the *prtT* gene, a regulator of secreted proteases, dramatically reduces the production of extracellular proteases. In characterizing gene regulators involved in development, Krijgsheld et al. (2013b) have demonstrated that deletion of *flbA*, the gene encoding a regulator of asexual development, leads to enhanced protein secretion. Only a handful of the hundreds of putative gene regulators have so far been characterized. Uncovering the function and interactions of this plethora of gene regulators should provide ample tools for altering the genome through synthetic biology approaches.

20.6 Conclusion

Robust genetic transformation systems have been developed for *Aspergillus niger* over the past 30 years. These systems have already been used to produce many homologous and heterologous proteins as well as to regulate the production of metabolites. The release of the genome sequence has substantially enhanced our understanding of its metabolic network and the regulation of its genes. Furthermore, genome-based studies combined with genome manipulation through genetic transformation have led to the rational design of strains for improved production of proteins and metabolites. The knowledge gained from these studies should prepare *A. niger* for a new role as a platform organism for synthetic biology.

Acknowledgements This work was supported by Genome Canada, Génome Québec, and the NSERC Strategic Bioconversion Network. We thank Dr. Ian Reid for critical reading of the manuscript.

References

Adav SS, Li AA, Manavalan A, Punt P, Sze SK (2010) Quantitative iTRAQ secretome analysis of Aspergillus niger reveals novel hydrolytic enzymes. J Proteome Res 9:3932–3940. doi:10.1021/pr100148j

Ahuja M, Punekar NS (2008) Phosphinothricin resistance in Aspergillus niger and its utility as a selectable transformation marker. Fungal Genet Biol 45:1103–1110, DOI: http://dx.doi.org/10.1016/j.fgb.2008.04.002

Andersen MR, Lehmann L, Nielsen J (2009) Systemic analysis of the response of Aspergillus niger to ambient pH. Genome Biol 10:R47. doi:10.1186/gb-2009-10-5-r47

Andersen MR, Nielsen ML, Nielsen J (2008) Metabolic model integration of the bibliome, genome, metabolome and reactome of Aspergillus niger. Mol Syst Biol 4:178. doi:10.1038/msb.2008.12

Andersen MR, Salazar MP, Schaap PJ, van de Vondervoort PJ, Culley D, Thykaer J, Frisvad JC, Nielsen KF, Albang R, Albermann K, Berka RM, Braus GH, Braus-Stromeyer SA, Corrochano LM, Dai Z, van Dijck PW, Hofmann G, Lasure LL, Magnuson JK, Menke H, Meijer M, Meijer SL, Nielsen JB, Nielsen ML, van Ooyen AJ, Pel HJ, Poulsen L, Samson RA, Stam H, Tsang A, van den Brink JM, Atkins A, Aerts A, Shapiro H, Pangilinan J, Salamov A, Lou Y, Lindquist E, Lucas S, Grimwood J, Grigoriev IV, Kubicek CP, Martinez D, van Peij NN, Roubos JA, Nielsen J, Baker SE (2011) Comparative genomics of citric-acid-producing Aspergillus niger ATCC 1015 versus enzyme-producing CBS 513.88. Genome Res 21:885–897. doi:10.1101/gr.112169.110

Archer DB, Jeenes DJ, MacKenzie DA, Brightwell G, Lambert N, Lowe G, Radford SE, Dobson CM (1990) Hen egg white lysozyme expressed in, and secreted from, Aspergillus niger is correctly processed and folded. Biotechnology (N Y) 8:741–745

Azizi M, Yakhchali B, Ghamarian A, Enayati S, Khodabandeh M, Khalaj V (2013) Cloning and expression of Gumboro VP2 antigen in Aspergillus niger. Avicenna J Med Biotechnol 5:35–41

Bartling S, van den Hombergh JP, Olsen O, von Wettstein D, Visser J (1996) Expression of an Erwinia pectate lyase in three species of Aspergillus. Curr Genet 29:474–481

Battaglia E, Hansen SF, Leendertse A, Madrid S, Mulder H, Nikolaev I, de Vries RP (2011a) Regulation of pentose utilisation by AraR, but not XlnR, differs in Aspergillus nidulans and Aspergillus niger. Appl Microbiol Biotechnol 91:387–397. doi:10.1007/s00253-011-3242-2

Battaglia E, Visser L, Nijssen A, van Veluw GJ, Wosten HA, de Vries RP (2011b) Analysis of regulation of pentose utilisation in Aspergillus niger reveals evolutionary adaptations in Eurotiales. Stud Mycol 69: 31–38. doi:10.3114/sim.2011.69.03

Benech R-O, Li X, Patton D, Powlowski J, Storms R, Bourbonnais R, Paice M, Tsang A (2007) Recombinant expression, characterization, and pulp prebleaching property of a Phanerochaete chrysosporium endo-β-1,4-mannanase. Enzym Microb Tech 41:740–747, DOI: http://dx.doi.org/10.1016/j.enzmictec.2007.06.012

Benghazi L, Record E, Suarez A, Gomez-Vidal JA, Martinez J, de la Rubia T (2014) Production of the Phanerochaete flavido-alba laccase in Aspergillus niger for synthetic dyes decolorization and biotransformation. World J Microbiol Biotechnol 30:201–211

Blumhoff ML, Steiger MG, Mattanovich D, Sauer M (2013) Targeting enzymes to the right compartment: metabolic engineering for itaconic acid production by Aspergillus niger. Metab Eng 19:26–32. doi:10.1016/j.ymben.2013.05.003

Bohlin C, Jonsson LJ, Roth R, van Zyl WH (2006) Heterologous expression of Trametes versicolor laccase in Pichia pastoris and Aspergillus niger. Appl Biochem Biotechnol 129–132:195–214

Bojsen K, Yu S, Kragh KM, Marcussen J (1999) A group of α-1,4-glucan lyases and their genes from the red alga Gracilariopsis lemaneiformis: purification, cloning, and heterologous expression. Biochim Biophys Acta 1430:396–402, DOI: http://dx.doi.org/10.1016/S0167-4838(99)00017-5

Braaksma M, Martens-Uzunova ES, Punt PJ, Schaap PJ (2010) An inventory of the Aspergillus niger secretome by combining in silico predictions with shotgun proteomics data. BMC Genomics 11:584. doi:10.1186/1471-2164-11-584

Broekhuijsen MP, Mattern IE, Contreras R, Kinghorn JR, van den Hondel CA (1993) Secretion of heterologous proteins by Aspergillus niger: production of active human interleukin-6 in a protease-deficient mutant by KEX2-like processing of a glucoamylase-hIL6 fusion protein. J Biotechnol 31:135–145

Bussink HJ, Buxton FP, Visser J (1991) Expression and sequence comparison of the Aspergillus niger and Aspergillus tubigensis genes encoding polygalacturonase II. Curr Genet 19:467–474

Buxton FP, Gwynne DI, Davies RW (1985) Transformation of Aspergillus niger using the argB gene of Aspergillus nidulans. Gene 37:207–214

Buxton FP, Gwynne DI, Davies RW (1989) Cloning of a new bidirectionally selectable marker for Aspergillus strains. Gene 84:329–334, DOI: http://dx.doi.org/10.1016/0378-1119(89)90507-6

Camarero S, Pardo I, Canas AI, Molina P, Record E, Martinez AT, Martinez MJ, Alcalde M (2012) Engineering platforms for directed evolution of Laccase from Pycnoporus cinnabarinus. Appl Environ Microbiol 78:1370–1384. doi:10.1128/AEM.07530-11

Carvalho ND, Jorgensen TR, Arentshorst M, Nitsche BM, van den Hondel CA, Archer DB, Ram AF (2012) Genome-wide expression analysis upon constitutive activation of the HacA bZIP transcription factor in

Aspergillus niger reveals a coordinated cellular response to counteract ER stress. BMC Genomics 13:350. doi:10.1186/1471-2164-13-350

Carvalho NDSP, Arentshorst M, Jin KM, Meyer V, Ram AFJ (2010) Expanding the ku70 toolbox for filamentous fungi: establishment of complementation vectors and recipient strains for advanced gene analyses. Appl Microbiol Biotechnol 87:1463–1473. doi:10.1007/s00253-010-2588-1

Chiang YM, Meyer KM, Praseuth M, Baker SE, Bruno KS, Wang CC (2011) Characterization of a polyketide synthase in Aspergillus niger whose product is a precursor for both dihydroxynaphthalene (DHN) melanin and naphtho-gamma-pyrone. Fungal Genet Biol 48:430–437. doi:10.1016/j.fgb.2010.12.001

Conesa A, Jeenes D, Archer DB, van den Hondel CA, Punt PJ (2002) Calnexin overexpression increases manganese peroxidase production in Aspergillus niger. Appl Environ Microbiol 68:846–851

Conesa A, van De Velde F, van Rantwijk F, Sheldon RA, van Den Hondel CA, Punt PJ (2001) Expression of the Caldariomyces fumago chloroperoxidase in Aspergillus niger and characterization of the recombinant enzyme. J Biol Chem 276(21):17635–17640

Conesa A, van den Hondel CA, Punt PJ (2000) Studies on the production of fungal peroxidases in Aspergillus niger. Appl Environ Microbiol 66:3016–3023

Cortes-Espinosa DV, Absalon AE, Sanchez N, Loera O, Rodriguez-Vazquez R, Fernandez FJ (2011) Heterologous expression of manganese peroxidase in Aspergillus niger and its effect on phenanthrene removal from soil. J Mol Microbiol Biotechnol 21:120–129. doi:10.1159/000331563

Dave K, Ahuja M, Jayashri TN, Sirola RB, Punekar NS (2012) A novel selectable marker based on Aspergillus niger arginase expression. Enzyme Microb Technol 51:53–58, DOI: http://dx.doi.org/10.1016/j.enzmictec.2012.04.001

Dave K, Punekar NS (2011) Utility of Aspergillus niger citrate synthase promoter for heterologous expression. J Biotechnol 155:173–177, DOI: http://dx.doi.org/10.1016/j.jbiotec.2011.06.012

Davies RW (1994) Heterologous gene expression and protein secretion in Aspergillus. Prog Ind Microbiol 29:527–560

de Bekker C, Bruning O, Jonker MJ, Breit TM, Wosten HA (2011) Single cell transcriptomics of neighboring hyphae of Aspergillus niger. Genome Biol 12:R71. doi:10.1186/gb-2011-12-8-r71

de Graaff LK, van den Broeck HC, van Ooijen AJJ, Visser J (1994) Regulation of the xylanase-encoding xlnA gene of Aspergilius tubigensis. Mol Microbiol 12:479–490.doi:10.1111/j.1365-2958.1994.tb01036.x

de Graaff LK, van den Broeck HC, Visser J (1992) Isolation and characterization of the Aspergillus niger. Curr Genet 22:21–27

de Groot MJA, Bundock P, Hooykaas PJJ, Beijersbergen AGM (1998) Agrobacterium tumefaciens-mediated transformation of filamentous fungi. Nat Biotech 16:839–842

de Jongh WA, Nielsen J (2008) Enhanced citrate production through gene insertion in Aspergillus niger. Metab Eng 10:87–96. doi:10.1016/j.ymben.2007.11.002

de Souza WR, de Gouvea PF, Savoldi M, Malavazi I, de Souza Bernardes LA, Goldman MH, de Vries RP, de Castro Oliveira JV, Goldman GH (2011) Transcriptome analysis of Aspergillus niger grown on sugarcane bagasse. Biotechnol Biofuels 4:40. doi:10.1186/1754-6834-4-40

de Souza WR, Maitan-Alfenas GP, de Gouvêa PF, Brown NA, Savoldi M, Battaglia E, Goldman MH, de Vries RP, Goldman GH (2013) The influence of Aspergillus niger transcription factors AraR and XlnR in the gene expression during growth in D-xylose, L-arabinose and steam-exploded sugarcane bagasse. Fungal Genet Biol. doi:10.1016/j.fgb.2013.07.007

de Vries RP (2003) Regulation of Aspergillus genes encoding plant cell wall polysaccharide-degrading enzymes; relevance for industrial production. Appl Microbiol Biotechnol 61:10–20. doi:10.1007/s00253-002-1171-9

de Vries RP, Visser J, de Graaff LH (1999) CreA modulates the XlnR-induced expression on xylose of Aspergillus niger genes involved in xylan degradation. Res Microbiol 150:281–285

Debets F, Swart K, Hoekstra RF, Bos CJ (1993) Genetic maps of eight linkage groups of Aspergillus niger based on mitotic mapping. Curr Genet 23:47–53

Decelle B, Tsang A, Storms RK (2004) Cloning, functional expression and characterization of three Phanerochaete chrysosporium endo-1,4-beta-xylanases. Curr Genet 46:166–175. doi:10.1007/s00294-004-0520-x

Delmas S, Pullan ST, Gaddipati S, Kokolski M, Malla S, Blythe MJ, Ibbett R, Campbell M, Liddell S, Aboobaker A, Tucker GA, Archer DB (2012) Uncovering the genome-wide transcriptional responses of the filamentous fungus Aspergillus niger to lignocellulose using RNA sequencing. PLoS Genet 8:e1002875. doi:10.1371/journal.pgen.1002875

Denison SH (2000) pH regulation of gene expression in fungi. Fungal Genet Biol 29:61–71. doi:10.1006/fgbi.2000.1188

Eibes GM, Lú-Chau TA, Ruiz-Dueñas FJ, Feijoo G, Martínez MJ, Martínez AT, Lema JM (2009) Effect of culture temperature on the heterologous expression of Pleurotus eryngii versatile peroxidase in Aspergillus hosts. Bioprocess Biosyst Eng 32:129–134. doi:10.1007/s00449-008-0231-7

Ferracin LM, Fier CB, Vieira ML, Monteiro-Vitorello CB, Varani Ade M, Rossi MM, Muller-Santos M, Taniwaki MH, Thie Iamanaka B, Fungaro MH (2012) Strain-specific polyketide synthase genes of Aspergillus niger. Int J Food Microbiol 155:137–145. doi:10.1016/j.ijfoodmicro.2012.01.020

Ferreira de Oliveira JM, van Passel MW, Schaap PJ, de Graaff LH (2010) Shotgun proteomics of Aspergillus niger microsomes upon D-xylose induction. Appl Environ Microbiol 76:4421–4429. doi:10.1128/AEM.00482-10

Fleissner A, Dersch P (2010) Expression and export: recombinant protein production systems for Aspergillus. Appl Microbiol Biotechnol 87:1255–1270. doi:10.1007/s00253-010-2672-6

Flipphi MJA, Visser J, van der Veen P, de Graaff LH (1994) Arabinase gene expression in Aspergillus niger: indications for coordinated regulation. Microbiology 140:2673–2682. doi:10.1099/00221287-140-10-2673

Fowler T, Berka RM, Ward M (1990) Regulation of the glaA gene of Aspergillus niger. Curr Genet 18:537–545

Franken AC, Lokman BC, Ram AF, Punt PJ, van den Hondel CA, de Weert S (2011) Heme biosynthesis and its regulation: towards understanding and improvement of heme biosynthesis in filamentous fungi. Appl Microbiol Biotechnol 91:447–460. doi:10.1007/s00253-011-3391-3

Franken AC, Lokman BC, Ram AF, van den Hondel CA, de Weert S, Punt PJ (2012) Analysis of the role of the Aspergillus niger aminolevulinic acid synthase (hemA) gene illustrates the difference between regulation of yeast and fungal haem- and sirohaem-dependent pathways. FEMS Microbiol Lett 335:104–112. doi:10.1111/j.1574-6968.2012.02655.x

Gems D, Johnstone IL, Clutterbuck AJ (1991) An autonomously replicating plasmid transforms Aspergillus nidulans at high frequency. Gene 98:61–67

Gielesen B, van den Berg M (2013) Transformation of filamentous fungi in microtiter plate. V.K.Gupta et al. (eds), Laboratory Protocols in Fungal Biology: Current Methods in Fungal Biology, pp 343–348, Springer, New York

Gielkens MM, Visser J, de Graaff LH (1997) Arabinoxylan degradation by fungi: characterization of the arabinoxylan-arabinofuranohydrolase encoding genes from Aspergillus niger and Aspergillus tubingensis. Curr Genet 31:22–29

Gil Girol C, Fisch KM, Heinekamp T, Günther S, Hüttel W, Piel J, Brakhage AA, Müller M (2012) Regio- and stereoselective oxidative phenol coupling in Aspergillus niger. Angew Chem Int Ed Engl 51:9788–9791. doi:10.1002/anie.201203603

Goedegebuur F, Fowler T, Phillips J, van der Kley P, van Solingen P, Dankmeyer L, Power S (2002) Cloning and relational analysis of 15 novel fungal endoglucanases from family 12 glycosyl hydrolase. Curr Genet 41:89–98. doi:10.1007/s00294-002-0290-2

Goosen T, Bloemheuvel G, Christoph G, de Bie DA, Henk WJ, van Den B, Klaas S (1987) Transformation of Aspergillus niger using the homologous orotidine-5″-phosphate-decarboxylase gene. Curr Genet 11:499–503

Gouka RJ, Punt PJ, van den Hondel CAMJJ (1997) Efficient production of secreted proteins by Aspergillus: progress, limitations and prospects. Appl Microbiol Biotechnol 47:1–11

Gruben BS, Zhou M, de Vries RP (2012) GalX regulates the D-galactose oxido-reductive pathway in Aspergillus niger. FEBS Lett 586:3980–3985. doi:10.1016/j.febslet.2012.09.029

Guillemette T, Ram AFJ, Carvalho NDSP, Joubert A, Simoneau P, Archer DB (2011) Methods for investigating the UPR in filamentous fungi. Methods Enzymol 490:1–29. doi:10.1016/B978-0-12-385114-7.00001-5

Guillemette T, van Peij N, Goosen T, Lanthaler K, Robson GD, van den Hondel CA, Stam H, Archer DB (2007) Genomic analysis of the secretion stress response in the enzyme-producing cell factory Aspergillus niger. BMC Genomics 8:158. doi:10.1186/1471-2164-8-158

Halaouli S, Record E, Casalot L, Hamdi M, Sigoillot JC, Asther M, Lomascolo A (2006) Cloning and characterization of a tyrosinase gene from the white-rot fungus Pycnoporus sanguineus, and overproduction of the recombinant protein in Aspergillus niger. Appl Microbiol Biotechnol 70:580–589. doi:10.1007/s00253-005-0109-4

Hartingsveldt WV, Mattern IE, van Zeijl CMJ, Pouwels PH, van den Hondel CAMJJ (1987) Development of a homologous transformation system for Aspergillus niger based on the pyrG gene. Mol Gen Genet 206:71–75

Harvey AR, Ward M, Archer DB (2010) Identification and characterisation of eroA and ervA, encoding two putative thiol oxidases from Aspergillus niger. Gene 461:32–41. doi:10.1016/j.gene.2010.04.011

Hasper AA (2004) Functional analysis of the transcriptional activator XlnR from Aspergillus niger. Microbiology 150:1367–1375. doi:10.1099/mic.0.26557-0

Hasper AA, Visser J, de Graaff LH (2000) The Aspergillus niger transcriptional activator XlnR, which is involved in the degradation of the polysaccharides xylan and cellulose, also regulates D-xylose reductase gene expression. Mol Microbiol 36:193–200

Hynes MJ (1996) Genetic transformation of filamentous fungi. J Genet 75:297–311

Jacobs DI, Olsthoorn MM, Maillet I, Akeroyd M, Breestraat S, Donkers S, van der Hoeven RA, van den Hondel CA, Kooistra R, Lapointe T, Menke H, Meulenberg R, Misset M, Muller WH, van Peij NN, Ram A, Rodriguez S, Roelofs MS, Roubos JA, van Tilborg MW, Verkleij AJ, Pel HJ, Stam H, Sagt CM (2009) Effective lead selection for improved protein production in Aspergillus niger based on integrated genomics. Fungal Genet Biol 46(Suppl 1):S141–S152. doi:10.1016/j.fgb.2008.08.012

James E, van Zyl W, van Zyl P, Gorgens J (2012) Recombinant hepatitis B surface antigen production in Aspergillus niger: evaluating the strategy of gene fusion to native glucoamylase. Appl Microbiol Biotechnol 96:385–394. doi:10.1007/s00253-012-4191-0

Jeenes DJ, Mackenzie DA, Archer DB (1994) Transcriptional and post-transcriptional events affect the production of secreted hen egg white lysozyme by Aspergillus niger. Transgenic Res 3:297–303

Jørgensen TR, Park J, Arentshorst M, van Welzen AM, Lamers G, Vankuyk PA, Damveld RA, van den Hondel CAM, Nielsen KF, Frisvad JC, Ram AFJ (2011) The molecular and genetic basis of conidial pigmentation in Aspergillus niger. Fungal Genet Biol 48:544–553. doi:10.1016/j.fgb.2011.01.005

Juge N, Svensson B, Williamson G (1998) Secretion, purification, and characterisation of barley alpha-amylase produced by heterologous gene expression in Aspergillus niger. Appl Microbiol Biotechnol 49:385–392

Kappeler SR, van den Brink HJ, Rahbek-Nielsen H, Farah Z, Puhan Z, Hansen EB, Johansen E (2006) Characterization of recombinant camel chymosin reveals superior properties for the coagulation of bovine and camel milk. Biochem Biophys Res Commun 342:647–654. doi:10.1016/j.bbrc.2006.02.014

Karnaukhova E, Ophir Y, Trinh L, Dalal N, Punt PJ, Golding B, Shiloach J (2007) Expression of human alpha1-proteinase inhibitor in Aspergillus niger. Microb Cell Fact 6:34. doi:10.1186/1475-2859-6-34

Kelly JM, Hynes MJ (1985) Transformation of Aspergillus niger by the amdS gene of Aspergillus nidulans. EMBO J 4:475–479

Koda A, Bogaki T, Minetoki T, Hirotsune M (2005) High expression of a synthetic gene encoding potato alpha-glucan phosphorylase in Aspergillus niger. J Biosci Bioeng 100:531–537. doi:10.1263/jbb.100.531

Koivistoinen OM, Richard P, Penttila M, Ruohonen L, Mojzita D (2012) Sorbitol dehydrogenase of Aspergillus niger, SdhA, is part of the oxido-reductive D-galactose pathway and essential for D-sorbitol catabolism. FEBS Lett 586:378–383. doi:10.1016/j.febslet.2012.01.004

Krasevec N, van de Hondel CA, Komel R (2000a) Expression of human lymphotoxin alpha in Aspergillus niger. Pflugers Arch 440:R83–R85

Krasevec N, van den Hondel CA, Komel R (2000b) Can hTNF-alpha be successfully produced and secreted in filamentous fungus Aspergillus niger? Pflugers Arch 439:R84–R86

Krijgsheld P, Bleichrodt R, van Veluw GJ, Wang F, Muller WH, Dijksterhuis J, Wosten HA (2013a) Development in Aspergillus. Stud Mycol 74:1–29. doi:10.3114/sim0006

Krijgsheld P, Nitsche BM, Post H, Levin AM, Müller WH, Heck AJR, Ram AFJ, Altelaar AFM, Wösten HAB (2013b) Deletion of flbA results in increased secretome complexity and reduced secretion heterogeneity in colonies of Aspergillus niger. J Proteome Res 12:1808–1819. doi:10.1021/pr301154w

Kwon MJ, Arentshorst M, Roos ED, van den Hondel CA, Meyer V, Ram AF (2011) Functional characterization of Rho GTPases in Aspergillus niger uncovers conserved and diverged roles of Rho proteins within filamentous fungi. Mol Microbiol 79:1151–1167. doi:10.1111/j.1365-2958.2010.07524.x

Lee BN, Adams TH (1994) Overexpression of flbA, an early regulator of Aspergillus asexual sporulation, leads to activation of brlA and premature initiation of development. Mol Microbiol 14:323–334

Lee BN, Adams TH (1996) FluG and flbA function interdependently to initiate conidiophore development in Aspergillus nidulans through brlA beta activation. EMBO J 15:299–309

Lenouvel F, Fraissinet-Tachet L, van de Vondervoort PJ, Visser J (2001) Isolation of UV-induced mutations in the areA nitrogen regulatory gene of Aspergillus niger, and construction of a disruption mutant. Mol Genet Genomics 266:42–47

Li A, Pfelzer N, Zuijderwijk R, Punt P (2012) Enhanced itaconic acid production in Aspergillus niger using genetic modification and medium optimization. BMC Biotechnol 12:57. doi:10.1186/1472-6750-12-57

Li A, van Luijk N, ter Beek M, Caspers M, Punt P, van der Werf M (2011a) A clone-based transcriptomics approach for the identification of genes relevant for itaconic acid production in Aspergillus. Fungal Genet Biol 48:602–611, DOI: http://dx.doi.org/10.1016/j.fgb.2011.01.013

Li J, Luo Y, Lee JK, Zhao H (2011b) Cloning and characterization of a type III polyketide synthase from Aspergillus niger. Bioorg Med Chem Lett 21:6085–6089. doi:10.1016/j.bmcl.2011.08.058

Li W, Chen G, Gu L, Zeng W, Liang Z (2013) Genome shuffling of Aspergillus niger for improving transglycosylation activity. Appl Biochem Biotechnol. doi:10.1007/s12010-013-0421-x

Lu X, Sun J, Nimtz M, Wissing J, Zeng AP, Rinas U (2010) The intra- and extracellular proteome of Aspergillus niger growing on defined medium with xylose or maltose as carbon substrate. Microb Cell Fact 9:23. doi:10.1186/1475-2859-9-23

Mach-Aigner AR, Omony J, Jovanovic B, van Boxtel AJ, de Graaff LH (2012) D-Xylose concentration-dependent hydrolase expression profiles and the function of CreA and XlnR in Aspergillus niger. Appl Environ Microbiol 78:3145–3155. doi:10.1128/AEM.07772-11

MacKenzie DA, Kraunsoe JA, Chesshyre JA, Lowe G, Komiyama T, Fuller RS, Archer DB (1998) Aberrant processing of wild-type and mutant bovine pancreatic trypsin inhibitor secreted by Aspergillus niger. J Biotechnol 63:137–146

Magnuson JK, Lasure LL (2004) Organic acid production by filamentous fungi. Adv Fungal Biotechnol Ind Agr Med 24:307–340

Martens-Uzunova ES, Schaap PJ (2009) Assessment of the pectin degrading enzyme network of Aspergillus niger by functional genomics. Fungal Genet Biol 46(Suppl 1):S170–S179

Martens-Uzunova ES, Zandleven JS, Benen JA, Awad H, Kools HJ, Beldman G, Voragen AG, Van den Berg JA, Schaap PJ (2006) A new group of exo-acting family 28 glycoside hydrolases of Aspergillus niger that are involved in pectin degradation. Biochem J 400:43–52. doi:10.1042/BJ20060703

Master ER, Zheng Y, Storms R, Tsang A, Powlowski J (2008) A xyloglucan-specific family 12 glycosyl hydrolase from Aspergillus niger: recombinant expression, purification and characterization. Biochem J 411:161–170. doi:10.1042/BJ20070819

Mattern JE, Punt PJ, van den Hondel CAMJJ (1988) A vector of Aspergillus transformation conferring phleomycin resistance. Fungal Genet Newslett 35:25

Meijer S, de Jongh WA, Olsson L, Nielsen J (2009) Physiological characterisation of acuB deletion in

Aspergillus niger. Appl Microbiol Biotechnol 84:157–167. doi:10.1007/s00253-009-2027-3

Meyer V (2008) Genetic engineering of filamentous fungi – progress, obstacles and future trends. Biotechnol Adv 26:177–185, DOI: http://dx.doi.org/10.1016/j.biotechadv.2007.12.001

Meyer V, Arentshorst M, El-Ghezal A, Drews AC, Kooistra R, van den Hondel CA, Ram AF (2007) Highly efficient gene targeting in the Aspergillus niger kusA mutant. J Biotechnol 128:770–775. doi:10.1016/j.jbiotec.2006.12.021

Michielse CB, Hooykaas PJ, van den Hondel CA, Ram AF (2005) Agrobacterium-mediated transformation as a tool for functional genomics in fungi. Curr Genet 48:1–17. doi:10.1007/s00294-005-0578-0

Mikosch T, Klemm P, Gassen HG, van den Hondel CA, Kemme M (1996) Secretion of active human mucus proteinase inhibitor by Aspergillus niger after KEX2-like processing of a glucoamylase-inhibitor fusion protein. J Biotechnol 52:97–106

Mojzita D, Koivistoinen OM, Maaheimo H, Penttila M, Ruohonen L, Richard P (2012) Identification of the galactitol dehydrogenase, LadB, that is part of the oxido-reductive D-galactose catabolic pathway in Aspergillus niger. Fungal Genet Biol 49:152–159. doi:10.1016/j.fgb.2011.11.005

Mulder HJ, Saloheimo M, Penttila M, Madrid SM (2004) The transcription factor HACA mediates the unfolded protein response in Aspergillus niger, and up-regulates its own transcription. Mol Genet Genomics 271:130–140. doi:10.1007/s00438-003-0965-5

Murphy C, Powlowski J, Wu M, Butler G, Tsang A (2011) Curation of characterized glycoside hydrolases of fungal origin. Database (Oxford) 2011:bar020. DOI: 10.1093/database/bar020

Nielsen BR, Lehmbeck J, Frandsen TP (2002) Cloning, heterologous expression, and enzymatic characterization of a thermostable glucoamylase from Talaromyces emersonii. Protein Expr Purif 26:1–8

Nikolaev I, Mathieu M, van de Vondervoort P, Visser J, Felenbok B (2002) Heterologous expression of the Aspergillus nidulans alcR-alcA system in Aspergillus niger. Fungal Genet Biol 37:89–97

Nitsche BM, Jorgensen TR, Akeroyd M, Meyer V, Ram AF (2012) The carbon starvation response of Aspergillus niger during submerged cultivation: insights from the transcriptome and secretome. BMC Genomics 13:380. doi:10.1186/1471-2164-13-380

Novodvorska M, Hayer K, Pullan ST, Wilson R, Blythe MJ, Stam H, Stratford M, Archer DB (2013) Trancriptional landscape of Aspergillus niger at breaking of conidial dormancy revealed by RNA-sequencing. BMC Genomics 14:246. doi:10.1186/1471-2164-14-246

Omony J, de Graaff LH, van Straten G, van Boxtel AJ (2011) Modeling and analysis of the dynamic behavior of the XlnR regulon in Aspergillus niger. BMC Syst Biol 5(Suppl 1):S14. doi:10.1186/1752-0509-5-S1-S14

Ozeki K, Kyoya F, Hizume K, Kanda A, Hamachi M (1994) Transformation of Intact Aspergillus niger by electroporation. Biosci Biotechnol Biochem 58:2224–2227

Pachlinger R, Mitterbauer R, Adam G, Strauss J (2005) Metabolically independent and accurately adjustable Aspergillus sp. expression system. Appl Environ Microbiol 71:672–678. doi:10.1128/AEM.71.2.672-678.2005

Papadopoulou S, Sealy-Lewis HM (1999) The Aspergillus niger acuA and acuB genes correspond to the facA and facB genes in Aspergillus nidulans. FEMS Microbiol Lett 178:35–37

Pel HJ, de Winde JH, Archer DB, Dyer PS, Hofmann G, Schaap PJ, Turner G, de Vries RP, Albang R, Albermann K, Andersen MR, Bendtsen JD, Benen JA, van den Berg M, Breestraat S, Caddick MX, Contreras R, Cornell M, Coutinho PM, Danchin EG, Debets AJ, Dekker P, van Dijck PW, van Dijk A, Dijkhuizen L, Driessen AJ, d'Enfert C, Geysens S, Goosen C, Groot GS, de Groot PW, Guillemette T, Henrissat B, Herweijer M, van den Hombergh JP, van den Hondel CA, van der Heijden RT, van der Kaaij RM, Klis FM, Kools HJ, Kubicek CP, van Kuyk PA, Lauber J, Lu X, van der Maarel MJ, Meulenberg R, Menke H, Mortimer MA, Nielsen J, Oliver SG, Olsthoorn M, Pal K, van Peij NN, Ram AF, Rinas U, Roubos JA, Sagt CM, Schmoll M, Sun J, Ussery D, Varga J, Vervecken W, van de Vondervoort PJ, Wedler H, Wosten HA, Zeng AP, van Ooyen AJ, Visser J, Stam H (2007) Genome sequencing and analysis of the versatile cell factory Aspergillus niger CBS 513.88. Nat Biotechnol 25:221–231. doi:10.1038/nbt1282

Piddington CS, Houston CS, Paloheimo M, Cantrell M, Miettinen-Oinonen A, Nevalainen H, Rambosek J (1993) The cloning and sequencing of the genes encoding phytase (phy) and pH 2.5-optimum acid phosphatase (aph) from Aspergillus niger var. awamori. Gene 133:55–62

Pisanelli I, Kujawa M, Gschnitzer D, Spadiut O, Seiboth B, Peterbauer C (2010) Heterologous expression of an Agaricus meleagris pyranose dehydrogenase-encoding gene in Aspergillus spp. and characterization of the recombinant enzyme. Appl Microbiol Biotechnol 86:599–606. doi:10.1007/s00253-009-2308-x

Pluddemann A, Van Zyl WH (2003) Evaluation of Aspergillus niger as host for virus-like particle production, using the hepatitis B surface antigen as a model. Curr Genet 43:439–446. doi:10.1007/s00294-003-0409-0

Poulsen L, Andersen MR, Lantz AE, Thykaer J (2012) Identification of a transcription factor controlling pH-dependent organic acid response in Aspergillus niger. PLoS One 7:e50596. doi:10.1371/journal.pone.0050596

Prathumpai W, Flitter S, McIntyre M, Nielsen J (2004) Lipase production by recombinant strains of Aspergillus niger expressing a lipase-encoding gene from

Thermomyces lanuginosus. Appl Microbiol Biotechnol 65:714–719. doi:10.1007/s00253-004-1699-y

Punt PJ, Oliver RP, Dingemanse MA, Pouwels PH, van den Hondel CA (1987) Transformation of Aspergillus based on the hygromycin B resistance marker from Escherichia coli. Gene 56:117–124

Punt PJ, Schuren FH, Lehmbeck J, Christensen T, Hjort C, van den Hondel CA (2008) Characterization of the Aspergillus niger prtT, a unique regulator of extracellular protease encoding genes. Fungal Genet Biol 45:1591–1599. doi:10.1016/j.fgb.2008.09.007

Record E, Punt PJ, Chamkha M, Labat M, van Den Hondel CA, Asther M (2002) Expression of the Pycnoporus cinnabarinus laccase gene in Aspergillus niger and characterization of the recombinant enzyme. Eur J Biochem 269:602–609

Roberts IN, Jeenes DJ, MacKenzie DA, Wilkinson AP, Sumner IG, Archer DB (1992) Heterologous gene expression in Aspergillus niger: a glucoamylase-porcine pancreatic prophospholipase A2 fusion protein is secreted and processed to yield mature enzyme. Gene 122:155–161

Roberts IN, Oliver RP, Punt PJ, van den Hondel CA (1989) Expression of the Escherichia coli beta-glucuronidase gene in industrial and phytopathogenic filamentous fungi. Curr Genet 15:177–180

Rodríguez E, Ruiz-Dueñas FJ, Kooistra R, Ram A, Martínez ÁT, Martínez MJ (2008) Isolation of two laccase genes from the white-rot fungus Pleurotus eryngii and heterologous expression of the pel3 encoded protein. J Biotechnol 134:9–19, DOI: http://dx.doi.org/10.1016/j.jbiotec.2007.12.008

Rose SH, van Zyl WH (2002) Constitutive expression of the Trichoderma reesei beta-1,4-xylanase gene (xyn2) and the beta-1,4-endoglucanase gene (egl) in Aspergillus niger in molasses and defined glucose media. Appl Microbiol Biotechnol 58:461–468

Roth AFJ, Dersch P (2010) A novel expression system for intracellular production and purification of recombinant affinity-tagged proteins in Aspergillus niger. Appl Microbiol Biotechnol 86:659–670. doi:10.1007/s00253-009-2252-9

Sagt CM, ten Haaft PJ, Minneboo IM, Hartog MP, Damveld RA, van der Laan JM, Akeroyd M, Wenzel TJ, Luesken FA, Veenhuis M, van der Klei I, de Winde JH (2009) Peroxicretion: a novel secretion pathway in the eukaryotic cell. BMC Biotechnol 9:48. doi:10.1186/1472-6750-9-48

Salovuori I, Makarow M, Rauvala H, Knowles J, Kaariainen L (1987) Low molecular weight high-mannose type glycans in a secreted protein of the filamentous fungus Trichoderma reesei. Nat Biotech 5:152–156

Sandgren M, Gualfetti PJ, Shaw A, Gross LS, Saldajeno M, Day AG, Jones TA, Mitchinson C (2003) Comparison of family 12 glycoside hydrolases and recruited substitutions important for thermal stability. Protein Sci 12:848–860. doi:10.1110/ps.0237703

Schachtschabel D, Arentshorst M, Lagendijk EL, Ram AF (2012) Vacuolar H(+)-ATPase plays a key role in cell wall biosynthesis of Aspergillus niger. Fungal Genet Biol 49:284–293. doi:10.1016/j.fgb.2011.12.008

Schuster E, Dunn-Coleman N, Frisvad JC, Van Dijck PW (2002) On the safety of Aspergillus niger – a review. Appl Microbiol Biotechnol 59:426–435. doi:10.1007/s00253-002-1032-6

Scott-Craig JS, Borrusch MS, Banerjee G, Harvey CM, Walton JD (2011) Biochemical and molecular characterization of secreted alpha-xylosidase from Aspergillus niger. J Biol Chem 286:42848–42854. doi:10.1074/jbc.M111.307397

Sealy-Lewis HM, Fairhurst V (1998) Isolation of mutants deficient in acetyl-CoA synthetase and a possible regulator of acetate induction in Aspergillus niger. Microbiology 144(Pt 7):1895–1900

Semova N, Storms R, John T, Gaudet P, Ulycznyj P, Min XJ, Sun J, Butler G, Tsang A (2006) Generation, annotation, and analysis of an extensive Aspergillus niger EST collection. BMC Microbiol 6:7. doi:10.1186/1471-2180-6-7

Smith DJ, Burnham MK, Edwards J, Earl AJ, Turner G (1990) Cloning and heterologous expression of the penicillin biosynthetic gene cluster from penicillum chrysogenum. Biotechnology (N Y) 8:39–41

Spencer A, Morozov-Roche LA, Noppe W, MacKenzie DA, Jeenes DJ, Joniau M, Dobson CM, Archer DB (1999) Expression, purification, and characterization of the recombinant calcium-binding equine lysozyme secreted by the filamentous fungus Aspergillus niger: comparisons with the production of hen and human lysozymes. Protein Expr Purif 16:171–180. doi:10.1006/prep.1999.1036

Srivastava S, Luqman S, Khan F, Chanotiya CS, Darokar MP (2010) Metabolic pathway reconstruction of eugenol to vanillin bioconversion in Aspergillus niger. Bioinformation 4:320–325

Storms R, Zheng Y, Li H, Sillaots S, Martinez-Perez A, Tsang A (2005) Plasmid vectors for protein production, gene expression and molecular manipulations in Aspergillus niger. Plasmid 53:191–204, DOI: http://dx.doi.org/10.1016/j.plasmid.2004.10.001

Stricker AR, Mach RL, de Graaff LH (2008) Regulation of transcription of cellulases- and hemicellulases-encoding genes in Aspergillus niger and Hypocrea jecorina (Trichoderma reesei). Appl Microbiol Biotechnol 78:211–220. doi:10.1007/s00253-007-1322-0

Svetina M, Krasevec N, Gaberc-Porekar V, Komel R (2000) Expression of catalytic subunit of bovine enterokinase in the filamentous fungus Aspergillus niger. J Biotechnol 76:245–251. doi:10.1016/S0168-1656(99)00191-1

Tamayo-Ramos JA, van Berkel WJ, de Graaff LH (2012) Biocatalytic potential of laccase-like multicopper oxidases from Aspergillus niger. Microb Cell Fact 11:165. doi:10.1186/1475-2859-11-165

Tamayo Ramos JA, Barends S, Verhaert RM, de Graaff LH (2011) The Aspergillus niger multicopper oxidase family: analysis and overexpression of laccase-like encoding genes. Microb Cell Fact 10:78

Tambor JH, Ren H, Ushinsky S, Zheng Y, Riemens A, St-Francois C, Tsang A, Powlowski J, Storms R (2012) Recombinant expression, activity screening and functional characterization identifies three novel endo-1,4-beta-glucanases that efficiently hydrolyse cellulosic substrates. Appl Microbiol Biotechnol 93:203–214. doi:10.1007/s00253-011-3419-8

Tsang A, Butler G, Powlowski J, Panisko EA, Baker SE (2009) Analytical and computational approaches to define the Aspergillus niger secretome. Fungal Genet Biol 46(Suppl 1):S153–S160

Turbe-Doan A, Arfi Y, Record E, Estrada-Alvarado I, Levasseur A (2013) Heterologous production of cellobiose dehydrogenases from the basidiomycete Coprinopsis cinerea and the ascomycete Podospora anserina and their effect on saccharification of wheat straw. Appl Microbiol Biotechnol 97:4873–4885. doi:10.1007/s00253-012-4355-y

Turnbull IF, Smith DR, Sharp PJ, Cobon GS, Hynes MJ (1990) Expression and secretion in Aspergillus nidulans and Aspergillus niger of a cell surface glycoprotein from the cattle tick, Boophilus microplus, by using the fungal amdS promoter system. Appl Environ Microbiol 56:2847–2852

Unkles SE, Campbell EI, Carrez D, Grieve C, Contreras R, Fiers W, Van den Hondel CAMJJ, Kinghorn JR (1989) Transformation of Aspergillus niger with the homologous nitrate reductase gene. Gene 78:157–166, DOI: http://dx.doi.org/10.1016/0378-1119(89)90323-5

Van Den Hombergh JP, MacCabe AP, Van De Vondervoort PJ, Visser J (1996) Regulation of acid phosphatases in an Aspergillus niger pacC disruption strain. Mol Gen Genet 251:542–550

van Leeuwen MR, Krijgsheld P, Bleichrodt R, Menke H, Stam H, Stark J, Wosten HA, Dijksterhuis J (2013) Germination of conidia of Aspergillus niger is accompanied by major changes in RNA profiles. Stud Mycol 74:59–70. doi:10.3114/sim0009

van Munster JM, van der Kaaij RM, Dijkhuizen L, van der Maarel MJ (2012) Biochemical characterization of Aspergillus niger CfcI, a glycoside hydrolase family 18 chitinase that releases monomers during substrate hydrolysis. Microbiology 158:2168–2179. doi:10.1099/mic.0.054650-0

van Peij NN, Gielkens MM, de Vries RP, Visser J, de Graaff LH (1998a) The transcriptional activator XlnR regulates both xylanolytic and endoglucanase gene expression in Aspergillus niger. Appl Environ Microbiol 64:3615–3619

van Peij NN, Visser J, de Graaff LH (1998b) Isolation and analysis of xlnR, encoding a transcriptional activator co-ordinating xylanolytic expression in Aspergillus niger. Mol Microbiol 27:131–142

van Zyl PJ, Moodley V, Rose SH, Roth RL, van Zyl WH (2009) Production of the Aspergillus aculeatus endo-1,4-beta-mannanase in A. niger. J Ind Microbiol Biotechnol 36:611–617. doi:10.1007/s10295-009-0551-x

vanKuyk PA, Benen JA, Wosten HA, Visser J, de Vries RP (2012) A broader role for AmyR in Aspergillus niger: regulation of the utilisation of D-glucose or D-galactose containing oligo- and polysaccharides. Appl Microbiol Biotechnol 93:285–293. doi:10.1007/s00253-011-3550-6

Verdoes JC, Calil MR, Punt PJ, Debets F, Swart K, Stouthamer AH, van den Hondel CA (1994a) The complete karyotype of Aspergillus niger: the use of introduced electrophoretic mobility variation of chromosomes for gene assignment studies. Mol Gen Genet 244:75–80

Verdoes JC, Punt PJ, van der Berg P, Debets F, Stouthamer AH, van den Hondel CA (1994b) Characterization of an efficient gene cloning strategy for Aspergillus niger based on an autonomously replicating plasmid: cloning of the nicB gene of A. niger. Gene 146:159–165

Ward M, Kodama KH, Wilson LJ (1989) Transformation of Aspergillus awamori and A. niger by electroporation. Exp Mycol 13:289–293, DOI: http://dx.doi.org/10.1016/0147-5975(89)90050-9

Ward M, Lin C, Victoria DC, Fox BP, Fox JA, Wong DL, Meerman HJ, Pucci JP, Fong RB, Heng MH, Tsurushita N, Gieswein C, Park M, Wang H (2004) Characterization of humanized antibodies secreted by Aspergillus niger. Appl Environ Microbiol 70:2567–2576

Ward M, Wilson LJ, Carmona ICL, Turner G (1988) The oliC3 gene of Aspergillus niger: isolation, sequence and use as a selectable marker for transformation. Curr Genet 14:37–42

Ward M, Wilson LJ, Kodama KH, Rey MW, Berka RM (1990) Improved production of chymosin in Aspergillus by expression as a glucoamylase-chymosin fusion. Nat Biotech 8:435–440

Wiebe MG, Karandikar A, Robson GD, Trinci AP, Candia JL, Trappe S, Wallis G, Rinas U, Derkx PM, Madrid SM, Sisniega H, Faus I, Montijn R, van den Hondel CA, Punt PJ (2001) Production of tissue plasminogen activator (t-PA) in Aspergillus niger. Biotechnol Bioeng 76:164–174

Wright JC, Sugden D, Francis-McIntyre S, Riba-Garcia I, Gaskell SJ, Grigoriev IV, Baker SE, Beynon RJ, Hubbard SJ (2009) Exploiting proteomic data for genome annotation and gene model validation in Aspergillus niger. BMC Genomics 10:61. doi:10.1186/1471-2164-10-61

Yuan XL, Goosen C, Kools H, van der Maarel MJ, van den Hondel CA, Dijkhuizen L, Ram AF (2006) Database mining and transcriptional analysis of genes encoding inulin-modifying enzymes of Aspergillus niger. Microbiology 152:3061–3073. doi:10.1099/mic.0.29051-0

Yuan XL, Roubos JA, van den Hondel CA, Ram AF (2008a) Identification of InuR, a new Zn(II)2Cys6 transcriptional activator involved in the regulation of inulinolytic genes in Aspergillus niger. Mol Genet Genomics 279:11–26. doi:10.1007/s00438-007-0290-5

Yuan XL, van der Kaaij RM, van den Hondel CA, Punt PJ, van der Maarel MJ, Dijkhuizen L, Ram AF (2008b) Aspergillus niger genome-wide analysis reveals a large number of novel alpha-glucan acting enzymes with unexpected expression profiles. Mol Genet Genomics 279:545–561. doi:10.1007/s00438-008-0332-7

Zabala AO, Xu W, Chooi Y-H, Tang Y (2012) Characterization of a silent azaphilone gene cluster from Aspergillus niger ATCC 1015 reveals a hydroxylation-mediated pyran-ring formation. Chem Biol 19:1049–1059. doi:10.1016/j.chembiol.2012.07.004

Zhang JX, Pan J, Guan GH, Li Y, Xue W, Tang GM, Wang AQ, Wang HM (2008) Expression and high-yield production of extremely thermostable bacterial xylanaseB in Aspergillus niger. Enzyme Microb Technol 43:513–516, DOI: 10.1016/j.enzmictec.2008.07.010

Zhao W, Zheng J, Zhou HB (2011) A thermotolerant and cold-active mannan endo-1,4-beta-mannosidase from Aspergillus niger CBS 513.88: Constitutive overexpression and high-density fermentation in Pichia pastoris. Bioresour Technol 102:7538–7547. doi:10.1016/j.biortech.2011.04.070

Genetic Manipulation of *Meyerozyma guilliermondii*

21

Nicolas Papon, Yuriy R. Boretsky,
Vincent Courdavault, Marc Clastre,
and Andriy A. Sibirny

21.1 Introduction

21.1.1 Classification and Characterization of *Meyerozyma (Candida) guilliermondii* Strains

Pichia guilliermondii (anamorph *Candida guilliermondii*; since 2010 *Meyerozyma guilliermondii*) is an ascomycetous yeast widely distributed in the natural environment and also a part of the human saprophyte microflora. At the beginning of the twentieth century this yeast was described by Castellani as *Endomyces guilliermondii*; it

---

N. Papon (✉) • V. Courdavault, Ph.D.
M. Clastre, Ph.D.
Department EA2106 Biomolécules et
Biotechnologies Végétales, Faculté de Pharmacie,
Université François-Rabelais de Tours,
31 Avenue Monge, Tours F-37200, France
e-mail: nicolas.papon@univ-tours.fr

Y.R. Boretsky, Ph.D.
Department of Molecular Genetics and
Biotechnology, Institute of Cell Biology, National
Academy of Sciences of Ukraine, Lviv, Ukraine

A.A. Sibirny, Ph.D., Dr.Sc. (✉)
Department of Molecular Genetics and
Biotechnology, Institute of Cell Biology, National
Academy of Sciences of Ukraine, Lviv, Ukraine

Department of Biotechnology and Microbiology,
University of Rzeszow, Lviv, Ukraine
e-mail: sibirny@cellbiol.lviv.ua

was isolated in Sri Lanka from bronchomycosis patients (Castellani 1912). For a long time, strains classified as *C. guilliermondii* and its teleomorph *P. guilliermondii* were considered to be a genetically heterogeneous complex comprising numerous phenotypically undistinguishable taxa including notably *Candida fermentati* (*Pichia caribbica*) and *Candida carpophila* (Bai 1996; San Millan et al. 1997). Electrophoretic karyotyping as well as internal transcribed spacer (ITS)/26S rRNA sequence comparisons carried out during the past two decades provided a clear separation of taxa in the *P. guilliermondii* clade (Vaughan-Martini et al. 2005; Desnos-Ollivier et al. 2008; Yamamura et al. 2009; Savini et al. 2011). *P. guilliermondii* currently represents a collection of sporogenous strains formerly belonging to the asporogenous species *Candida guilliermondii* (Cast.) Langeron and Guerra (Wickerham and Burton 1954; Wickerham 1966; Kreger van-Rij 1970). This also means that each strain of *C. guilliermondii* which has a sexual life cycle should be moved to the species *P. guilliermondii*. Close relations between known strains are also seen at nucleotide level. Alignment of several nucleotide the sequences cloned from the well-mating strain *P. guilliermondii* L2 revealed 99.4–99.8 % homology with those derived from the ATCC6260 strain assigned as *C. guilliermondii* (Boretsky and Sibirny, unpublished). Both belong to the so-called "fungal CTG clade": yeast species that translate CUG as serine instead of leucine (Fitzpatrick et al. 2006; Sibirny and

---

M.A. van den Berg and K. Maruthachalam (eds.), *Genetic Transformation Systems in Fungi, Volume 2*, Fungal Biology, DOI 10.1007/978-3-319-10503-1_21,
© Springer International Publishing Switzerland 2015

Boretsky 2009; Butler et al. 2009). The newly assigned teleomorph species name of *C. guilliermondii* is *Meyerozyma guilliermondii* (http://www.uniprot.org/taxonomy/4929; Kurtzman and Suzuki 2010). For clarity, the name *M. (C.) guilliermondii* is used throughout the chapter. Recent identification and characterization of the *M. (C.) guilliermondii MAT* locus demonstrated that the *a1* gene is missing within the *MATa* allele and that the *MATα* allele lacks the *α2* gene; however, *M. (C.) guilliermondii* has retained a complete sexual cycle including hybridization, meiosis, and sporulation (Reedy et al. 2009; Butler et al. 2009), controllable by environmental conditions (Sibirny and Boretsky 2009), thus allowing the use of formal genetics tools.

The sequencing of the *M. (C.) guilliermondii* ATCC6260 reference strain genome clearly reflects the increasing interest of the scientific community in this species. The *M. (C.) guilliermondii* has a relatively small genome (10.6 Mb); yet, the number of predicted genes (total of 5920) is similar to other fungal CTG clade species (Butler et al. 2009).

As mentioned above, the natural habitat of this species is highly diverse. In most cases, strains were isolated from oil-containing soil, plant leaves, lake water, and cow paunch (Zharova et al. 1977, 1980; Sibirny 1996). Some clinical isolates from immunocompromised patients also are reported (Pfaller and Diekema 2007; Pfaller et al. 2010).

M. (C.) guilliermondii is a typical representative of aerobic yeasts and cannot grow under strictly anaerobic conditions. The standard growth temperature for *M. (C.) guilliermondii* is 30 °C and the upper limit is near 42 °C. Standard yeast media are used for *M. (C.) guilliermondii* cultivation (Sibirny 1996). Cells of *M. (C.) guilliermondii* are heterogeneous, mostly elongated in shape (approximately 2×10 µm). In contrast to *C. albicans*, *M. (C.) guilliermondii* is unable to produce true hyphae. More details concerning peculiarities of this yeast species can be found in several reviews (Sibirny 1996; Sibirny and Boretsky 2009; Papon et al. 2013a, b).

Despite some relations to *C. albicans*, strains of *M. (C.) guilliermondii* are of great interest due to their potential for several industrial applications. *M. (C.) guilliermondii* is a rarely observed pathogen, accounting for 1–3 % of all candidemias. Notably, most cases of infection are associated with oncology patients ((http://www.broad.mit.edu/, Pfaller and Diekema 2007; Savini et al. 2011). It should be emphasized that no cases of candidiasis caused by laboratory strains of *M. (C.) guilliermondii* were noted during decades of work with this species at the Institute of Cell Biology, NAS of Ukraine, Lviv. All recorded strains *M. (C.) guilliermondii* that were selected at this institute were properly checked, and none of them was found to be pathogenic to laboratory animals. Thus, it can be stated that most strains of *M. (C.) guilliermondii* (except some clinical isolates) could be considered as safe and useful model for different purposes.

M. (C.) guilliermondii is considered as a model organism of the so-called "flavinogenic yeasts", organisms capable of over-producing riboflavin (vitamin B_2) under iron limitation (Tanner et al. 1945; Boretsky et al. 2005; Abbas and Sibirny 2011). Besides, this organism is the only one that is known to be capable of riboflavin uptake through active transport catalyzed by riboflavin permease leading to hyperaccumulation of the vitamin in the cells (Sibirnyi et al. 1977; Sibirny and Shavlovsky 1984). *M. (C.) guilliermondii* belongs to relatively rare yeast species capable of growing on n-alkanes as sole carbon and energy source (Shchelokova et al. 1974) and thus to produce single-cell protein from hydrocarbons. This yeast species appears to be one of the most effective organisms for bioconversion of xylose into xylitol, the anti-caries sweetener, and is able to utilize even hemicellulosic hydrolysates obtained by acid hydrolysis as an energy source (Canettieri et al. 2001; Carvalho et al. 2002; Rodrigues et al. 2006). In addition, *M. (C.) guilliermondii* is a promising source for enzymes like inulinase (Guo et al. 2009), biofuel (Schirmer-Michel et al. 2008), and aromas (Wah et al. 2013).

Some *M. (C.) guilliermondii* isolates exhibit an important potential in post-harvest biological control of spoilage fungi during storage of fruits and vegetables. This yeast-mediated biological control has notably emerged as a potential alternative technique to fungicide treatments to prevent and control decay loss of harvested commodities (Zhang et al. 2011; Coda et al. 2013).

To obtain an insight into the mechanisms of pathogenesis as well as its ability to overproduce valuable biocompounds, a robust molecular toolbox for *M. (C.) guilliermondii* genetics has been developed over the last decade. This involves construction of recipient strains, vectors with recessive and dominant markers, optimization of transformation methods, isolation of knock out strains, and the cloning of strong promoters. These are prerequisites for further development of *M. (C.) guilliermondii* research and practical use of strains of biotechnological interest. Here, we report on protocols providing efficient transformation of *M. (C.) guilliermondii*, available strains, markers and vectors, as well as some basic molecular biology techniques adapted for this yeast.

21.2 Selective Media and Markers

In most cases yeast cells are cultured in YPS medium (0.5 % yeast extract, 1 % peptone, and 2 % sucrose) supplemented with special additives (riboflavin, uridine, and/or antibiotics) if required. Solid medium contained 2.0 % agar. Synthetic defined medium (SD: 0.67 %, yeast nitrogen base, supplemented with casamino acid (Difco) and 2 % glucose but without uridine) or YPS medium without riboflavin added (or with antibiotics) were used for yeast transformants selection. Available strains and markers are listed in Table 21.1. More details concerning composition of media and peculiarities of recipient strains growth can be found in the provided references. Fortunately, it is not necessary to select a specific auxotrophic recipient strains for transformation as dominant markers are available and are more preferable for initial experiments despite the relatively high price of the corresponding antibiotics (e.g. phleomycin, hygromycin B, nourseothricin). This approach can be used to construct a suitable recipient strain for subsequent scaled up transformation experiments. In addition, the use of heterologous dominant markers may result in increased percentage of homologous recombination at the target loci during gene knockout experiments in *M.(C) guilliermondii*.

As the major part of ascomycetous yeast species, *M. (C.) guilliermondii* preferentially uses the non-homologous end-joining (NHEJ) pathway over the homologous recombination pathway for

Table 21.1 Strains of *M.(C) guilliermondii* and selective markers available for transformation

Recipient	Selective agent	Gene	Origin	Reference
rib1-21	Riboflavin +/− 200 mg/L	*RIB1*	*M.(C) guilliermondii*	Liauta-Teglivets et al. (1995)
rib7-162	Riboflavin +/− 200 mg/L	*RIB7*	*M.(C) guilliermondii*	Boretsky et al. (1999); Boretskii et al. (2002)
M3, R-66	Uridine +/− 400 mg/L	*URA3*	*S. cerevisiae*	Boretsky et al. (2007); Pynyaha et al. (2009)
U312	Uridine +/− 400 mg/L	*URA3*	*M.(C) guilliermondii*	Foureau et al. (2012b)
NP566U	Uridine +/− 400 mg/L	*URA5*	*M.(C) guilliermondii*	Millerioux et al. (2011b)
leu2 [REP]	Leucine +/− 100 mg/L	*LEU2*	*M.(C) guilliermondii*	Courdavault et al. (2011)
	Uridine +/− 400 mg/L	*URA5*		
All strains	Nourseothricin −/+ 150 mg/L	*SAT-1*	*E. coli*	Millerioux et al. (2011a)
All strains	hygromycin B −/+ 500 mg/L	*HPH#*	*E. coli*	Millerioux et al. (2011a)
trp5 [HPH#]	Tryptophan +/− 100 mg/L	*TRP5*	*M.(C) guilliermondii*	Foureau et al. (2012a)
trp5 [REP]	Tryptophan +/− 100 mg/L	*TRP5*	*M.(C) guilliermondii*	Foureau et al. (2013a)
	Uridine +/− 400 mg/L	*URA5*		
ade2 [Sable]	Adenine +/− 40 mg/L	*ADE2*	*M.(C) guilliermondii*	Foureau et al. (2013c)
All strains	Phleomycin −/+ 600 mg/mL	*ble*	*Staphylococcus aureus MRSA252*	Foureau et al. (2013c)
KU141F1	Uridine +/− 400 mg/L	*URA5*	*M.(C) guilliermondii*	Foureau et al. (2013b)
ura5, lig4	Uridine +/− 400 mg/L	*URA5*	*M.(C) guilliermondii*	Foureau et al. (2013a)

+/− Means that the recommended quantity of a substance should be supplemented to the media when growing a recipient strain/or omitted to select transformants in the case of an auxotrophic marker
−/+ Correspondingly recommended quantity of a substance should be omitted when growing a recipient strain/or supplemented to the media to select transformants in the case of a dominant marker

DNA double-strand break (DSB) repair. This implies that DNA cassettes are frequently integrated ectopically into genomes following transformation experiments, as previously described for a large number of model yeasts. This phenomenon is favourable to random insertion mutagenesis experiments in *M. (C.) guilliermondii* (Piniaga et al. 2002; Pynyaha et al. 2009; Boretsky et al. 2011; Foureau et al. 2013b). However, this remains highly deleterious for targeted gene disruption in this species. To circumvent the natural low frequency of homologous integration of transforming DNA, some NHEJ-deficient recipient strains were recently engineered (KU141F1 and *ura5, lig4*).

21.3 Transformation of *M.(C.) guilliermondii*

21.3.1 Reagents

1. Double distilled water (dd-water), or equivalent.
2. Peptone.
3. Yeast extract.
4. Sucrose.
5. Agar.
6. Yeast Nitrogen Base w/o A.A. (DIFCO).
7. Uridine.
8. Riboflavin.
9. Leucine.
10. L-tryptophan.
11. Nourseothricin.
12. Hygromycin B.
13. Phleomycin.
14. 1,4-Dithiothreitol (DTT).
15. Lithium acetate.
16. Potassium phosphate dibasic.
17. Potassium phosphate monobasic.
18. Polyethylene glycol PEG 3350.
19. Tris-base.
20. Lyticase from *Arthrobacter luteus*.
21. Ethylenediaminetetraacetic acid.
22. Dithiothreitol.
23. Sodium citrate.

21.3.2 Equipments

1. 1.5 mL-microcentrifuge tubes.
2. 50 mL centrifuge tubes.
3. Graduated cylinders and beakers.
4. Lockable storage bottles.
5. 20 mL glass tubes.
6. 500 mL flasks.
7. 1,000 mL flasks.
8. Petri dishes.
9. Vortexer.
10. Filters for sterilization.
11. Micropipettors and tips.
12. Shaker at 28 °C.
13. Microcentrifuge.
14. Centrifuge.
15. Incubator at 28–30 °C.
16. pH meter.
17. UV–Vis spectrophotometer.
18. Analytical balance.
19. Laboratory balance.
20. Electroporator and 1-mm electroporation cuvettes.
21. Microscope.

21.4 Special Precautions

- Ensure that cells are harvested at appropriate growth phase.
- Ensure that all steps (including centrifugation) are done at the recommended temperature.
- During centrifugation steps traces of media and buffers should be removed completely since they affect subsequent steps of the transformation procedure significantly.
- Ensure that the PEG solution is added and the sample is mixed immediately after that.
- Efficiency of transformation strongly depends on quality of PEG.
- Ensure that the cells are not over-treated with Zymolyase.
- Solution of 0.3 M sucrose in YPS medium is added immediately after electroporation.

21.5 Transformation of *M. (C.) guilliermondii* Using Lithium Acetate Treatment

Principle: The cells are growing in a rich medium until early logarithmic phase, harvested, treated with lithium acetate, followed by incubation with DNA, and PEG, heat shocked and plated on a selective medium.

1. Inoculate a colony of freshly subcloned yeast strain in 2 mL of liquid YPS medium to be transformed. Incubate for 24 h at 30 °C with vigorous shaking. Measure A_{600}.

2. Inoculate 100 mL of liquid YPS medium in a 500 mL flask with an aliquot (3–10 µL) of the pre-grown culture. It is better to start three cultures simultanenously with inocula of different volumes in order to guarantee one culture at the proper density at a reasonable time in the next morning. Optical density measured at 600 nm should be 0.6–0.8.

3. Pellet cells at 3,000 g for 10 min (room temperature). Discard the supernatant. Resuspend cells in 50 mL of sterile double-distilled water.

4. Pellet cells at 3,000 g for 10 min. Discard the supernatant.

5. Resuspend cells in 2 mL of 0.1 M lithium acetate prepared on TE buffer pH 7.5 (LiAc/TE buffer). Incubate for 1 h at 30 °C; periodic gentle agitation is recommended.

6. Pellet the cells and resuspend them in an aliquot of fresh LiAc/TE buffer to obtain an optical density $A_{600} = 100$.

7. Tranfer 50 µL aliquots of culture into 1.5 mL tubes. Add to each tube plasmid DNA (1–10 µg in 1–10 µL of TE buffer) and mix well.

8. Add to each tube 250 µL 50 % PEG in LiAc/TE, mix vigorously and incubate for 30 min at 30 °C.

9. Make heat-shock for 15 min at 42 °C. Chill samples in ice for 1 min.

10. Spin down samples at 3,000 g for 10 min, resuspend samples in 1 mL YPS and incubated for 1 h at 30 °C.

11. Spin down samples for 10 min at 3,000 g and resuspend the cells in 150 µL of 1 M sucrose.

12. Plate cells on a selective medium and incubate at 30 °C for 3–5 days.

13. Note: Steps 10 and 11 can be omitted when using dominant selective marker.

21.6 Transformation of *M. (C.) guilliermondii* Using Spheroplasting Method

Principle: The cells are growing in a rich medium until early logarithmic phase, harvested, treated with zymolyase, incubated with DNA, followed by incubation in PEG, washed and plated on a selective medium. This is a modification of protocol for *P. pastoris* transformation (Cregg et al. 1985).

1. Inoculate a colony of a freshly subcloned yeast strain in 2 mL of liquid YPS medium to be transformed. Incubate for 24 h at 30 °C with vigorous shaking. Measure A_{600}.

2. Inoculate 150–200 mL of liquid YPS medium in a 1,000 mL flask with an aliquot (usually 3–10 µL) of the pre-grown culture. We usually start 2 cultures simultanenously (2 flasks for each) with inocula of different volumes in order to guarantee one culture at the proper density at a reasonable time in the next morning. Optical density measured at 600 nm should be ≤ 0.3.

3. Pellet cells at 3,000 g for 15 min (room temperature). Discard the supernatant. Resuspend cells in 50 mL of sterile double-distilled water.

4. Pellet cells again and resuspended them in 10 mL of 1 M sucrose, 50 mM dithiothreitol, 25 mM EDTA, pH 8.0.

5. Incubate the suspension at room temperature for 15 min, again pellet cells, wash them twice with 1 M sucrose and resuspended in 1 M sucrose, 25 mM EDTA, 100 mM sodium citrate pH 5.8.

6. Add zymolyase (or lyticase), incubate for 10–40 min (30 °C with slow agitation) in order to get approximately 5–10 % of

spheroplasted cells. Quantity of an enzyme used for the treatment of cells should be adjusted before.

7. Cells are pelleted at 2,000 g for 5 min, washed twice with 1 M sucrose, once with 1 M sucrose, and 10 mM $CaCl_2$ and resuspended in the last solution to a final density 2×10^8 cells/mL. To resuspend the pelleted cells and spheroplasts slowly and repeatedly draw them into a pipette and transfer them into the tube.

8. Dispense 0.1 mL aliquots of the cells in 1.5 mL tubes and add 0.1–1 µg of transforming DNA together with 5 µg of single strand carrier DNA.

9. Incubate at 25 °C for 20 min, add 1 mL of 20 % PEG 3350, 10 mM Tris–HCl, pH 7.5, 10 mM $CaCl_2$ and immediately (gently, avoid wortexing) mix suspension.

10. Incubate the mixture for additional 15 min at 25 °C.

11. Pellet spheroplasts and cells (1,000 g for 10 min), resuspend them in 1 M sucrose, 10 mM $CaCl_2$, and incubate for 30 min at 25 °C.

12. Plate the suspension on a selective medium containing 1 M sucrose and incubate at 30 °C for 3–5 days.

21.7 Electroporation Transformation of *M. (C.) guilliermondii*

Principle: Penetration of transforming DNA is facilitated by changes in membrane structure that occured under short-time applied electrical field impulse. For yeast electroporation, a modified protocol of Becker and Guarente (1991) was used.

1. Inoculate 2 mL of liquid YPS medium with a colony of a freshly subcloned yeast strain to be transformed. Incubate for 24 h at 30 °C with vigorous shaking. Measure A_{600}.

2. Inoculate 100 mL of liquid YPS medium in a 500 mL flask with an aliquot (3–10 µL) of the pre-grown culture. We usually start three cultures simultanenously with inocula of different volumes in order to guarantee one

culture at the proper density at a reasonable time the next morning. Optical density measured at 600 nm should be ≤ 0.5.

3. Chill cell suspension in ice for 10–15 min. All subsequent manipulations should be done at 2–4 °C.

4. Pellet cells at 3,000 g for 10 min. Discard the supernatant. Resuspend cells in 2 mL of 0.1 M LiAc, 10 mM Tris–HCl, 1 mM EDTA (pH 7.5), incubate at 30 °C for 60 min.

5. Pellet cells at $3,000 \times g$ for 10 min, resuspend in 50 mL of ice-cold water, and pellet them again.

6. Resuspend cells in ice-cold 1 M sucrose (5 mL), incubate for 5–10 min and pellet them.

7. Resuspend cells in fresh ice-cold 1 M sucrose to a final density at 600 nm around 100.

8. Dispense 0.05 mL aliquots of the cells 1.5 mL tubes, add transforming DNA (0.05–0.50 µg in 1–2 µL of TE buffer) and mix gently.

9. Transfer the mixture into prechilled 1 or 2 mm electroporation cuvettes. Electroporation was performed as follows: resistance—200 Ω; capacitance—25 µF; voltage—1.8–2.5 kV.

10. Wash out the cells from the cuvettes with 1 mL of 0.3 M sucrose in YPS medium and incubate for 1 h at 30 °C.

11. Pellet the cells and resuspend in 1 M sucrose.

12. Plate on selective medium and incubate at 30 °C for 3–5 days.

Note: Steps 11 can be omitted when using dominant selective marker.

21.8 Insertional Mutagenesis of *M. (C.) guilliermondii*

Principle: As mentioned above, the linear DNA fragments integrate into the genome of transformed *M. (C.) guilliermondii* cells mostly via non-homologous recombination. This phenomenon has been applied to generate desired mutants subsequently used to facilitate identification of genes influencing the branch of metabolism studied (Boretsky et al. 2011). This approach involves several distinct steps: transformation of an appropriate strain with linear DNA fragment bearing a selective marker; screening for the desired mutants among the obtained transformants; isolation of total DNA from the selected mutants and cloning of the integrated cassette. Since there are substantial collections of selective markers and appropriate recipient strains reported (see above), the main difficulty in this approach is the design of the corresponding screening system that allows easy selection of the desired mutants.

1. Transform an appropriate *M. (C.) guilliermondii* strain with linear DNA fragment bearing a suitable selective marker. Transformation protocols are given above.

2. Screen transformants for the desired phenotype mutants. Check the selected mutants (e.g., enzyme activity, or an accumulated product, etc).

3. Isolate chromosomal DNA from the chosen mutant using protocol for yeast NucleoSpin Tissue Kit (e.g. Macherey-Nagel).

4. Determine quality and quantity of prepared samples of chromosomal DNA by agarose gel electrophoresis.

5. Digest 5 µg of the purified DNA with appropriate restriction enzymes that do not cleave the cassette in total volume of 50 µL.

6. Purify digested DNA with DNeasy Blood & Tissue Kit (e.g. Qiagen).

7. Self-ligate digested and purified DNA with 10U of T4 DNA ligase in total volume of 50 µL overnight at room temperature.

8. Purify self-ligated mixture with DNeasy Blood & Tissue Kit (e.g. Qiagen).

9. Transform 0.2 volume of purified ligation mixture into *E. coli* using electrocompetent cells.

10. Isolate plasmid DNA from *E. coli* with the Wizard® *Plus* SV Minipreps DNA Purification System (e.g. Promega, USA).

11. Perform sequencing of the isolated plasmids for identification of a locus of the insertion with primers homologous to the sequence of the insertion cassette.

Note: The method selected for *M.(C.) guilliermondii* transformation (see above) can influence on percentage of homologous integration events slightly but not significantly for this approach (Boretsky et al. 2007). Sometimes the desired mutants could not be found among the large number of transformants. In such cases it is recommended to switch to another selective marker.

21.9 Multiple Gene Disruption Using Blaster Systems in *M. (C.) guilliermondii*

Principle: Blaster transformation systems are applicable for multiple gene disruption since the selection markers used, flanked by two short repeated sequences, can easily pop-out from the target locus following a counter-selection on an antimetabolite-containing medium (Fig. 21.1). Three different blaster systems are currently available in *M. (C.) guilliermondii*: the *URA5-blaster* system (Millerioux et al. 2011b) (Fig. 21.1a), the *URA3*-blaster system (Foureau et al. 2012b) (Fig. 21.1b), and the *TRP5*-blaster system (Foureau et al. 2013a) (Fig. 21.1c). The *URA5*-blaster system was originally developed for the recipient strain NP566U (genotype *ura5*) (Fig. 21.1a). It is important to note that the *URA5* copy contained in REP-URA5-REP blaster cassette described in Millerioux et al. (2011b) naturally includes an autonomously replicating sequence (ARS) (Foureau et al. 2012a, see next section) which strongly reduces the frequency of integration of this blaster cassette the in *M. (C.) guilliermondii* genome. For gene disruption in *M. (C.) guilliermondii* it is thus better to use (i) the *URA5$^{\Delta ARS}$*-blaster cassette described in (Foureau et al. 2013b) and (ii) the KU141F1 recipient strain (genotype *ku70, ura5*) (Table 21.1) (Fig. 21.1a). This strategy allows high target gene disruption frequencies (40–100 % of transformants), but results also depend on the length of 5′ and 3′ homologous arms that flank the REP-URA5-REP cassette (500–2,000 bp each are convenient).

The two other blaster systems, the *URA3-blaster* system (Foureau et al. 2012b) (Fig. 21.1b) and the *TRP5*-blaster system (Foureau et al. 2013a) (Fig. 21.1c), are also convenient but the gene targeting frequencies do not exceed

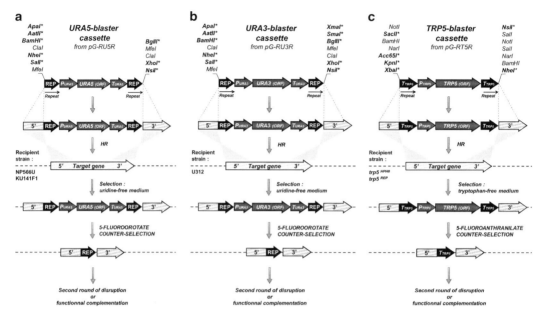

Fig. 21.1 The different blaster systems are currently available in *M. (C.) guilliermondii*: (**a**) the *URA5*-blaster system, (**b**) the *URA3*-blaster system, and (**c**) the *TRP5*-blaster system. The *URA5*-blaster system was originally developed for the recipient strain NP566U (genotype *ura5*). It is important to note that the *URA5* copy contained in REP-URA5-REP blaster cassette described in Millerioux et al. (2011b) naturally includes an autonomously replicating sequence (ARS) which strongly reduces the frequency of integration of this blaster cassette the *M. (C.) guilliermondii* genome. It is thus better for gene disruption in *M. (C.) guilliermondii* to use (1) the *URA5*$^{\Delta ARS}$-blaster cassette described in (Foureau et al. 2013b) and (2) the KU141F1 recipient strain (genotype *ku70, ura5*). This strategy allows high target gene disruption frequencies (40–100 % of transformants), but results also depend on the length of 5′ and 3′ homologous arms that flank the REP-URA5-REP cassette (500–2,000 bp each are convenient). The two remaining blaster systems, the *URA3*-blaster system and the *TRP5*-blaster system, are also convenient but the gene targeting frequencies do not exceed 15–25 %. The loss of the *URA5* or the *URA3* marker at the target loci can be obtained after a counter-selection on 5-fluoroorotate/uridine containing minimal medium plates. The loss of the *TRP5* marker at the target loci can be obtained after a counter-selection on 5-fluoroanthranilate/tryptophan containing minimal medium plates. *HR* Homologous recombination

15–25 % due to the lack of *KU70* mutation in the proposed recipient strains (Table 21.1).

21.9.1 Reagents

1. Sterile distilled water.
2. dNTPs—a mixture of dATP, dCTP, dGTP, and dTTP (10 mM each, stored at −20 °C).
3. DNA polymerase kit (e.g. Phusion from Fermentas). All reagents are stored at −20 °C.
4. Oligonucleotide primers (custom-made). Stored at −20 °C.
5. DNA clean-up kit of your choice.
6. Yeast genomic DNA purification kit of your choice.
7. Alkaline phosphatase kit of your choice.
8. T4 DNA ligase kit of your choice.
9. *E coli* (TOP10 or XL1Blue for example) competent cells.
10. Ampicillin (100 μg/mL) containing Luria Bertani plates.
11. Yeast nitrogen base powder.
12. L-tryptophan.
13. Uridine.
14. Tryptophan.

21 Genetic Manipulation of *Meyerozyma guilliermondii*

15. 5-Fluoroorotate.
16. 5-Fluoroanthranilate.
17. Toyn medium (Toyn et al. 2000).
18. Plasmid miniprep extraction kit of your choice.
19. Agarose.
20. TAE buffer—50 mM Tris, 50 mM Acetic acid, 1 mM EDTA.
21. Ethidium bromide—0.5 mg/mL stock.
22. Gel loading mixture—40 % (w/v) sucrose, 0.1 M EDTA, 0.15 mg/mL bromophenol blue.

21.9.2 Equipments

1. Vortexer.
2. Thermoregulatable water bath.
3. Microcentrifuge.
4. Disposable polypropylene microcentrifuge tubes 1.5 mL conical.
5. PCR tubes (thin walled).
6. Thermal cycler.
7. Horizontal electrophoresis equipment.

21.9.3 Protocol

1. Extract and purify genomic DNA of *M. (C.) guilliermondii* ATCC6260 strain using a standard yeast genomic DNA purification kit.
2. Check the quantity of extracted genomic DNA by loading a 5 µL of the purified sample on a standard agarose/ethidium bromide gel electrophoresis.
3. Design and synthesize couple of primers allowing the amplification by PCR of a 5′ and a 3′ 0.5–2 kb fragment overlapping the promoter and the terminator sequences of the target gene. Include a suitable restriction endonuclease adaptor in these primers for subsequent cloning of the 5′ and 3′ fragments on both side of the selected blaster plasmid (i.e. pG-blaster: pG-RU5R, pG-RU3R or pGRT5R, Fig. 21.1).
4. Amplify following a standard PCR protocol the 5′ and 3′ 0.5–2 kb fragments overlapping the promoter and the terminator sequences of the target gene in a final reaction volume of 50 µL. Use *M. (C.) guilliermondii* ATCC6260 strain's genomic DNA as PCR matrix.
5. Check the PCR amplification by loading a 2 µL of each sample on a standard agarose/ethidium bromide gel electrophoresis.
6. Purify the two PCR samples (5′ PCR fragment and 3′ PCR fragment of the target gene) using a standard DNA purification kit.
7. Use appropriate endonucleases to digest each adaptor at both sides of the 5′ fragment of the target gene. Use the same (or compatible) appropriate endonucleases to digest the selected blaster plasmid (pG-RU5R, pG-RU3R or pGRT5R, Fig. 21.1). If a unique restriction site is used in the plasmid for cloning, dephosphorylate the digested plasmid with standard alkaline phosphatase kit to prevent self-ligation of the vector.
8. Purify the digested 5′ fragment and the digested blaster plasmid using a standard DNA purification kit.
9. Check the DNA digestion by loading a 2 µL of each sample on a standard agarose/ethidium bromide gel electrophoresis.
10. Ligate the digested 5′ fragment and the digested blaster plasmid using a standard T4 DNA ligase kit.
11. Transform *E. coli* (TOP10 or XL1Blue for example) competent cells with the ligation sample. Pour cells onto Ampicillin (100 µg/mL) containing Luria Bertani plates. Incubate plates at 37 °C for 16 h.
12. Screen correct insertion of the 5′ fragment in the blaster plasmid from a set of bacterial clones using standard colony PCR or differential endonuclease restriction after plasmid extraction.
13. Select a positive bacterial clone. Use a plasmid miniprep extraction kit of your choice to obtain 10–20 µg of this purified plasmid named pG-5′gene-blaster.
14. Use appropriate endonucleases to digest each adaptor at both sides of the 3′ fragment of the target gene. Use the same (or compatible) appropriate endonucleases to digest the

selected pG-5′gene-blaster. If a unique restriction site is used in the plasmid for cloning, dephosphorylate the digested pG-5′gene-blaster with standard alkaline phosphatase kit to prevent self-ligation of the vector.

15. Purify the digested 3′ fragment and the digested pG-5′gene-blaster plasmid using a standard DNA purification kit.

16. Check the digestion by loading a 2 μL of each sample on a standard agarose/ethidium bromide gel electrophoresis.

17. Ligate the digested 3′ fragment and the digested pG-5′gene-blaster plasmid using a standard T4 ligase kit.

18. Transform *E. coli* (TOP10 or XL1Blue for example) competent cells with the ligation sample. Pour cells onto Ampicillin (100 μg/mL) containing Luria Bertani plates. Incubate plates 16 h at 37 °C.

19. Screen correct insertion of the 3′ fragment in the pG-5′gene-blaster plasmid from a set of bacterial clones using standard colony PCR or differential endonuclease restriction after plasmid extraction.

20. Select a positive bacterial clone. Use a plasmid miniprep extraction kit of your choice to obtain 10–20 μg of this purified plasmid named pG-5′gene-blaster-3′gene.

21. Amplify following a standard PCR protocol the whole 5′gene-blaster-3′gene disruption cassette in a final reaction volume of 50 μL. Use pG-5′gene-blaster-3′gene as PCR matrix, the forward primer of the 5′ fragment and the reverse primer of the 3′ fragment.

22. Check the PCR amplification by loading a 2 μL of the sample on a standard agarose/ethidium bromide gel electrophoresis.

23. Purify the PCR sample using a standard DNA purification kit.

24. Use 5–10 μg of this purified disruption cassette to transform the appropriate *M. (C.) guilliermondii* recipient strain using one of the protocols described in this chapter. After transformation, pour cells onto selective medium (CSM-URA if *URA5* or *URA3* markers are used and CSM-TRP if *TRP5* is used).

25. Screen correct insertion by gene replacement of the disruption cassette at the target locus using standard yeast colony PCR.

26. To select Ura⁻ cells derived from *URA3* or *URA5* markers loss, allow to grow a previously selected target gene knocked-out clone overnight in YPS liquid medium. Collect cells (3,800 g, 10 min), wash with sterile water, serially dilute in water and finally spot cells onto YNB petri dishes supplemented with 200 μg/mL uridine and 1 mg/mL 5-fluoroorotate.

27. To recover Trp⁻ cells derived from the *TRP5* cassette loop-out, allow to grow a previously selected target gene knocked-out clone overnight in YPS liquid medium supplemented with 200 μg/mL tryptophan. Dilute the culture (1:20) in the same medium and allow to grow overnight. Collect cells by centrifugation at $3,800 \times g$ for 10 min, wash with sterile water, serially dilute in water and finally spread cells on Toyn medium plates (Toyn et al. 2000) supplemented with 600 μg/mL 5-fluoroanthranilate and 40 μg/mL tryptophan. After 5 days of growth, isolate colonies by streaking onto the same selection medium and allow growing for 2 days to confirm FAA resistance. All *5-fluoroanthranilate*-resistant clones have to be replicated on YNB alone and YNB supplemented with 200 μg/mL tryptophan to screen for tryptophan auxotrophy.

21.10 A 60-BP DNA Tag That Confers Autonomous Replication Ability to DNAS in *M. (C.) guilliermondii*

Principle: Autonomously replicating sequences (ARS) correspond to fungal replication origins. They found a broad range of application in yeast genetics by supporting plasmid maintenance in the cells. Recent advances allowed the identification of a 60-bp ARS within an A/T rich region located upstream of the *URA5* open reading frame (ORF) (Foureau et al. 2012a) (Fig. 21.2). Linear double-strand DNAs containing this putative ARS are circularized in the endogenous

21 Genetic Manipulation of *Meyerozyma guilliermondii*

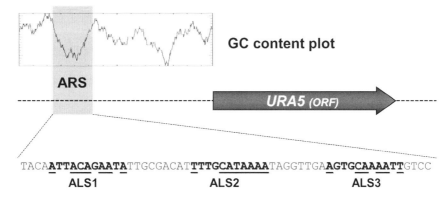

Fig. 21.2 Identification of an ARS located upstream of the *M.(C.) guilliermondii* URA5 ORF. Schematic representation of the *URA5* locus. The GC content plot was obtained with the ISOCHORE program. The minimal sequence allowing circularization/autonomous replication of DNAs in *M. (C.) guilliermondii*. This includes a combination of three ARS-like sequences (ALS1, ALS2, and ALS3)

M. (C.) guilliermondii Lig4p ligase and then autonomously replicated in transformed cells. In addition, simple integration of this ARS in a backbone plasmid confers autonomous replication ability to this latter.

This ARS sequence represents a powerful DNA tag to easily and efficiently confer autonomous replication ability to all *M. (C.) guilliermondii* molecular tools. The simplest methods to integrate this ARS in molecular constructs of *M. (C.) guilliermondii* remains the addition of the 60-bp as a flag (full sequence provided in Fig. 21.2) in one of the primers used for the construction of the plasmid.

21.10.1 Reagents

1. Sterile distilled water.
2. dNTPs—a mixture of dATP, dCTP, dGTP, and dTTP (10 mM each, stored at −20 °C).
3. DNA polymerase kit (e.g. Phusion from Fermentas). All reagents are stored at −20 °C.
4. Oligonucleotide primers (custom-made). Stored at −20 °C.
5. DNA clean-up kit of your choice.
6. Yeast genomic DNA purification kit of your choice.
7. Alkaline phosphatase kit of your choice.
8. T4 ligase kit of your choice.
9. *E. coli* (TOP10 or XL1Blue for example) competent cells.
10. Ampicillin (100 µg/mL) containing Luria Bertani plates.
11. Yeast nitrogen base powder.
12. Plasmid miniprep extraction kit of your choice.
13. Agarose.
14. TAE buffer—50 mM Tris, 50 mM Acetic acid, 1 mM EDTA.
15. Ethidium bromide—0.5 mg/mL stock.
16. Gel loading mixture—40 % (w/v) sucrose, 0.1 M EDTA, 0.15 mg/mL bromophenol blue.
17. Sorbitol.
18. Sodium phosphate 0.1 M, pH 6.5.
19. β-Mercaptoethanol.
20. Zymolase.
21. NaCl.
22. Tris–HCl.
23. EDTA.
24. SDS.
25. Glass beads.
26. Phenol.
27. Chloroform.

21.10.2 Equipments

1. Microcentrifuge.
2. Vortexer.
3. Disposable polypropylene microcentrifuge tubes 1.5 mL conical flask.
4. Thermal cycler.
5. Horizontal electrophoresis equipment.

21.10.3 Protocol

1. Design and synthesize couple of primers allowing the amplification by PCR of one of the elements of the plasmid you want to construct. Do not omit to include in 5′ end of one of your primer the whole 60-bp sequence given in Fig. 21.2. In addition, do not omit to include at the extremities of both primers suitable restriction endonuclease adaptors for subsequent cloning of the ARS-containing DNA fragment in the target plasmid.

2. Amplify following a standard PCR protocol the DNA fragments using an appropriate DNA matrix in a final reaction volume of 50 μL.

3. Check the PCR amplification by loading a 2 μL of each sample on a standard agarose/ethidium bromide gel electrophoresis.

4. Purify the PCR sample using a standard DNA purification kit.

5. Use appropriate endonucleases to digest each adaptor on both sides of the DNA fragment. Use the same (or compatible) appropriate endonucleases to digest the selected recipient plasmid. If a unique restriction site is used in the plasmid for cloning, dephosphorylate the digested plasmid with standard alkaline phosphatase kit to prevent self-ligation of the vector.

6. Purify the digested DNA fragment and the digested recipient plasmid using a standard DNA purification kit.

7. Check the digestion occurred by loading a 2 μL of each sample on a standard agarose/ethidium bromide gel electrophoresis.

8. Ligate the digested DNA fragment and the digested recipient plasmid using a standard T4 DNA ligase kit.

9. Transform *E. coli* (TOP10 or XL1Blue for example) competent cells with the ligation sample. Pour cells onto antibiotic-containing (compatible with the selection of the plasmid used) Luria Bertani plates. Incubate plates at 37 °C for 16 h.

10. Screen correct insertion of the DNA fragment in the recipient plasmid from a set of bacterial clones using standard colony PCR or differential endonuclease restriction after plasmid extraction.

11. Select a positive bacterial clone. Use a plasmid miniprep extraction kit of your choice to obtain 10–20 μg of this purified ARS-containing plasmid.

12. The efficiency of transformation of *M. (C.) guilliermondii* cells with this ARS-containing plasmid must be 50–1,000-fold higher compared to the efficiency of transformation of *M. (C.) guilliermondii* cells with the same backbone plasmid that do not contain the ARS sequence.

13. Extraction of circular DNA from these *M. (C.) guilliermondii* transformed strains can be performed with the following procedure.

14. Allow growing transformed yeast cells overnight in liquid selective medium (150 rpm, 30 °C) to stationary phase.

15. Harvest cells by centrifugation at $3,800 \times g$ for 10 min and suspend in 1 mL of YCWDB buffer (sorbitol 1 M, sodium phosphate 0.1 M, pH 6.5, β-mercaptoethanol 1 %, zymolase 0.5 mg/mL) and incubate for 1 h at 37 °C.

16. Spheroplasts are harvested by centrifugation at $3,800 \times g$ for 10 min and resuspended in 600 μL of lysis buffer (100 mM NaCl, 10 mM Tris–HCl, pH 8.0, 1 mM EDTA, 0.1 % SDS) to which is added twelve glass beads.

17. The mixture is vortexed vigorously for 1 min and 600 μL of phenol chloroform (1:1) solution is added.

18. The mixture is vigorously vortexed 1 min and centrifuged. The 400 μL upper aqueous phase was purified using a standard plasmid miniprep extraction kit.

19. Check the quantity of extracted autonomously replicated plasmid by loading a 5 μL of the purified sample on a standard agarose/ethidium bromide gel electrophoresis.

21.11 Fluorescent Protein Fusion in *M. (C.) guilliermondii*

Principle: Reporter genes such as β-glucuronidase, β-galactosidase, chloramphenicol acetyltransferases, luciferases, or fluorescent proteins are now widely used to study many aspects of gene regulation, signal-transduction pathways, and other cellular processes.

Fluorescent proteins are notably powerful tools in cell biology since they are often used for monitoring both protein localization/interaction and promoter activity in a large range of fungal species. With the goal to provide practical molecular tools usable in cell biology studies in *M. (C.) guilliermondii*, a recipient strain (leu2[REP]) (Table 21.1) as well as a complete and practical set of plasmids (Fig. 21.3a) for construction of fluorescent protein fusions was recently developed (Courdavault et al. 2011). In this set of plasmids, the strong *PGK* transcription-regulating sequences allow a constitutive expression of codon-optimized GFP, YFP, CFP,

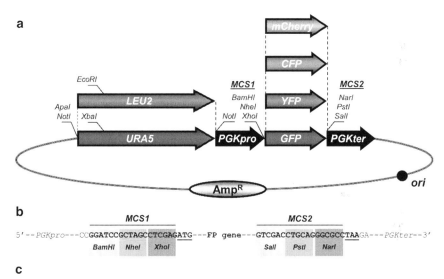

Fig. 21.3 Global strategy for fluorescent protein fusion in *M. (C.) guilliermondii*. (**a**) Representation of the set of plasmids available for fluorescent protein fusions. (**b**) To facilitate molecular manipulations, each fluorescent protein gene is bordered by multiple cloning sites: *BamHI, NheI, and XhoI* for C-terminal fusion or *SalI, PstI,* and *NarI* for N-terminal fusion. The start codon ATG of the fluorescent protein and the stop codon TAA are underlined. The selectable marker (here *URA5* or *LEU2*) is bordered by the rare restriction site *NotI* that allows easy exchange. (**c**) Fluorescent markers of various subcellular compartments (plasma membrane, peroxisomes, mitochondria, vacuole, nucleus, and cytosol) available in *M. (C.) guilliermondii* for dual labeling co-expression and co-localization experiments with YFP/CFP and GFP/mCherry pairs

and mCherry genes in *M. (C.) guilliermondii* cells (Fig. 21.3a). In addition, the gene fusion can be made in N-terminus or C-terminus depending on the presence of multiple cloning sites that border the fluorescent protein gene sequence (Fig. 21.3b). Finally, a series of fluorescent markers of various subcellular compartments (plasma membrane, peroxisomes, mitochondria, vacuole, nucleus, and cytosol) of *C. guilliermondii* is also available (Fig. 21.3c). In this way, the co-expression and co-localization experiments of YFP/CFP and GFP/mCherry pairs can be used for dual labeling in *M. (C.) guilliermondii*.

21.11.1 Reagents

1. Sterile distilled water.
2. dNTPs—a mixture of dATP, dCTP, dGTP, and dTTP (10 mM each, stored at −20 °C).
3. DNA polymerase kit (e.g. Phusion from Fermentas). All reagents are stored at −20 °C.
4. Oligonucleotide primers (custom-made). Stored at −20 °C.
5. DNA clean-up kit of your choice.
6. Yeast genomic DNA purification kit of your choice.
7. Alkaline phosphatase kit of your choice.
8. T4 DNA ligase kit of your choice.
9. *E. coli* (TOP10 or XL1Blue for example) competent cells.
10. Ampicillin (100 µg/mL) containing Luria Bertani plates.
11. Yeast nitrogen base powder.
12. L-leucine.
13. Uridine.
14. Plasmid miniprep extraction kit of your choice.
15. Agarose.
16. TAE buffer—50 mM Tris, 50 mM Acetic acid, 1 mM EDTA.
17. Ethidium bromide—0.5 mg/mL stock.
18. Gel loading mixture—40 % (w/v) sucrose, 0.1 M EDTA, 0.15 mg/mL bromophenol blue.

21.11.2 Equipments

1. Vortexer.
2. Thermoregulatable water bath.
3. Microcentrifuge.
4. Disposable polypropylene microcentrifuge tubes 1.5 mL conical.
5. PCR tubes (thin walled).
6. Thermal cycler.
7. Horizontal electrophoresis equipment for agarose gels.
8. Epifluorescence microscope with set of filters.

21.11.3 Protocol

21.11.3.1 Fusion of the Fluorescent Protein in C-terminus of the Target Gene

1. Extract and purify genomic DNA of *M. (C.) guilliermondii* ATCC6260 strain using a standard yeast genomic DNA purification kit.
2. Check the quality the and quantity of extracted genomic DNA by loading a 5 µL of the purified sample on a standard agarose/ethidium bromide gel electrophoresis.
3. Design and synthesize couple of primers allowing the amplification by PCR of the target gene. To avoid the disruption of the open reading frame of the fusion protein, the forward primer must contain in 5′ a six nucleotide restriction site (*Bam*HI, *Nhe*I, *Xho*I or compatible sites, Fig. 21.3b) followed by the 24th first nucleotide (including the ATG start codon) of the target gene. The reverse primer must contain in 5′ a six nucleotide restriction site (*Bam*HI, *Nhe*I, *Xho*I or compatible sites, Fig. 21.3b) followed by the 24th ultimate nucleotide (do not include the stop codon) of the target gene. This will make easier the subsequent cloning of the target gene in the selected fluorescent plasmid (Fig. 21.3a).
4. Amplify following a standard PCR protocol the target gene DNA in a final reaction volume of 50 µL. Use *M. (C.) guilliermondii* ATCC6260 strain genomic DNA as PCR matrix.

21 Genetic Manipulation of *Meyerozyma guilliermondii*

5. Check the PCR amplification by loading a 2 µL of each sample on a standard agarose/ethidium bromide gel electrophoresis.

6. Purify the PCR sample using a standard DNA purification kit.

7. Use appropriate endonucleases to digest each adaptor at both sides of the target gene. Use the same (or compatible) appropriate endonucleases to digest the selected fluorescent plasmid (Fig. 21.3a). If a unique restriction site is used in the plasmid for cloning, dephosphorylate the digested plasmid with standard alkaline phosphatase kit to prevent self-ligation of the vector.

8. Purify the digested target gene DNA and the digested fluorescent plasmid using a standard DNA purification kit.

9. Check the DNA digestion by loading a 2 µL of each sample on a standard agarose/ethidium bromide gel electrophoresis.

10. Ligate the digested 5′ fragment and the digested plasmid using a standard T4 ligase kit.

11. Transform *E. coli* (TOP10 or XL1Blue for example) competent cells with the ligation sample. Pour cells onto Ampicillin (100 µg/mL) containing Luria Bertani plates. Incubate plates at 37 °C for 16 h.

12. Screen correct insertion of the target DNA in the fluorescent plasmid from a set of bacterial clones using standard colony PCR or differential endonuclease restriction after plasmid extraction.

13. Select a positive bacterial clone. Use a plasmid miniprep extraction kit of your choice to obtain 10–20 µg of this purified plasmid.

14. Use 2–5 µg of this purified plasmid to transform the *leu2^REP* (Table 21.1) *M. (C.) guilliermondii* recipient strain using one of the protocols described in this chapter. After transformation, transfer cells onto selective medium (YNB + leucine 100 µg/mL if *URA5* marker is used and YNB + uridine 200 µg/mL if *LEU2* is used). Incubate plates at 30 °C for 3 days.

15. After 3 days of growth on selective medium, fluorescent protein expressing could be directly identified by epifluorescence microcopy using the appropriate filter.

21.11.3.2 Fusion of the Fluorescent Protein in N-terminus of the Target Gene

1. Extract and purify genomic DNA of *M. (C.) guilliermondii* ATCC6260 strain using a standard yeast genomic DNA purification kit.

2. Check the quality and quantity of extracted genomic DNA by loading a 5 µL of the purified sample on a standard agarose/ethidium bromide gel electrophoresis.

3. Design and synthesize couple of primers allowing the amplification by PCR of the target gene. To avoid the disruption of the open reading frame of the fusion protein, the forward primer must contain in 5′ a six nucleotide restriction site (*Sal*I, *Pst*I, *Nar*I or compatible sites, Fig. 21.3b) followed by the 24th first nucleotide (including the ATG start codon) of the target gene. The reverse primer must contain in 5′ a six nucleotide restriction site (*Sal*I, *Pst*I, *Nar*I or compatible sites, Fig. 21.3b) followed by the 24th ultimate nucleotide (do not include the stop codon) of the target gene. This will make easier the subsequent cloning of the target gene in the selected fluorescent plasmid (Fig. 21.3a).

4. Amplify following a standard PCR protocol the target gene DNA in a final reaction volume of 50 µL. Use *M. (C.) guilliermondii* ATCC6260 strain genomic DNA as PCR matrix.

5. Check the PCR amplification by loading a 2 µL of each sample on a standard agarose/ethidium bromide gel electrophoresis.

6. Purify the PCR sample using a standard DNA purification kit.

7. Use appropriate endonucleases to digest each adaptor at both sides of the target gene. Use the same (or compatible) appropriate endonucleases to digest the selected fluorescent plasmid (Fig. 21.3a). If a unique restriction site is used in the plasmid for cloning, dephosphorylate the digested plasmid with standard alkaline phosphatase kit to prevent self-ligation of the vector.

8. Purify the digested target gene DNA and the digested fluorescent plasmid using a standard DNA purification kit.

9. Check the DNA digestion by loading a 2 μL of each sample on a standard agarose/ethidium bromide gel electrophoresis.
10. Ligate the digested 5′ fragment and the digested plasmid using a standard T4 ligase kit.
11. Transform *E. coli* (TOP10 or XL1Blue for example) competent cells with the ligation sample. Pour cells onto Ampicillin (100 μg/mL) containing Luria Bertani plates. Incubate plates 16 h at 37 °C.
12. Screen correct insertion of the target DNA in the fluorescent plasmid from a set of bacterial clones using standard colony PCR or differential endonuclease restriction after plasmid extraction.
13. Select a positive bacterial clone. Use a plasmid miniprep extraction kit of your choice to obtain 10–20 μg of this purified plasmid.
14. Use 2–5 μg of this purified plasmid to transform the *leu2^REP* (Table 21.1) *M. (C.) guilliermondii* recipient strain using one of the protocols described in this chapter. After transformation, pour cells onto selective medium (YNB + leucine 100 μg/mL if *URA5* marker is used and YNB + uridine 200 μg/mL if *LEU2* is used). Incubate plates at 30 °C for 3 days.
15. After 3 days of growth on selective medium, fluorescent protein expressing could be directly identified by epifluorescence microcopy using the appropriate filter.

21.11.3.3 Dual Fluorescent Labeling

1. Select a proper marker for a particular compartment as listed in Fig. 21.3c. Make sure to use the complementary metabolic marker (URA5/LEU2) as well as the opposite fluorescent protein: the co-expression and co-localization experiments must use YFP/CFP or GFP/mCherry pairs when using an epifluorescence microspcope. For example, if your preliminary experiments show that the studied protein-YFP fusion (constructed in the pG-LEU2-YFP) seems to be localized in the nucleus, use the pG-URA5-CFP-SKN7 as compartment marker plasmid.

2. Use 2–5 μg of the plasmid with appropriate marker for the complementation to transform one previously obtained yeast fluorescent clone of *M. (C.) guilliermondii* (obtained following section A or B) using one of the protocols described in this chapter. After transformation, pour cells onto selective medium (YNB alone). Incubate plates at 30 °C for 3 days.

3. After 3 days of growth on selective medium, fluorescent protein expressing could be directly identified by epifluorescence microcopy using the appropriate filter.

References

Abbas CA, Sibirny AA (2011) Genetic control of biosynthesis and transport of riboflavin and flavin nucleotides and construction of robust biotechnological producers. Microbiol Mol Biol Rev 75:321–360

Bai FY (1996) Separation of *Candida fermentati* comb nov from *Candida guilliermondii* by DNA base composition and electrophoretic karyotyping. Syst Appl Microbiol 19:178–181

Becker DM, Guarente L (1991) High-efficiency transformation of yeast by electroporation. Methods Enzymol 194:182–187

Boretskii IR, Petrishin AV, Kriger K, Rikhter G, Fedorovich DV, Bakher A (2002) Cloning and expression of a gene encoding riboflavin synthase of the yeast *Pichia guilliermondii*. Tsitol Genet 36:3–7, Russian

Boretsky YR, Kapustyak KY, Fayura LR, Stasyk OV, Stenchuk MM, Bobak YP, Drobot LB, Sibirny AA (2005) Positive selection of mutants defective in transcriptional repression of riboflavin synthesis by iron in the flavinogenic yeast *Pichia guilliermondii*. FEMS Yeast Res 5:829–837

Boretsky YR, Pynyaha YV, Boretsky VY, Kutsyaba VI, Protchenko OV, Philpott CC, Sibirny AA (2007) Development of a transformation system for gene knock-out in the flavinogenic yeast *Pichia guilliermondii*. J Microbiol Meth 70:13–19

Boretsky Y, Voronovsky A, Liuta-Tehlivets O, Hasslacher M, Kohlwein SD, Shavlovsky GM (1999) Identification of an ARS element and development of a high efficiency transformation system for *Pichia guilliermondii*. Curr Genet 36:215–221

Boretsky YR, Pynyaha YV, Boretsky VY, Fedorovych DV, Fayura LR, Protchenko O, Philpott CC, Sibirny AA (2011) Identification of the genes affecting the regulation of riboflavin synthesis in the flavinogenic yeast *Pichia guilliermondii* using insertion mutagenesis. FEMS Yeast Res 11:307–314

Butler G, Rasmussen MD, Lin MF, Santos MA, Sakthikumar S, Munro CA, Rheinbay E, Grabherr M, Forche A, Reedy JL, Agrafioti I, Arnaud MB, Bates S, Brown AJ, Brunke S, Costanzo MC, Fitzpatrick DA, de Groot PW, Harris D, Hoyer LL, Hube B, Klis FM, Kodira C, Lennard N, Logue ME, Martin R, Neiman AM, Nikolaou E, Quail MA, Quinn J, Santos MC, Schmitzberger FF, Sherlock G, Shah P, Silverstein KA, Skrzypek MS, Soll D, Staggs R, Stansfield I, Stumpf MP, Sudbery PE, Srikantha T, Zeng Q, Berman J, Berriman M, Heitman J, Gow NA, Lorenz MC, Birren BW, Kellis M, Cuomo CA (2009) Evolution of pathogenicity and sexual reproduction in eight Candida genomes. Nature 459:657–662

Canettieri EV, Almeida e Silva JB, Felipe MG (2001) Application of factorial design to the study of xylitol production from eucalyptus hemicellulosic hydrolysate. Appl Biochem Biotechnol 94:159–168

Carvalho W, Silva SS, Converti A, Vitolo M (2002) Metabolic behavior of immobilized *Candida guilliermondii* cells during batch xylitol production from sugarcane bagasse acid hydrolyzate. Biotechnol Bioeng 79:165–169

Castellani A (1912) Observations on the fungi found in tropical bronchomycosis. Lancet 1:13–15

Coda R, Rizzello CG, Di Cagno R, Trani A, Cardinali G, Gobbetti M (2013) Antifungal activity of *Meyerozyma guilliermondii*: identification of active compounds synthesized during dough fermentation and their effect on long-term storage of wheat bread. Food Microbiol 33:243–251

Courdavault V, Millerioux Y, Clastre M, Simkin AJ, Marais E, Creche J, Giglioli-Guivarc'h N, Papon N (2011) Fluorescent protein fusions in *Candida guilliermondii*. Fungal Genet Biol 48:1004–1011

Cregg JM, Barringer KJ, Hessler AY, Madden KR (1985) Pichia pastoris as a host system for transformations. Mol Cell Biol 5:3376–3385

Desnos-Ollivier M, Ragon M, Robert V, Raoux D, Gantier JC, Dromer F (2008) *Debaryomyces hansenii* (*Candida famata*), a rare human fungal pathogen often misidentified as *Pichia guilliermondii* (*Candida guilliermondii*). J Clin Microbiol 46:3237–3242

Fitzpatrick DA, Logue ME, Stajich JE, Butler G (2006) A fungal phylogeny based on 42 complete genomes derived from supertree and combined gene analysis. BMC Evol Biol 6:99

Foureau E, Clastre M, Millerioux Y, Simkin AJ, Cornet L, Dutilleul C, Besseau S, Marais E, Melin C, Guillard J, Creche J, Giglioli-Guivarc'h N, Courdavault V, Papon N (2012a) A TRP5/5-fluoroanthranilic acid counterselection system for gene disruption in *Candida guilliermondii*. Curr Genet 58:245–254

Foureau E, Courdavault V, Navarro Gallón SM, Besseau S, Simkin AJ, Crèche J, Atehortùa L, Giglioli-Guivarc'h N, Clastre M, Papon N (2013a) Characterization of an autonomously replicating sequence in *Candida guilliermondii*. Microbiol Res 168:580–588

Foureau E, Courdavault V, Rojas LF, Dutilleul C, Simkin AJ, Crèche J, Atehortùa L, Giglioli-Guivarc'h N, Clastre M, Papon N (2013b) Efficient gene targeting in a Candida guilliermondii non-homologous end-joining pathway-deficient strain. Biotechnol Lett 35:1035–1043

Foureau E, Courdavault V, Simkin AJ, Pichon O, Creche J, Giglioli-Guivarc'h N, Clastre M, Papon N (2012b) Optimization of the URA-blaster disruption system in *Candida guilliermondii*: efficient gene targeting using the *URA3* marker. J Microbiol Meth 91:117–120

Foureau E, Courdavault V, Simkin AJ, Sibirny AA, Crèche J, Giglioli-Guivarc'h N, Clastre M, Papon N (2013c) Transformation of *Candida guilliermondii* wild-type strains using the Staphylococcus aureus MRSA 252 *ble* gene as a phleomycin-resistant marker. FEMS Yeast Res 13:354–358

Guo N, Gong F, Chi Z, Sheng J, Li J (2009) Enhanced inulinase production in solid state fermentation by a mutant of the marine yeast *Pichia guilliermondii* using surface response methodology and inulin hydrolysis. J Ind Microbiol Biotechnol 36:499–507

Kreger van Rij NJW (1970) In: Lodder J (ed) The yeasts. A taxonomic study, 2nd edn. North-Holland Publ, Amsterdam, pp 455–458

Kurtzman CP, Suzuki M (2010) Phylogenetic analysis of ascomycete yeasts that form coenzyme Q-9 and the proposal of the new genera Babjeviella, Meyerozyma, Millerozyma, Priceomyces, and Scheffersomyces. Mycoscience 51:2–14

Liauta-Teglivets O, Hasslacher M, Boretskii IR, Kohlwein SD, Shavlovskii GM (1995) Molecular cloning of the GTP-cyclohydrolase structural gene RIB1 of Pichia guilliermondii involved in riboflavin biosynthesis. Yeast 11:945–952

Millerioux Y, Clastre M, Simkin AJ, Courdavault V, Marais E, Sibirny AA, Noel T, Creche J, Giglioli-Guivarc'h N, Papon N (2011a) Drug-resistant cassettes for the efficient transformation of *Candida guilliermondii* wild-type strains. FEMS Yeast Res 11:457–463

Millerioux Y, Clastre M, Simkin AJ, Marais E, Sibirny AA, Noel T, Creche J, Giglioli-Guivarc'h N, Papon N (2011b) Development of a URA5 integrative cassette for gene disruption in the Candida guilliermondii ATCC 6260 strain. J Microbiol Meth 84:355–358

Papon N, Courdavault V, Clastre M, Bennett RJ (2013a) Emerging and emerged pathogenic Candida species: beyond the Candida albicans paradigm. PLoS Pathog 9(9):e1003550

Papon N, Savini V, Lanoue A, Simkin AJ, Crèche J, Giglioli-Guivarc'h N, Clastre M, Courdavault V, Sibirny AA (2013b) *Candida guilliermondii*: biotechnological applications, perspectives for biological control, emerging clinical importance and recent advances in genetics. Curr Genet 59:73–90

Pfaller MA, Diekema DJ (2007) Epidemiology of invasive candidiasis: a persistent public health problem. Clin Microbiol Rev 20:133–163

Pfaller MA, Diekema DJ, Gibbs DL, Newell VA, Ellis D et al (2010) Results from the ARTEMIS DISK Global Antifungal Surveillance Study, 1997 to 2007: a 10.5-year analysis of susceptibilities of Candida Species to fluconazole and voriconazole as determined by CLSI standardized disk diffusion. J Clin Microbiol 48:1366–1377

Piniaga IV, Prokopiv TM, Petrishin AV, Khalimonchuk OV, Protchenko OV, Fedorovich DV, Boretskiĭ IR (2002) The reversion of *Pichia guilliermondii* transformants to the wild-type phenotype. Mikrobiologiia 71:368–372, Russian

Pynyaha YV, Boretsky YR, Fedorovych DV, Fayura LR, Levkiv AI, Ubiyvovk VM, Protchenko OV, Philpott CC, Sibirny AA (2009) Deficiency in frataxin homologue YFH1 in the yeast *Pichia guilliermondii* leads to missregulation of iron acquisition and riboflavin biosynthesis and affects sulfate assimilation. Biometals 22:1051–1061

Reedy JL, Floyd AM, Heitman J (2009) Mechanistic plasticity of sexual reproduction and meiosis in the *Candida* pathogenic species complex. Curr Biol 19:891–899

Rodrigues RC, Sene L, Matos GS, Roberto IC, Pessoa A Jr, Felipe MG (2006) Enhanced xylitol production by precultivation of Candida guilliermondii cells in sugarcane bagasse hemicellulosic hydrolysate. Curr Microbiol 53:53–59

San Millan RM, Wu LC, Salkin IF, Lehmann PF (1997) Clinical isolates of Candida guilliermondii include Candida fermentati. Int J Syst Bacteriol 47:385–393

Savini V, Catavitello C, Onofrillo D, Masciarelli G, Astolfi D, Balbinot A, Febbo F, D'Amario C, D'Antonio D (2011) What do we know about *Candida guilliermondii*? A voyage throughout past and current literature about this emerging yeast. Mycoses 54:434–441

Schirmer-Michel AC, Flôres SH, Hertz PF, Matos GS, Ayub MA (2008) Production of ethanol from soybean hull hydrolysate by osmotolerant *Candida guilliermondii* NRRL Y-2075. Bioresour Technol 99:2898–2904

Sibirny AA (1996) Chapter VII. *Pichia guilliermondii*. In: Wolf K (ed) Nonconvential yeasts in biotechnology. Springer, Berlin, pp 255–272

Sibirnyi AA, Shavlovskii GM, Ksheminskaya GP, Orlovskaya AG (1977) Active transport of riboflavin in the yeast *Pichia guilliermondii*. Detection and some properties of the cryptic riboflavin permease. Biochemistry 42:1851–1860

Sibirny AA, Boretsky YR (2009) *Pichia guilliermondii*. In: Satyanarayana T, Kunze G (eds) Yeast biotechnology:

diversity and applications. Springer Science+Business Media B.V., Dordrecht, pp 113–134

Sibirny AA, Shavlovsky GM (1984) Identification of regulatory genes of riboflavin permease and alpha-glucosidase in the yeast *Pichia guilliermondii*. Curr Gen 8:107–114

Shchelokova IP, Zharova VP, Kvasnikov EI (1974) Obtaining hybrids of haploid strains of *Pichia guilliermondii* Wickerham that assimilate petroleum hydrocarbons. Mikrobiol Zh 36:275–278

Tanner FW, Vojnovich C, Vanlanen JM (1945) Riboflavin production by Candida species. Science 101:180–181

Toyn JH, Gunyuzlu PL, White WH, Thompson LA, Hollis GF (2000) A counterselection for the tryptophan pathway in yeast: 5-fluoroanthranilic acid resistance. Yeast 16:553–560

Vaughan-Martini A, Kurtzman CP, Meyer SA, O'Neill EB (2005) Two new species in the *Pichia guilliermondii* clade: *Pichia caribbica* sp. nov., the ascosporic state of *Candida fermentati*, and *Candida carpophila* comb. nov. FEMS Yeast Res 5:463–469

Yamamura M, Makimura K, Fujisaki R, Satoh K, Kawakami S, Nishiya H, Ota Y (2009) Polymerase chain reaction assay for specific identification of *Candida guilliermondii* (*Pichia guilliermondii*). J Infect Chemother 15:214–218

Wah TT, Walaisri S, Assavanig A, Niamsiri N, Lertsiri S (2013) Co-culturing of *Pichia guilliermondii* enhanced volatile flavor compound formation by *Zygosaccharomyces rouxii* in the model system of Thai soy sauce fermentation. Int J Food Microbiol 160:282–289

Wickerham LJ (1966) Validation of the species *Pichia guilliermondii*. J Bacteriol 92:1269

Wickerham LJ, Burton KA (1954) A clarification of the relationship of *Candida guilliermondii* to other yeasts by a study of their mating types. J Bacteriol 68:594–597

Zhang DP, Spadaro D, Valente S, Garibaldi A, Gullino ML (2011) Cloning, characterization and expression of an exo-1,3-beta-glucanase gene from the antagonistic yeast, *Pichia guilliermondii* strain M8 against grey mold on apples. Biol Contr 59:284–293

Zharova VP, Kvasnikov EI, Naumov GI (1980) Production and genetic analysis of *Pichia guilliermondii* Wicherham mutants that do not assimilate hexadecane. Mikrobiol Zh 42:167–171

Zharova VP, Schelokova IF, Kvasnikov EI (1977) Genetic study of alkane utilization in yeast *Pichia guilliermondii* Wickerham. 1. Identification of haploid cultures by mating type and obtaining their hybrids. Genetika 13:309–313

Index

A

AASM. *See* Amino acid scanning mutagenesis (AASM)

Adiopodoumé, 56

Agrobacterium-mediated transformation (AMT), 217

Agrobacterium tumefaciens
- crown gall disease, 217
- T-DNA, 217

Alternaria alternata, 175, 178

AMA1. *See* Autonomous maintenance in *Aspergillus* (AMA1)

Amino acid scanning mutagenesis (AASM), 187, 189, 190

4-amino-2,6-dibromophenol/3,5-dimethylaniline (ADBP/ DMA), 147

Anastomosis formation
- calcium, 14
- genes, *N. crassa*, 7–10
- *ham-11* mutant, 15
- and heterokaryons, 15–17
- MAP kinase signaling, 10–12
- mathematic modeling, 7
- pheromones, 14
- plasma membrane merger, 15
- ROS signaling, 12, 13
- SO protein, 7, 10
- STRIPAK complex, 12, 14

Aqueous droplet

Arabidopsis thaliana, 88

Arginase *(agaA)*
- *A. niger* D-42 strain, 159
- arginine to ornithine, 155
- chemicals and supplies, 155–156
- media components and reagents, 156–157
- MM + S plates, 159
- nutritional markers, 155
- organism, 156
- preparation of protoplasts, 157–158
- regeneration frequency, 159
- transformation and selection, 158–159
- transformation methods, 155
- vector construction, 157

ARS. *See* Autonomously replicating sequence (ARS)

Ascomycete fungi
- filaments, 7
- haploid, 16

N. crassa, 125

null mutants, 125

Ascomycetes. *See* Sexual reproduction, filamentous ascomycete fungi

Ascomycetous fungi. *See* Autonomously replicating vector

Ascospores
- *asm*-1, 8, 50, 51, 108
- definition, 47
- dikaryotic, 48
- dispersal, 47
- *erg-3* mutations, 49
- mitotic divisions, 40
- 1:1 segregation, mating types, 40
- single ascospore cultures, 38–40

Aspergillus fumigatus
- amino acid biosynthesis pathway, 136
- conditional gene expression, 132, 136
- conditional promoter replacement, 134
- doxycycline-dependent growth, 133
- doxycycline-responsive systems, 133
- fungal pathogenicity, 131
- gene regulation modules, 132
- human immune system, 131
- mutants, definition, 131
- prime application, 133
- recombinant Tet-ON isolation, 135–136
- saprobe and ubiquitous mould, 131
- strains, 133
- Tet-ON and Tet-OFF systems, 132, 133
- tetracycline resistance-conferring operon, 132
- transformation, 134–135

Aspergillus niger, 89
- *Impala* element from *Fusarium oxysporum*, 89
- parasexual recombination, 15
- proto-heterothallic species, 42

Aspergillus transformation
- *A. nidulans*, 141
- DNA delivery methods, 141
- genome sequences, 141
- molecular tools, 142
- nutrition (*see* Nutritional markers)
- resistance (*see* Resistance markers)
- saprophytic filamentous fungi, 141
- selectable markers, 142

M.A. van den Berg and K. Maruthachalam (eds.), *Genetic Transformation Systems in Fungi, Volume 2*, Fungal Biology, DOI 10.1007/978-3-319-10503-1,
© Springer International Publishing Switzerland 2015

Index

Automated morphological analysis, 205
Autonomously replicating sequence (ARS)
  A/T rich region, 254
  M. (C.) guilliermondii genome, 251, 255, 256
  principle, 254
Autonomously replicating vector
  AMA1 (*see* Autonomous maintenance
    in *Aspergillus* (AMA1))
  application, 166
  Aspergillus nidulans, 161
  eurotiomycetes, 161
  protoplast-PEG/CaCl$_2$ method, 162
  Rosellinia necatrix, 161–162
  transformation
    fungal cells, 164
    protoplast production, 162–164
    screening, 164
  transformation, fungi, 161
  validation
    AMA1-bearing vectors, 164, 165
    transformants, 165
Autonomous maintenance in *Aspergillus* (AMA1)
  A. nidulans, transformation efficiency, 161
  autonomous extrachromosomal replication, 161, 162
  co-transformation, 162, 166
  fungal genome, 164
  genome-integrating vectors, 161
  isolation, *A. nidulans*, 227
  protoplast-PEG/CaCl$_2$ method, 162
  R. necatrix, 162
  transformants
    distribution, 165
    stability, 165
  transformation system, 162, 163
2,2-azino-di-3-ethylbenzthiazoline sulfonate
  (ABTS), 147

B

Bacillus subtilis, 172
Barrage zone method, 36
Beta-lactam
  characterization, 128
  class III chitin synthase gene silencing, 129
  gene silencing (*see* Gene silencing)
  *Nco*I sites, 126
  Pc22g22150/*penV*, 128
  P. chrysogenum laeA gene, 127
  PCR analysis, 127
  PcRFX1, 128
  phleomycin resistance, 126
  plasmid pJL43-RNAi, 126
  RNAi-mediated gene silencing, 128
  roquefortine C biosynthesis, 128
  Southern hybridization analysis, 127
  transformation, 127
Bidirectional markers
  acetamide by *amdS*, 144
  agaA, 146
  Aspergilli, 146, 147

  pyrG and *amdS*, 144
  sB and *sC* gene, 146, 147
Bidirectional selection
  hph/amdS marker system, 170
  loxP marker cassette, 173
  materials, 170
  Neurospora, 169
  NHEJ, 169
  pMS-5loxP3 construction, 171–172
  pyr4 gene, 169
  strain and cultivation conditions, 170
  T. reesei, 169
  Trichoderma, protoplast transformation, 172–173
  xyn1 promoter, 169

C

*Candida guilliermondii. See Meyerozyma (Candida)
  guilliermondii*
CATs. *See* Conidial anastomosis tubes (CATs)
Cellulosic biofuel
  cost-effectiveness, 192
  production, 193
Cellulosic biomass
  biofuel production, 183, 193
  metabolic pathway gene ORFs, 190
Cercospora nicotianae, 175, 178
Conditional promoter replacement (CPR)
  A. fumigatus, 133, 136
  functional characterization, 133
  PCR amplicons, 134
  purification, 134
Conidia
  definition, 47
  protoperithecia fertilization, 47
  Te-4 conidia, 49
Conidial anastomosis tubes (CATs), 4
Convergent transcription, 112, 115
  Histoplasma capsulatum, 112
  pSilent-1 and pSilent-dual1, 112, 115
  RNAi vectors, 112
  silencing vectors, 112, 115
Cre/loxP system, 169, 170
Cre recombinase ass, 172

D

Deoxynucleoside triphosphates (dNTPs), 170, 219, 252,
  255, 258
Directed evolution, 185
Dominant
  eight-spore (E) mutation, 48
  in heterokaryotic strains, 107
  markers, 51
  selection marker (*see* Selection markers)

E

Engineering improved fungal strains, 173, 179, 186
Ergosterol mutants, 48

Index

265

Error-prone homologous recombination, 98, 99
Escherichia coli
  amplification and selection, 227
  arginine decarboxylase, 144
  Cre recombinase reaction, 172
  electrocompetent cells, 251
  EZ10 electrocompetent cells, 170
  tetracycline resistance-conferring operon, 132
  transformation, 253, 254, 256, 259, 260
  typical AMA1-bearing vector, 162
  Wizard® *Plus* SV minipreps DNA purification
      system, 251
  XL1 Blue, 157
Extrachromosomal
  AMA1-bearing vector, 162
  replication in fungal cells, 161, 162
  transformants, 164–165

F
Filamentous fungi
  advantage as experimental organism, 16
  anastomosis formation (*see* Anastomosis formation)
  enzyme solution, 176
  foreign genes, 175
  genetic and nutritional exchange, 5
  genetic transformation and targeted gene disruption,
      175
  growth conditions, 176
  heterokaryon formation, 16–17
  homologous integration, 175, 178–179
  intercellular communication, 13
  Neurospora crassa (*see Neurospora crassa*)
  PEG solution, 175
  protoplasts preparation, 176
  regeneration medium, 176
  RNA silencing (*see* RNA silencing)
  ROS and MAP kinase signaling, 12
  Saccharomyces cerevisiae, 179
  sexual reproduction, 29–30
  solution A and B, 176
  splitmarker-based transformation, 175
  splitmarker fragments, 176–178
  STC solution, 175
  TEs (*see* Transposable elements (TEs))
  transformation, 178
  wash solution, 175
Fot1 element, 60, 85
Functional genomics
  gene knockout approaches, 111
  research, 111
  RNA silencing (*see* RNA silencing)
Fungal genomes
  analysis (*see* RIPCAL analysis, fungal species)
  Crypton elements, 84
  regional variability, RIP-activity, 60
  RIP mutation calculation (*see* RIPCAL analysis,
      fungal species)
  selection markers, 149
  sequenced fungal genome, 111

Fusarium oxysporum
  Fot1 elements, 60, 85
  Impala element, 89
  MAK-2 homologous kinase, 10
  Puccinia striiformis f. sp. *tritici*, 118
  TEs, 88
  tomatinase, 118

G
Gene complementation, 23, 41–42
Gene essentiality
  direct gene replacement, 133
  genetic approaches, 111–112
  HIS3 locus, 100
  mutational load, 97
  rid-1 and *sad-1*, 49
Gene family expansion, 60, 63
Gene replacement
  CBS 513.88 and NRRL3 strains, 231
  Trichoderma, 169
  yeast colony PCR, 254
Gene silencing
  Acremonium chrysogenum, 125
  description, 125
  dsRNA molecules, 125
  filamentous fungi, 127
  functional genomic analysis, 125
  Penicillium chrysogenum, 125
  pJL43-RNAi vector, 126
Genetic transformation
  A. niger, 225, 235
  genome manipulation, 235
Genome comparison, 161, 233
Genome evolution, 56, 65, 79, 90
Genome manipulation. *See* Selection markers
Genome projects with TE content, 86–87
Genome sequence
  addition and rearrangement, 221
  A. niger strain, 225–226, 230–231
  completion, 184
  deletion and duplication, 221
  genome-wide gene expression, 232
  metabolic network, 233
  strain ATCC 1015 genome, 231
Germling fusion
  CATs, 4
  conidia/conidial germlings, 3
  description, 4
  Neurospora crassa, 11
  SO protein, 7
Green fluorescent protein (GFP), 190, 192, 258, 260

H
HCS. *See* High-content screening (HCS)
Heterokaryons
  formation, 16–17
  genetic analysis, 15–16
  strain features combination, 16

266 Index

Heterologous expression, 190, 230
Heterothallic fungus. *See Neurospora crassa*
Heterothallism *vs.* pseudohomothallism, 23–24
HGU. *See* Hyphal growth unit (HGU)
High-content screening (HCS), 195
High fidelity polymerase chain reaction (HF-PCR)
    assembly, 190, 192
High-throughput screening (HTS), 185, 193–195
Homologous integration
    Alternaria alternata and *Cercospora* spp., 175
    analytical PCR, 178
    A.niger genomic region, 228
    Cre/loxP-based marker excision system, 169
    dominant marker gene, 178–179
    filamentous fungi, 175
    flanking sequence, 178
    fungi, 169
    marker gene, 178
    M.(C.) guilliermondii transformation, 251
    NHEJ, 169
    non-homologous integration, 175
    phytopathogenic fungi, split-marker fragments, 178
    targeted gene disruption, 178, 179
    Trichoderma, 169
HTS. *See* High-throughput screening (HTS)
Hygromycin-sensitive phenotype, 48, 49
Hyphal fusion
    colony establishment and development, 5–7
    fusion-competent hyphal branches, 4–5
    pore formation, 5
Hyphal growth unit (HGU), 207–208

I
Imaging flow cytometry. *See also* Filamentous fungi
    acquisition platform, 202, 204
    advantages, 208
    Cytosense software, 206
    hyphal element, 208, 209
    in-flow image acquisition method, 208, 209
    Matlab script (m-file), 206
    pixel, 206
    skeletonization, 208, 209
Impala element, 89
Industrial fungi
    Aspergillus niger (*see Aspergillus niger*)
    M. (C.) guilliermondii, 246
    Pezizomycotina, 23, 25, 55, 57, 59, 64, 69, 73
    Trichoderma reesei, 28, 29, 34, 169, 170, 230
    yeast, 183
Inoculations, filamentous ascomycete fungi
    "barrage zone" method, 36
    fertilization method, 38
    mixed culture method, 36–38
Insertional mutagenesis, *M. (C.) guilliermondii*
    chromosomal DNA, 251
    DNeasy blood and tissue kit, 251
    E. coli, 251
    principle, 250–251
    screen transformants, 251

transformation protocols, 251
Wizard® *Plus* SV minipreps DNA purification
    system, 251
Intron-containing hairpin RNA (ihpRNA), 112
In vivo targeted mutagenesis. *See* Targeting Glycosylases
    to Embedded Arrays for Mutagenesis
    (TaGTEAM) in yeast

K
Knock-down
    dsRNA molecules, 126
    glandicoline B, 128
    mRNA targeting, 125
    penicillin/cephalosporin C production, 127
    transformants, 128

L
Large retrotransposon derivative (LARD), 84
L-arginine
    D-42 mutant, 144
    and L-ornithine, 144
    nitrogen source, 156
    PcitA-agaA expression, 159
loxP marker cassette, 173

M
Mandels–Andreotti (MA) agar medium, 170
MAP kinase signaling
    A. nidulans, 10
    COT-1, 11–12
    formin Bni1, 11
    HYM-1, 12
    N. crassa, 10, 11
    PP-1, transcription factor, 10, 11
Marker recycling, 169, 170
Mating-type (MAT) genes
    'asexual' fungal species, 29
    detection, 32
    idiomorphs, 24, 25
    MAT1-1 and *MAT1-2* types, 24
    PCR-based
        detection method, 25
        diagnostic test, 29
    PCR primers, diagnostic tests, 26–28
    RAPD-PCR DNA fingerprinting study, 29
Meiotic silencing by unpaired DNA (MSUD)
    genome defense mechanism, 49
    N. tetrasperma, 49–51
    QIP, 109
    RNA silencing, 108
Methylation
    5-methylcytosine DNA-methylation, 61–62
    RdDm, 62
5-Methylcytosine DNA-methylation, 61–62
Meyerozyma guilliermondii. See also Meyerozyma
    (Candida) guilliermondii
    and *C. guilliermondii*, 245–246

Index 267

and *E. guilliermondii*, 245
and *P. guilliermondii*, 245
Meyerozyma (Candida) guilliermondii
blaster systems
equipments, 253
gene disruption, 251
principle, 251
protocol, 253–254
reagents, 252–253
REP-URA5-REP blaster cassette, 251
TRP5-blaster system, 251–252
URA3-blaster system, 251, 252
60-bp ARS with A/T rich region
autonomous replication, 255
equipments, 255
identification, 254, 255
protocol, 256
reagents, 255
classification and characterization, 245–247
electroporation, 250
fluorescent protein fusion
cell biology studies, 257
C-terminus, target gene, 258–259
dual fluorescent labeling, 260
equipments, 258
global strategy, 257
N-terminus, target gene, 259–260
principle, 257
reagents, 258
subcellular compartments, 258
insertional mutagenesis, 250–251 (*see also* Insertional
mutagenesis, *M. (C.) guilliermondii*)
lithium acetate treatment, 249
media and markers, 247–248
precautions, 248
spheroplasting method, 249–250
transformation
equipments, 248
reagents, 248
Microbial biocatalysts
biocatalytic processes, 195
industrial traits, 186
petroleum-based fuels and chemicals, 183, 195
Microfluidic chip, droplet-based
description, 211
electroporation, 211
microelectrodes and PDMS, 212
microelectrodes fabrication, 214
PDMS, 214–215
photolithography (*see* Photolithography)
photomask design software, 211, 213
yeast cell, 212–213, 215–216
Microfluidic device, 211, 212, 215
Miniature inverted-repeat transposable element
(MITE), 84
Morphological characterization
acquisition, 201
fungi, 201, 205
MSUD. *See* Meiotic silencing by unpaired DNA (MSUD)
Multinucleate hyphae, 59, 64
Mycelial agar discs, 162, 165

N
Neurospora crassa
active transposon, 56
anastomosis formation (*see* Anastomosis formation)
centromeres, 62
cytoplasmic flow, 5
DICER enzymes, 108
hpRNAproducing constructs, 112
hyphal fusion, 5–7
LINE1-like retrotransposon Tad, 110
"methylation-leakage", 62
5-methylcytosine DNA methylation, 61
multiprotein complex, 12, 14
nutrients translocation, 5
quelling, 107
RdDm, 62
rid (RIP-defective) gene, 59
RIP (*see* Repeat-induced point mutation (RIP))
ROS signaling, 12
vegetative cell-to-cell fusion, 4, 8–10
Neurospora tetrasperm
haploid nuclei, 48
MSUD, 49–51
and *N. crassa* comparison, 47–48
NHEJ-defective, 51
RIP-defective mutants screening, 49
transformation, *ERG-3* mutant enabled, 48–49
N,N-dimethyl-*p*-phenylenediamine sulfate (DMPPDA), 147
Non-homologous end joining (NHEJ), 169, 247
A. nidulans, 150
defective *N. tetrasperma* strains, 51
Nutritional markers
agaA, 144, 146
A. niger D-42 strain, 144
in Aspergilli, 144, 145
Aspergillus strains, 144
auxotrophic recipient hosts, 226
and resistance markers, 155

O
Oligonucleotide primers, 170, 178, 188, 255

P
Pathogenicity, 201, 208
PDMS. *See* Polydimethylsiloxane (PDMS)
pGEM-T vector system, 170, 172
Photolithography
AZ® 50XT, 212
hot plate, 212
mask aligner with UV lamp, 212
microelectrodes and channels, 213–214
petri dish, 212
silicon wafers, 211
spin coater, 212
wafer tweezers, 212
Pichia guilliermondii, 245
pJL43-RNAi silencing vector, 126
Plant pathogen, 162, 217
pMS-5loxP3 construction, 171–172

268   Index

Polydimethylsiloxane (PDMS)
    microelectrodes, 212
    microfluidic channel fabrication, 214–215
Promoters
    fluorescent protein fusions, 257
    gene expression, 228
    ORFs, 190
    SUMO expression, 190, 191
    transcriptional regulatory sequences, *A. niger*, 227
    xyn1 promoter, 169, 173
Protein production, 225, 227, 228, 234
Protoplasts
    collection, 163
    concentration, 164
    filtration, 163
    PEG/CaCl$_2$ method, 162
    preparation, 176
    separation, 163
    suspension, 166
    transformation, *Trichoderma*, 172–173
    YCDA plates, 164

Q

Quelling
    in fungi, 107
    mutants, 110
    qde genes, 107
    RNA silencing (*see* RNA silencing)
    vegetative phase and MSUD, 109

R

Random mutagenesis, TEs
    Impala element, 89
    molecular tools, 80
    Plasmodium falciparum, 88
Reactive oxygen species (ROS), 12, 13
Recombinant progeny, 40
Regeneration medium (RM), 176
Renewable transportation fuels, 183
Repeat-induced point mutation (RIP)
    active transposon, 56
    Adiopodoumé, 56
    CpA dinucleotide, 55
    cytosine transitions, 55
    defective mutants, *N. tetrasperma*, 49
    gene family expansion, 63
    index scan mode, 73
    induced nonsense mutation
        *M. graminicol*a, 64
        N. crassa, 63–64
        Pezizomycotina, 64
        R. solani, 64
        SSPs, 64
    leakage, single copy genes, 63
    and methylation (*see* Methylation)
    molecular machinery
        Ascomycota, 59–60
        rid gene, 59

Sordaria macrospora, 60
    mutation, fungal genomes (*see* RIPCAL analysis,
        fungal species)
    Neurospora crassa, 55–56
    regional variability, fungal genome, 60
    repeated DNA segments, 56
    taxonomic range
        cytosine transition mutations, 59
        experimental and computational evidence, 57–58
        RIP-like CpG mutations, 59
        sub-phylum Pucciniomycotina, 59
    transformation and reverse genetics, 64–65
    transition mutations, 55
    transposable elements, 56
    variability, 60–61
Repetitive DNA, 55, 56, 59, 63, 69, 77
Resistance markers
    A. niger glutamate uptake system, 142
    antibiotic/antimetabolite, 143
    disadvantages, 143
    EGFP expression, 143
    PPT, 142
    wild type/natural isolates, 142
Reverse genetic methods, 64–65
RIP. *See* Repeat-induced point mutation (RIP)
RIPCAL analysis, fungal species
    alignment-based mode
        CLI, 75
        comparative "model" sequences, 71–72
        dominance metrics, 72–73
        GUI, 75
        pre-analysis considerations, 71
        *_RIPALIGN.TXT files, 75
    application, 70
    CLI, 74–75
    deRIP, 76
    dinucleotide mode, 73, 76
    GUI, 74, 75
    index scan mode, 76
    input formats, 71
    installation, 70
    output formats, 73
    RIP-index scan mode, 73
RIP in *Neurospora crassa*
    active transposon, 56
    Adiopodoumé, 56
    CpA dinucleotide, 55
    cytosine transitions, 55
    repeated DNA segments, 56
    transition mutations, 55
    transposable elements (TEs), 56
RNA interference
    C. albicans, 111
    histone modifications, 61
    and siRNA, 125
RNA silencing
    convergent transcription, 112, 115
    dsRNA/siRNA delivery, 116
    fungal genome sequencing, 108
    in fungal research

Index 269

Aspergillus oryzae, 117
"building blocks method", 117
H. capsulatum and *C. neoformans*, 118
pathogenicity, 117
P. infestans, 117
in fungi and fungus-like organisms, 113–114
genetic tool
Ascomycota, 111
EFG1 gene, 111
gene knockout strategies, 111
homologous recombination, 112
genome stability, 110
host-induced gene silencing, 118
long hairpin RNA, 112
molecular genetic approach, 107
molecular mechanisms
dsRNA and sRNA, 108
Neurospora crassa, 109
PTGS, 108
QDE-2, 109
QDE-3, 108
SAD-1, 108
single-stranded DNA-binding complex, 110
PAZ and PIWI domains, 107
PTGS and dsRNA, 107
simultaneous silencing, multiple genes, 116–117
against transposons, 110–111
as viral defense, 110
ROS. *See* Reactive oxygen species (ROS)
Rosellinia necatrix, 162, 164–166

S
Saccharomyces cerevisiae
conventional strains, 192
cytochrome c gene transcriptional terminator, 126
ethanol production, 192
fungal *MAT* genes, 25
GFP sequence, 192
heme biosynthesis, 233
linear double-stranded vector, 185
MAK-2 module, 10
non-*Saccharomyces* yeasts, 184
Pigpa1-silenced transformants, 117
recombinant cellulosic, 195
single-stranded oligonucleotides, 185
split marker technique, 148
synIXR and semi-synVIL, 185
TaGTEAM, 98
Ty1 and δ elements, 80, 85
in vivo, 97
xylose fermentation, 183
Second-division segregation, 48, 49
Selection markers. *See also Aspergillus* transformation
dual selection, 149
homologous recombination, 148
NHEJ pathway, 148
rescue, 149–150
split marker, 148–149
transformation, 148

Sexual cycle utilization
gene identification, 40–41
gene manipulation, 41–42
Sexual reproduction, filamentous ascomycete fungi
agar media, 32–33
gene identification, 40–41
gene manipulation, 41–42
heterothallic species, 24
homothallic fungal species, 23–24
idiomorphs, 24
incubation conditions, 33–36
induction methods
agar media, 30, 32–33
solutions, 30
inoculations, 36–38
MAT genes, 24–29, 32
pezizomycete species, heterothallic, 24
progeny analysis, 40
'pseudohomothallism', 24
single ascospore cultures, 38–40
strain selection
identification, 30–31
origin of strains, 31
strain typing, 31–32
Single ascospore cultures
direct isolation, 38
ejected ascospores, 39
heat treatment, 38–39
isolation, sclerotial fruiting bodies, 39–40
Small non-autonomous CACTA transposon
(SNAC), 84
SOE. *See* Splicing-by-overlapping-extension (SOE)
Southern hybridization, 164
Splicing-by-overlapping-extension (SOE), 171–172
Split-marker gene fragments, 176, 177
Split marker technique
A. fumigatus genes, 149
ectopic integration, 149
selection markers, 148, 149
Strain improvement, sexual reproduction
classical methodologies, 184
functional genomics, 184
and gene complementation, 41–42
heterokaryon formation, 17
STRIPAK complex, 12, 14

T
Tad1 element, 85
TAIL-PCR. *See* Thermal asymmetric interlaced PCR
(TAIL-PCR)
Targeting Glycosylases to Embedded Arrays for
Mutagenesis (TaGTEAM) in yeast
description, 97
evolutionary protocol, 104
fluctuation analysis, 102–103
materials, 100
mutagenesis, 98–99
mutant libraries, 103
non-recombinogenic tetO array, 97, 98

Targeting Glycosylases to Embedded Arrays for Mutagenesis (TaGTEAM) in yeast (*cont.*)
primers, 102
S. cerevisiae, 98
strain design
ChrI:197000, 99
distance-dependence, 99
generation, 100, 101
mutation rates and spectrum, 100
tetO array and target sequences, 99
tetO DNA binding sites, 97
T-DNA insertion site
bended structures, 221
computer program counts, 220
and integration, 217
TAIL-PCR, 217
type 1 sequences, 220
T-DNA integration
AMT, 217
A. tumefaciens, 217
in fungi, 221
identification, 220–221
large-scale TAIL-PCR, 218–219
materials, 218
TAIL-PCR, 217, 218
Telomere-like DNA sequences (TEL), 190
Terminal inverted repeat (TIR), 81, 83, 84
Terminal repeat retrotransposon in miniature (TRIM), 84
tetO array
gene-poor region in chromosome I, 99
integrate mutator, 100, 102
Mag1-sctetR-mediated targeted mutagenesis, 99
PCR, 103
TaGTEAM works, 98
target sequences, 99
Thermal asymmetric interlaced PCR (TAIL-PCR). *See also* T-DNA integration
agarose gels, 220
composition, 219
diluted PCR products, 219
dilute genomic DNA, 218
ExoSAP-IT®, 220
PCR cycling parameters, 219
primary and secondary PCR products, 219–220
TAILand inverse-PCRs, 217
Thermostable DNA polymerase, 170
TIR. *See* Terminal inverted repeat (TIR)
Tolypocladium inflatum, 85, 88
Transformation experiments, *N. crassa*
ergosterol-3 (erg-3) mutations, 48
Te-4 conidia, 48–49
Transformation of *M. (C.) guilliermondii*, 247, 249, 250
Transposable elements (TEs), 56
bioinformatical analyses, 85–88
bioinformatics tools, 90
classification, 80–84
epigenetical changes, 79
filamentous fungi (*see* Filamentous fungi)
genomic distribution, 79
gypsy elements, 85
history, 80
mechanisms, 79–80

molecular tools
Ac/Ds elements, 88
Aspergillus niger, 89
function, 89
Fusarium oxysporum, 88, 89
hAT-superfamily, 88
mus-51 and *mus-52* mutants, 88
PiggyBac superfamily, 88
Tolypocladium nitrogen regulator 1, 88
non-LTR retrotransposons, 85
origin, 84–85
protein-encoding genes, 79
Restless element, 85
RNA-(retro-) and DNA transposons, 90
Tad1 element, 85
Transposons
tool mechanism, 88
transposition mechanism, 81
Trichoderma species
heterologous proteins, 229–230
protoplast transformation, 172–173
TRIM. *See* Terminal repeat retrotransposon in miniature (TRIM)

V

Vader element, *Aspergillus niger*, 89
Vader vector pIB635, 89
Variable nucleotide tandem repeat (VNTR) analysis, 195
Vectors
AMA1 (*see* Autonomous maintenance in *Aspergillus* (AMA1))
co-transformation, 166
Cre recombinase assay, 172
expression, *A. niger*, 227–228
pGEM-T vector system, 170
self-ligation, 259
Vegetative fusion. *See also* Anastomosis formation
germling fusion, 3–4
hyphal fusion, 4–5
Visual marker systems, 146–148

W

Whole-slide microscopy
acquisition methods, 201
bioprocesses, 208
hyphal growth unit, 210
imaging flow cytometry, 205
P. chrysogenum, 204, 208

Y

Yeast artificial chromosome (YAC), 186, 191–193
Yeast cell. *See also* Microfluidic chip, droplet-based
bright field image, 216
electroporation, 212–213, 215–216
MAP kinase, 11
preparation, 212, 215
transformation, 256
YPS medium, 247